ENGINE EMISSIONS
Pollutant Formation and Measurement

CONTRIBUTORS

J. B. Edwards
Department of Chemical Engineering
The University of Detroit
and Chrysler Corporation
Highland Park, Michigan

N. A. Henein
Mechanical Engineering Sciences Department
Wayne State University
Detroit, Michigan

R. W. Hurn
Bureau of Mines
U.S. Department of the Interior
Bartlesville, Oklahoma

E. L. Knuth
School of Engineering and Applied Science
University of California
Los Angeles, California

R. A. Matula
Environmental Studies Institute
Drexel University
Philadelphia, Pennsylvania

P. S. Myers
Mechanical Engineering Department
University of Wisconsin
Madison, Wisconsin

H. K. Newhall
Mechanical Engineering Department
University of Wisconsin
Madison, Wisconsin

S. L. Soo
Department of Mechanical Engineering
University of Illinois
Urbana, Illinois

G. S. Springer
Department of Mechanical Engineering
University of Michigan
Ann Arbor, Michigan

O. A. Uyehara
Mechanical Engineering Department
University of Wisconsin
Madison, Wisconsin

ENGINE EMISSIONS

Pollutant Formation and Measurement

Edited by
George S. Springer
and
Donald J. Patterson

Department of Mechanical Engineering
University of Michigan
Ann Arbor, Michigan

⨮ PLENUM PRESS • NEW YORK–LONDON • 1973

Library of Congress Catalog Card Number 71-188716
ISBN 0-306-30585-2

© 1973 Plenum Press New York
A Division of Plenum Publishing Corporation
227 West 17th Street, New York, N.Y. 10011

United Kingdom edition published by Plenum Press, London
A Division of Plenum Publishing Company, Ltd.
Davis House (4th Floor), 8 Scrubs Lane, Harlesden, London,
NW10 6SE, England

Printed in the United States of America

Preface

In recent years, emissions from transportation engines have been studied widely because of the contribution of such engines to atmospheric pollution. During this period the amounts of pollutants emitted, the mechanism of their formation, and means of controlling emissions have been investigated in industrial and government laboratories, as well as at universities. The results of these investigations have generally been published as individual articles in journals, transactions, meeting proceedings, and, frequently, in company reports. This proliferation of technical information makes it difficult for workers in the field to keep abreast of all developments. For this reason, the editors felt the need for a book which would survey the existing state of knowledge in wide, albeit selected areas, and would provide a guide to the relevant literature. This book is intended to fulfill this function.

It is recognized that all aspects of transportation engine emissions cannot be explored in a single volume. In this book attention is focused primarily on sources and mechanisms of emission formation within the combustion process, and on measurement techniques. Beyond this objective, no restrictions were placed on the authors. Within the framework of the general theme each author has been free to treat his subject as he saw fit. The editors have not strived to replace by uniformity the highly personal and attractive divergences of style. Considerable efforts were made, however, to ensure clarity and minimum overlap between the chapters. It is hoped that the resulting book will be as useful a reference to specialists and practicing engineers as to workers coming fresh to this subject.

GEORGE S. SPRINGER
DONALD J. PATTERSON

Ann Arbor, Michigan
May 1972

v

Contents

Chapter 1.   Engine Exhaust Emissions

P. S. Myers, O. A. Uyehara, and H. K. Newhall

Chapter 2.   The Chemistry of Spark-Ignition Engine Combustion and Emission Formation

J. B. Edwards

Chapter 3.   Mechanism of Hydrocarbon Formation in Combustion Processes

R. A. Matula

Chapter 4. The Kinetics of Pollutant Formation in Spark-Ignition Engines

H. K. Newhall

Chapter 5. Particulate Emission from Spark-Ignition Engines

G. S. Springer

Chapter 6. Diesel Engines Combustion and Emissions

N. A. Henein

Chapter 7. Diffusion and Fallout of Pollutants Emitted by Aircraft Engines

S. L. Soo

Chapter 8. Instrumentation and Techniques for Measuring Emissions

R. W. Hurn

Chapter 9. Direct-Sampling Studies of Combustion Processes

E. L. Knuth

Chapter 1

Engine Exhaust Emissions

P. S. Myers, O. A. Uyehara, and H. K. Newhall

Mechanical Engineering Department
University of Wisconsin
Madison, Wisconsin

I. INTRODUCTION

The growth in population (Fig. 1) plus the growth in energy consumption per person[1] have combined together to give dramatic increases in both air and water pollution problems. The automotive vehicle has been a significant contributor to air pollution on a total mass basis (Fig. 2), although its contribution is considerably smaller if the toxicity of emissions as well as the quantity of emissions is considered (Fig. 3).[2] Thus, the purpose of this paper is to first of all indicate why certain exhaust constituents from automotive vehicles are considered as pollutants, to describe how these pollutants are formed in automotive vehicle operation, and finally, to describe

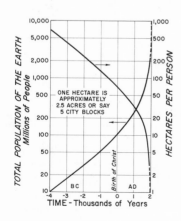

Fig. 1.   Plot of earths population versus time.

1

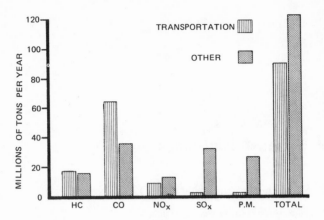

Fig. 2.   1968 mass emissions for the United States.

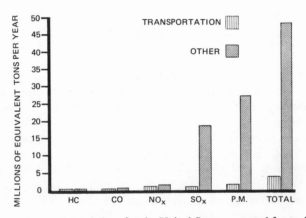

Fig. 3.   1968 mass emissions for the United States corrected for toxicity.

control techniques being used and considered. As indicated in the title, the paper is intended to educate by organizing and systematizing existing information rather than presenting primarily new information.

II. WHEN IS AN EMITTANT A POLLUTANT?

There are three sources of emittants from most automotive vehicles, as shown in Fig. 4. Emittants from the crankcase on all new vehicles have been led back to the carburetor and burned in the combustion chamber, starting in 1961 in California and in 1964 nationwide. Fuel evaporation controls were put on all new cars in 1970 in California and in 1971 nationwide. These fuel

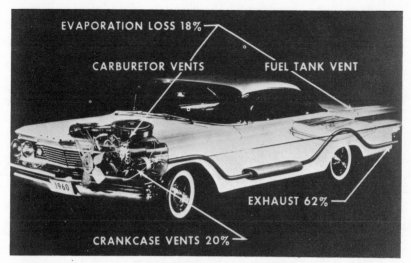

Fig. 4. Sources of pollution from automobiles.

evaporation controls store and then, later during subsequent engine operation, burn in the combustion chamber most of the fuel vapors from the fuel tank and carburetor. Thus the primary emittant source from automotive vehicles is the exhaust.

The primary energy source for our automotive vehicles is crude oil from underground which typically contains varying amounts of sulfur. Much (but not all) of the sulfur is removed during refining of automotive fuels. Thus, except for a limited number of engines, the final fuel is hydrocarbons with only a small amount of sulfur. If for the moment we neglect the sulfur and assume perfect and complete combustion only water and carbon dioxide would appear in the exhaust.

Except at the time of Noah, water is not generally considered undesirable and therefore is not considered a pollutant. Likewise, carbon dioxide is not normally considered a pollutant, although it appears[1] that about one-half of the carbon dioxide we are putting into the atmosphere via combustion is remaining in the atmosphere with consequent concentration increases. Since CO_2 in the atmosphere has the potential of affecting the weather it may ultimately be considered a pollutant.

Basically then, except for the sulfur dioxide which is a result of perfect combustion, the compounds currently considered as pollutants are the result of imperfect and/or incomplete combustion. It should be clear, however, that the energy involved in these exhaust pollutants is normally negligibly small.

Figure 5 shows those automotive vehicle emissions which are usually considered to be pollutants, and the basis on which they are considered

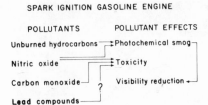

SPARK IGNITION GASOLINE ENGINE

Fig. 5. Pollutants and their harmful effects.

pollutants. Other emissions such as hydrogen, for example, are not considered as pollutants because they do not produce these objectionable effects. Let us look at each of the pollutants and their effect.

A. Unburned Hydrocarbons

There are some unburned or partially burned hydrocarbons in the exhaust in addition to the small amount that escapes even when evaporation controls are used. The amount we are talking about is insignificant from an energy standpoint, but can be objectionable from an odor, photochemical smog, or conceivably a carcinogenic standpoint.

As will be pointed out later, odor is observed primarily when extremely lean mixtures are present during combustion,[3] typically in diesel or stratified-charge engines. The odors are believed to be caused by incomplete oxidation. However, because the nose is at least several orders of magnitude more sensitive than our best chemical detection techniques, we do not know specifically which compounds cause odors. Likewise, we do not know if these compounds are toxic or carcinogenic.

While the details of the mechanism are not completely established, it is known that unburned or partially burned hydrocarbons in the gaseous form plus oxides of nitrogen and sunlight combine together to form what is called photochemical smog. The products of photochemical smog cause watering and burning of the eyes, and affect the respiratory system, especially when the respiratory system is marginal for other reasons. Hydrocarbons differ in their tendency to form photochemical smog. The relative tendency of a hydrocarbon to cause photochemical smog is currently described by a reactivity index.[4] The unsaturated hydrocarbons have the highest reactivity with the reactivity depending on the location of the double or triple bond.

Unburned or partially burned hydrocarbons may also show up as particulate matter. In the spark-ignition (SI) engine, using leaded fuel, the particulate matter will typically contain both hydrocarbon, lead, and lead scavenging compounds while in an SI engine using unleaded fuel, or in the compression-ignition (CI) engine, the particulate matter is primarily hydrocarbons. Some of the high molecular weight aromatic hydrocarbons have been shown to be carcinogenic in animals and thus are suspect from this standpoint.

B. Carbon Monoxide

Carbon monoxide is present particularly when there is a deficiency of air. When there is a considerable deficiency of air there can be considerable chemical energy present in the carbon monoxide.

The toxicity of carbon monoxide is well known. It occurs because the hemoglobin in the blood, which carries oxygen to the different parts of the body, has a higher affinity for carbon monoxide than for oxygen. When a human is exposed to an atmosphere containing carbon monoxide the percent carboxy hemoglobin (the percent of the hemoglobin carrying carbon monoxide rather than oxygen) gradually increases with time to an equilibrium value which depends upon the carbon monoxide concentration. Thus, the toxic effects of carbon monoxide are dependent upon both time and concentration, as shown in Fig. 6.[1]

Figure 6 indicates that carbon monoxide concentrations less than 70–80 ppm have but little effect. However, a recent study[5] has indicated that there may be some reduction in mental acuity even at concentrations as low as 15 to 30 ppm. It should also be noted that essentially no data on the effects of carbon monoxide at low concentrations and very long periods of time are available.

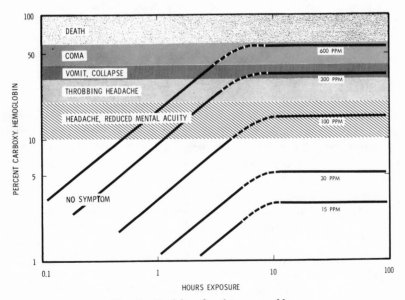

Fig. 6.   Toxicity of carbon monoxide.

C. Oxides of Nitrogen

Oxides of nitrogen (NO, NO_2, N_2O_2, etc.) are formed at the high combustion temperatures present in engines and, while not thermodynamically stable at room temperature, are destroyed quite slowly during engine exhaust and at atmospheric conditions.

Like carbon monoxide, the oxides of nitrogen tend to settle on the hemoglobin in the blood. However, probably the most undesirable toxic effect of oxides of nitrogen is their tendency to join with the moisture in the lungs to form dilute nitric acid. Because the amounts formed are minute and dilute, the effect is very small but over a long period of time can be cumulatively undesirable, especially when respiratory problems for other reasons are present.

As already mentioned, oxides of nitrogen also seem to be a necessary component of photochemical smog.

D. Sulfur Dioxide

Sulfur dioxide from automotive vehicles is small compared with the sulfur dioxide released by the burning of coal.[1] However, sulfur dioxide combined with fog were the primary ingredients responsible for the famous 1952 London Smog and the deaths at Donora, Pennsylvania in 1948. The combination of moisture and sulfur dioxide produces sulfuric acid particularly when high temperatures are present. It should be noted, as indicated in Fig. 5, that much of the sulfur dioxide combines with other materials in the atmosphere to form sulfates and consequently settles out as particulate matter.

E. Particulates

As previously indicated, particulate matter comes from hydrocarbons, lead additives, and sulfur dioxide. If lead is used with the fuel to control combustion about 70% of the lead is airborne with the exhaust gases. Almost half (30%) of the lead particulates are sufficiently large to rapidly settle to the ground while the remainder (40%) is smaller in size and remains in the atmosphere for an appreciable time.

Lead is well known as a toxic compound. It is not so well appreciated, however, that man continuously takes in significant quantities of lead via food and drink as shown in Fig. 7,[1] and that the ingested airborne lead is small in comparison. The evidence for detrimental health effects as a direct result of airborne lead from gasoline additives is inconclusive to nonexistent.[6] However, the increasing quantities of lead used, the fact that there does tend to be a concentrating effect in the soil, along roads, and in the runoff from

ADDITIONS

AIRBORNE -.015-.009 mg/day
 35-55% Exhaled
 35-50% Swallowed
 7-13 % Absorbed
FOOD and WATER
 5-10% Reaches Blood
 .25-.35 mg/day
SMOKING
 .5 mg/cigarette
STORAGE
 Bones - 200-4000 $\frac{ug}{100\,gm}$
 Soft Tissue - 10-280 $\frac{ug}{100\,gm}$

REMOVALS

EXHALED AIR
 35-55% of Inhaled Lead
URINE
 .01-.04 mg/day
FECES
 .1-.4 mg/day
PERSPIRATION
 Estimated same
 concentration as urine

Fig. 7. Mass rate balance for lead for an individual.

these roads plus the known toxicity of lead does lead to legitimate concern about the ultimate detrimental effects. Current concern for lead, however, should be based on its effect on emission control effects rather than its effect on humans.

III. FORMATION OF POLLUTANTS

A. Carbon Monoxide

The appearance of carbon monoxide in combustion processes is generally a simple result of oxygen insufficiency either on an overall or local basis. In principle, the concentration of carbon monoxide contained in exhaust products should correspond to a chemical equilibrium state represented by the water gas equation

$$H_2O + CO = H_2 + CO_2$$

At maximum flame temperatures this equilibrium yields significant quantities of CO relative to CO_2 — even for fuel-lean mixture ratios. However, as combustion gases cool from peak flame temperatures to the much lower temperatures characteristic of exhaust products, this equilibrium shifts in a direction favoring oxidation of CO to CO_2. Consequently, for fuel-lean or chemically correct mixture ratios, relatively small quantities of carbon monoxide ultimately appear in exhausted combustion products. For fuel-rich mixture ratios, however, due to the simple insufficiency of oxygen, significant concentrations of carbon monoxide persist even in cool exhaust products.

Figure 8 illustrates the behavior of carbon monoxide in the spark-ignition engine cycle. Here the variation of carbon monoxide concentration within the engine cylinder is plotted against time following completion of combustion.[7] As previously indicated, CO is formed in large quantities in the early high-temperature portions of the combustion process, and as the products expand and cool oxidation of CO to CO_2 occurs. The dashed line of Fig. 8 represents CO concentrations that would result if chemical equilibri-

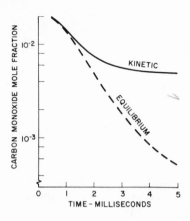

Fig. 8.   Carbon monoxide concentration as a function of time following combustion.

um prevailed throughout expansion. However, at the lower temperatures during expansion, chemical reaction rates lag behind equilibrium leading to the solid curve of Fig. 8 representing the actual situation.

The influence of fuel–air ratio on this behavior is demonstrated by Fig. 9 which compares theoretical calculations[7] with experimental exhaust measurements.[8] Here it is observed that real exhaust concentrations of CO fall between the peak temperature and exhaust temperature equilibrium values.

Diesels, gas turbines, steam engines, Stirling engines, etc., typically can or do operate at mixtures that may be locally rich in fuel, but are lean in fuel on an overall basis. As a consequence excess oxygen is ultimately available for oxidation of CO and the primary problem is to be certain that the excess oxygen reaches the CO before the temperature drops to a value too low for oxidation to proceed in the time available. Carbon monoxide emis-

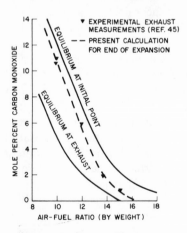

Fig. 9.   Carbon monoxide as a function of air–fuel ratio.

sions from these engines tend to be quite low while emissions from SI engines vary markedly with fuel–air ratio.

B. Nitric Oxide

The principal oxide of nitrogen formed in combustion processes is nitric oxide, NO. Nitric oxide is a high-enthalpy species relative to N_2 and O_2 from the standpoint of basic thermodynamics. Therefore, its presence is favored by the existence of high temperatures.

Chemical equilibrium calculation predicts that nitric oxide concentrations in high-temperature combustion products should reach levels ranging from several hundred parts per million to as much as one or two mole percent depending on temperature and fuel–air ratio. Such calculations implicitly assume that sufficient time is available for the combustion products (particularly nitric oxide) to reach the chemical equilibrium state during combustion. Recent investigations have shown that in reality the time required for nitric oxide to reach chemical equilibrium levels at typical flame temperatures is significant relative to the time scale of most combustion processes.[9] Furthermore, the rates of the chemical reactions directly responsible for nitric oxide formation are highly temperature sensitive. As a consequence, cooling of burned products immediately following combustion often occurs before peak-temperature chemical-equilibrium nitric oxides levels are achieved.

A second deviation from chemical equilibrium occurs during later stages of cooling of burned combustion products. Equilibrium calculations predict that at the relatively low temperatures at which combustion products exit from the combustion system, nitric oxide should have largely decomposed to N_2 and O_2. However, in reality the pertinent chemical reactions are much too slow at these temperatures to yield significant decomposition. Consequently, nitric oxide remains permanently fixed at levels formed earlier during higher-temperature portions of the combustion process.

This behavior is graphically summarized in Figs. 10 and 11 which show a hypothetical time–temperature plot for a combustion process and the resulting nitric oxide–time curve. Also displayed in Fig. 11 is the variation of nitric oxide that would result if chemical equilibrium existed at all times during the process. In Figs. 10 and 11 it is shown that during the earliest part of the process chemical reaction rates are not sufficiently rapid to bring the concentration of nitric oxide to the peak equilibrium level. Later at time τ, the temperature has dropped to a point where chemical reactions have effectively stopped and nitric oxide therefore remains fixed even though if chemical equilibrium existed it would drop to near zero levels at later times. The important point here is that the NO concentration exhausted is controlled primarily by the rate of NO formation during the high-temperature part of

Fig. 10. Temperature–time history for hypothetical combustion process.

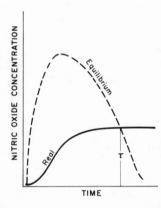

Fig. 11. Nitric oxide formation in hypothetical combustion process.

the process. Therefore, the ultimate quantity of NO formed will depend on the formation rate and the total time available.

In addition to being highly dependent upon temperature, the rate of formation of nitric oxide depends upon the concentration of oxygen present in combustion products. For this reason nitric oxide levels tend to be low for fuel-rich mixtures and increase to a maximum for mixture ratios having approximately 10% excess air. With further increases in excess air the attendant drop in combustion temperature offsets the increasing oxygen concentration leading to a reduction in nitric oxide formation. The interaction of oxygen availability and temperature is illustrated by Fig. 12 which presents plots of nitric oxide concentration versus equivalence ratio for a series of times following combustion.[10]

The formation of nitric oxide in real spark-ignition engines closely follows the preceding fundamental description. Following ignition, a flame propagates across the combustion chamber. In a given parcel of fuel–air mixture nitric oxide formation begins at the instant that the flame consumes that parcel. Thus nitric oxide formation begins first at those locations in the

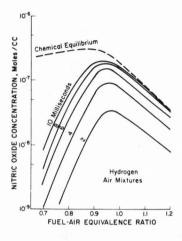

Fig. 12. Nitric oxide concentrations as a function of equivalence ratio and time after combustion.

chamber nearest the ignition point and last at those points farthest removed from the ignition source. As the combustion process nears completion the piston begins to move downward, rapidly expanding and cooling the combustion products and, as a consequence, stopping the NO formation reactions. Because the amount of NO formed at a given location depends on the time elapsed between flame arrival and cooling due to piston motion, nitric oxide will be formed in largest quantities at locations near the point of ignition. This phenomenon is accentuated by the fact that temperatures and hence nitric oxide formation rates are highest near the point of ignition. Thus at the end of the combustion process there exists a gradient in nitric oxide concentration across the chamber with highest concentrations near the point of ignition.

As mentioned previously, maximum nitric oxide formation occurs in slightly fuel-lean mixtures. In addition, ignition timing exerts a strong influence on NO levels due both to its influence on the time available for NO formation and on temperatures. These effects of fuel–air ratio and ignition timing are illustrated by Fig. 13.[11] Nitric oxide formation rates tend to increase with increasing pressure. Therefore engine emissions tend to increase with increasing load.

The preceding discussion of nitric oxide formation is implicitly based on the concept of a homogeneous combustion process, that is, one in which fuel and air are intimately and uniformly mixed before combustion. In the case of the heterogeneous combustion process typical of the diesel and gas-turbine engines in which fuel and air mix and burn simultaneously, the chemical reactions responsible for NO formation are unchanged, but the physical environment in which these reactions proceed is quite different from that of the homogeneous system.

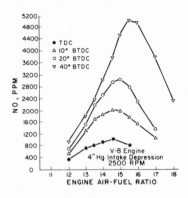

Fig. 13.   Effects of fuel–air ratio and timing on oxides of nitrogen.

Consider the case of a single droplet burning in surrounding high-temperature air, as shown in Fig. 14. Fuel vapor flows from the droplet into a combustion zone surrounding the droplet where fuel and oxygen burn to form combustion products. Based on the preceding discussion, nitric oxide formation should begin as products are formed. Following their formation these products diffuse outwardly and mix with the surrounding oxygen and nitrogen. As a consequence, the oxygen concentration of the products increases while temperature drops. Because nitric oxide formation depends critically on both oxygen concentration and temperature the progress of NO formation in the products is complex. Quantitatively, however, as products mix with surrounding air the rate of NO formation should reach a maximum and then diminish to essentially zero. The total quantity of NO generated in this process will depend on the many variables that determine the temperature and flow fields surrounding the droplet.

The combustion and hence nitric oxide formation processes in both the diesel and gas turbine are much more complex than the idealized process discussed above. In both processes it is probable that a certain portion of the fuel burns under premixed and hence more or less homogeneous conditions with the remainder burning as a true diffusion flame as described

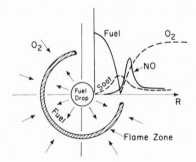

Fig. 14.   Diffusion flame around a burning
droplet.

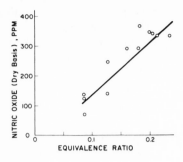

Fig. 15. Variation of NO_x with load for gas turbine.

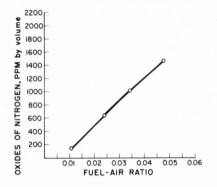

Fig. 16. Variation of NO_x with load for direct-injection diesel.

above. Generally the quantity of nitric oxide produced increases approximately in proportion to the quantity of fuel supplied. This is demonstrated by Figs. 15[12] and 16.[13] A notable exception to this behavior occurs in diesel engines of the prechamber type. Here, when the fueling rate reaches a certain point, further increases in fuel input yield diminishing quantities of nitric oxide. This is demonstrated by Fig. 17.[14]

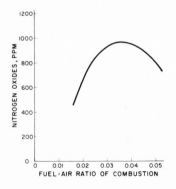

Fig. 17. Variation of NO_x with load for prechamber diesel.

As in the spark-ignition engine, nitric oxide formation in the diesel engine is influenced by combustion pressure and temperature, and by the time available for combustion. For this reason, nitric oxide levels tend to increase with advanced injection timing and with increased compression ratio or increased manifold temperature.

C. Hydrocarbons

Unlike carbon monoxide or nitric oxide, hydrocarbons are not substances that one expects to find in high-temperature combustion gases. Chemical equilibrium calculations readily show that the quantities of hydrocarbon gases that can exist in homogeneous high-temperature combustion products are immeasurable. Furthermore, the oxidation reactions for hydrocarbons under such conditions are among the most rapid observed. Thus, the appearance of unburned hydrocarbons in the exhaust of a combustion process must be associated with the existence of temperature or mixture heterogeneity at some point in the system and not with the homogeneous part of the system.

The appearance of unburned hydrocarbons in combustion products implies that these particular hydrocarbons were never successfully ignited. Thus, the effect of heterogeneity resulting in unburned hydrocarbons can be viewed as one of ignition inhibition.

The simplest criterion that can be used to determine whether or not ignition of a parcel of reactants occurs is that internal heat generation due to precombustion reactions within the mixture exceeds the rate of heat loss from the mixture to surroundings. There are at least three conceivable engine conditions that can markedly affect either the rate of internal heat generation or the rate of heat loss. These are:

1. The local mixture composition may be so rich or lean that the oxidation reactions are very slow and ignition cannot occur because of heat losses.
2. For very small isolated elements of fuel–air mixture the element surface-to-volume ratio may become so large and heat losses consequently so great that ignition cannot occur.
3. Heat losses from fuel–air mixture adjacent to a cool surface may be so great that ignition cannot occur.

One or more of these factors may play a role in each of the various types of combustion systems presently in use. In the case of the premixed flame, as employed in the spark-ignition engine, it is doubtful that item one or two above would be of major importance. Clearly, however, the presence of relatively cool combustion chamber walls can result in large heat losses from adjacent fuel–air mixture. As a consequence, a layer of unburned fuel–air mixture next to the walls will fail to ignite and will therefore persist throughout the combustion process. In a definitive study of this problem optical methods were used to experimentally confirm the role of combustion chamber surfaces in quenching combustion reactions.[15] It is generally considered that the thickness of this quench layer is a function of fuel–air ratio and pressure, among other things.

It is estimated[16] that a significant portion of those hydrocarbons quenched by the cold combustion-chamber walls are burned during the expansion process, and it has been shown that under many typical operating conditions a fraction of these hydrocarbons surviving the expansion are burned in the exhaust system. For this reason both temperature and oxygen concentration in the exhaust gases are important to the final concentrations of hydrocarbons emitted from the exhaust system. The importance of oxygen concentration is illustrated by Fig. 18 showing the variation of hydrocarbon emissions with air–fuel ratio.[4]

The thickness of the quench zone is probably minimum slightly on the rich side. However, as the engine mixture is made leaner more oxygen is available during the expansion and exhaust processes to help destroy the increased hydrocarbons present because of the increased quench thickness. Thus, for lean mixtures, even though the quench-zone thickness has increased, very few hydrocarbons survive the expansion and exhaust process until the mixture is so lean that erratic flame propagation occurs. High turbulence[17] plus turbulent heated inlet manifolds[18] will delay erratic flame propagation to very lean mixtures. On the rich side the quench thickness increases again with no offsetting increase in the destruction of hydrocarbons as on the lean side. The result is a very rapid increase in hydrocarbons in the exhaust when the mixture is made rich as shown in Fig. 18.

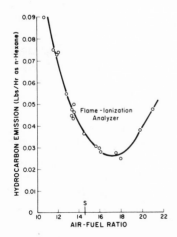

Fig. 18. Variation of hydrocarbon with air–fuel ratio.

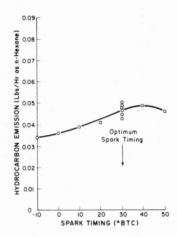

Fig. 19. Effect of ignition timing on hydrocarbon emissions.

Results presented in Fig. 19 indicate that retarded ignition timing decreases hydrocarbon emissions.[8] This is primarily a result of increased exhaust temperatures with retarded timing.

Diffusion combustion as employed in diesel and gas turbine engines can result in hydrocarbon emissions for any of the three fundamental causes listed previously. Impingement of a fuel spray on a cool surface can lead to quenching of hydrocarbons with subsequent emissions. Small parcels of fuel isolated from the main combustion regions can escape combustion and contribute to emissions. Similarly, development of very lean mixtures isolated from high-temperature combustion regions can result in unburned hydrocarbon emissions.

Generally, in well-designed diffusion combustion systems, the above problems can be minimized and total hydrocarbon emissions are not a serious problem, at least relative to the spark-ignition engine. This, however, does not preclude the possible emission of very small quantities of highly objectionable hydrocarbon types such as odorants or carcinogens. This is recognized as an important problem with regard to the diesel engine.

D. Particulates

Particulate material may be formed at a solid surface or in the gas phase. Let us first look at particulate matter formed in the gas phase. In general, in spark-ignition engines operating with a relatively homogeneous mixture, particulate matter in the gas phase is not observable unless extremely rich mixtures are used. Thus, when discussing gas-phase particulate matter, we will primarily be discussing particulate matter formed in the diffusion type of flame that exists in diesel, gas turbine, steam, Stirling, and other similar engines.

Particulate matter can occur in either a liquid or solid form. Liquid particulates are observed primarily in diesel engines at light load where ignition in very lean mixtures has failed to occur and occasionally in two-cycle spark-ignited engines when partially oxidized and cracked fuel in the condensed phase is carried out the exhaust. We shall be more concerned with the solid particulates.

A solid particulate consists of an agglomeration of carbon and hydrocarbon molecules with the hydrocarbon molecules having a higher C/H ratio than the original fuel. Thus the formation of the solid particulate must involve

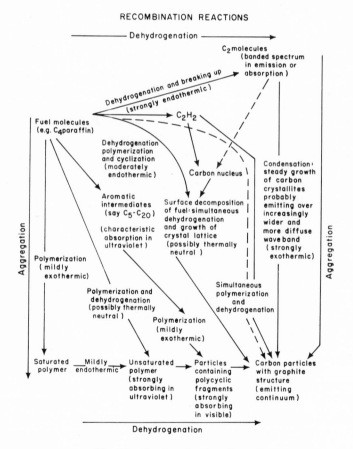

Fig. 20.   Routes of carbon formation.

the stripping off of some of the hydrocarbon molecules (dehydrogenation), polymerization (formation of higher molecular weight molecules), and finally agglomeration to form a particulate. Street,[19] after a detailed examination on past studies, concluded that the different carbon formation routes proposed in the literature differed mainly in the different emphasis put on the roles of dehydrogenation and polymerization. He prepared Fig. 20 which shows many different possible routes when going from a fuel molecule to a particulate. Note that all of the paths start with a molecule in the upper left-hand corner and finish in the lower right-hand corner by forming a solid particulate. It is not clear whether all these different paths actually exist and are followed in different physical situations or whether our knowledge is as yet insufficient to select between the many routes shown in Fig. 20.

The route that is currently considered most probable is shown as a dotted line in Fig. 20. It is considered most probable on two basis. The first is that when diffusion flames using different fuels are studied, C_2H_2 is found as an intermediate product in most cases, as is shown in Fig. 21, where all of the fuels, with the exception of methane, show high C_2H_2 concentrations. The second reason is results obtained by Porter,[20] who used a very high-intensity short-duration flash to provide almost instantaneous heating of the mixture followed by a much less intense flash at later times to study absorption spectra and thus chemical reactions. From his experiments Porter concluded that carbon formation occurs via a preliminary decomposition of the hydrocarbon as far as acetylene with carbon formation proceding directly from the acetylene without the formation of any stable intermediate. Electron micrographs of the carbon formed during these photolysis experiments showed it to be similar to that obtained from a smoky acetylene flame and to have a mean particle size of 500 Å.

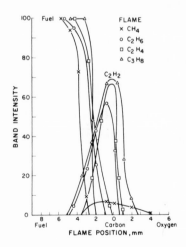

Fig. 21.   Formation of C_2H_2 when using different fuels.

Many studies of carbon formation in steady-state laboratory diffusion flames have been reported. A typical experimental setup is illustrated schematically in Fig. 22.[21] As can be seen in Fig. 22, gaseous fuel and air (or oxygen) are fed through two different channels with the oxygen (or air) being inserted between two "layers" of gaseous fuel. Nitrogen is fed on the outside of the two gaseous fuel layers so that a flat stable flame is formed. An idealized diagram of concentrations versus distance across the burner is shown in Fig. 22.

Figure 23 shows measured and estimated concentration gradients in an actual experiment using this burner with ethylene as a fuel. Experimental temperature measurements are also shown. Note that carbon is formed on the fuel side of the diffusion flame where the oxygen concentration is low and the temperature is relatively high (2000–2200 C).

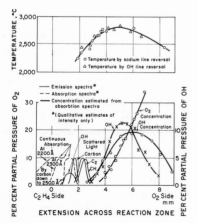

Fig. 22. Idealized concentrations versus distance in a diffusion burner.

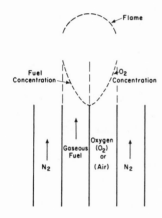

Fig. 23. Measured and estimated concentrations in a diffusion burner.

Let us apply the concepts of Fig. 23 to a fuel droplet burning in a combustion chamber. The vaporizing fuel droplet will be partially or completely surrounded by a diffusion flame as shown schematically in Fig. 14. Carbon particles should be formed on the droplet (fuel) side of the diffusion flame. As the drop vaporizes and the fuel vapor burns, there is a flow of gases through the flame. Thus these carbon particles should pass through the flame into a region of increasing oxygen concentration and at first increasing but soon decreasing temperature. Assuming the flame surface is not broken the amount of particulate finally escaping from the flame around the droplet will be the sum of the particulates formed minus the particulates oxidized in their passage through the flame front. However, if the flame surface is broken by turbulence or by impingement on a cold surface, the destruction mechanism may not exist with a consequent large increase in net carbon production. Flynn[22] found confirmation of the formation–destruction concept in his study of radiant-heat transfer from a diesel engine. He found that particularly during the first part of combustion the optical depth required to approach blackbody radiation conditions was very small (very high carbon concentrations) and that extrapolation of carbon particle densities during early combustion to exhaust conditions would give orders of magnitude more particles than are experimentally found in the exhaust, i.e., the rate of formation must be high to produce the observed optical densities and the rate of destruction must be high to reduce the carbon particle densities observed during early combustion to those observed in the exhaust with the exhaust concentration being the net difference between two large quantities.

In the diesel engine, assuming a proper match between the injector and combustion chamber, particulates are usually not a problem until a fuel–air ratio someplace between 75 and 100% stoichiometric is reached. Presumably until this load is reached the destruction mechanism is adequate to eliminate almost all of the carbon formed. As higher loads are reached and less oxygen is available the carbon formed is not so readily destroyed and net production of particulates (smoke) increases rapidly with each load.

In a gas turbine, net particulate matter is markedly affected by burner design but it is not clear whether the proper design is such that the flame around the droplet is not "broken" to let out the carbon formed without passing through the hot zone, or whether proper design includes sufficient turbulence and excess air to ensure that sufficient oxygen is always present to destroy the particulates as they pass through the diffusion flame. In steady-flow combustion engines, such as steam or Stirling engines which involve heat transfer to a solid surface, there is always the additional possibility of "breaking" of the flame by the relatively cold heat exchange surface if insufficient combustion chamber volume is provided.

In the last several years smoke-suppressing additives (usually containing barium) have been studied. It is not clear whether the additive affects the rate of formation or the rate of destruction of the carbon, or both. It would be very interesting to conduct radiation measurements using the technique employed by Flynn[22] on an engine using fuel with a smoke-suppressing additive.

It is observed experimentally in an SI engine, when using both leaded and unleaded fuels, that deposits (particulate matter) are formed on the cold chamber wall presumably from the heavy ends of the fuel and the lubricating oil being dehydrogenated and polymerized on the cold wall. Some of these deposits end up in the exhaust stream. Some of the same phenomena occur in a diesel engine, but their presence may not be observed because of the particulates formed in the gas phase. The less volatile fuel used in diesel engines also probably results in a different type and location of deposits.

The combustion deposits in the typical automotive SI engine reach an uneasy equilibrium between rate of formation and rate of destruction after 3000–10,000 miles, the exact mileage being highly dependent upon fuel, oil, and engine operating conditions as well as engine design. Cyclic engine operation will cause momentary changes in equilibrium thickness. Unfortunately particulate concentrations in the exhaust of SI engines are just now being seriously studied and not much data are available. Preliminary data indicate that the weight of particulates is greater when using leaded fuels versus unleaded fuels, but that the volume of particulate matter is larger when using unleaded fuels, i.e., leaded fuels give more dense deposits. Present test procedures also measure particulates emitted during transient conditions when deposit equilibrium thickness is changing because of these transient conditions. This raises a question as to whether the net amount of particulate matter formed or the stability of the deposits is the quantity being measured when using current test procedures.

Measurement and characterization of exhaust particulates is difficult. Comparison of light extinction from the sky by the smoke plume by comparison with a calibrated chart (Ringleman method) is commonly used. The percentage light extinction method where a photo cell is used is also a common technique. Filtering of the particles through either a stationary or moving filter followed by a measurement of light reflectivity to characterize particle density on the filter is also used. Traps using centrifugal action have been installed in SI engine exhausts. Much work remains to be done in this area yet.

E. Odor

Ever since the first diesel engine was developed, the odor from its exhaust has been recognized as undesirable. Determination of the cause of this odor has been slow and difficult because of the complexity of the heterogeneous combustion process, the lack of chemical instruments capable of detecting materials present in extremely low concentrations, and the extremely high sensitivity of the human nose. To illustrate the problem, in some cases under

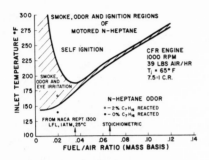

Fig. 24.   Odor and auto ignition limits as a function of fuel–air ratio.

ideal conditions the nose is capable of detecting concentrations of the order of magnitude of one part in 10^{18}–10^{21} while even today detection of concentrations of the order of magnitude of one part in 10^9 using chemical instruments requires extreme care. Thus it has been very difficult to isolate the odor-forming compounds.

Barnes[3] concluded that diesel odor resulted from partial oxidation reactions in the fuel-lean regions which are almost inevitable formed in heterogeneous combustion. Figure 24 from Trumpy[23] illustrates the relative odor-producing capabilities of different air–fuel regions and tends to confirm Barnes conclusions. The data shown in Fig. 24 were obtained using a motored engine and varying the air–fuel ratio and the inlet-air temperature to produce the results shown.

It would seem likely that fuel-lean regions existed in the heterogeneous combustion occurring in steady-flow combustion chambers, but odor in the exhaust from these engines does not seem to be as much of a problem as from a diesel.

In practice odor is most reproducibly measured by a specially-selected, specially-trained human panel working for only part of a day and for part of a week. Ideally the panel is trained to recognize both "quality" and "intensity" of odors. Exhaust samples are passed through systems carefully designed so as to not change the odor characteristic of the exhaust. These samples are then diluted in amounts ranging from 50:1 to 600:1 with the amount of dilution being chosen to correspond to the dilution found in the vicinity of engine exhausts. Dilution corresponding to field conditions is necessary since the response of the nose is highly dependent upon concentration — we are told that skunk oil in low concentrations is used in perfume.

IV. TRANSIENT OPERATION — ENGINES

In this section the effect of operating different types of engines, i.e., carburetor type spark-ignition; port fuel-injection spark-ignition; cylinder fuel-injection spark-ignition; naturally aspirated diesel; turbocharged diesel;

gas turbine; and external combustion under transient conditions, will be discussed.

A brief definition of steady-state engine operation will serve as a basis for comparison with transient operation. Under steady-state operation the temperatures of the component parts such as the manifold, cylinder walls, and the pressure and temperature of the gas in the cylinders, have reached an equilibrium value that varies only slightly and randomly from one cycle to the next. Dynamometer engine operating data are usually taken under these steady-state conditions. However, under transient engine operation the pressure and temperature of the mixture in the cylinder at the time of ignition as well as the temperature of the cylinder wall varies progressively rather than randomly from cycle to cycle.

The basic mechanism of formation of the different pollutants has already been presented. The quantity of different pollutants formed in the combustion chamber during transient operation will be presented in terms of the mechanisms described previously. Each type engine mentioned above will be presented under the following operating modes: cold start and warm-up; acceleration and deceleration.

A. Carbureted Spark Ignition

At the time the engine is started the manifold and cylinder walls are cold. In order to form a combustible mixture enough fuel must be vaporized so that the vapor–air mixture entering the cylinder is in the ignitable range. Since only part of the fuel is vaporized the fuel–air ratio formed in the carburetor must be made fuel-rich by choking. The mixture that enters the combustion chamber under this choked-carburetor condition is liquid fuel plus air–vapor mixture. During this period the exhaust will have high concentrations of HC and CO with some of the HC coming from fuel vaporized from the piston and walls late in the combustion process or during the exhaust stroke. Nitric oxide should be low since the combustion temperature is low due to the vaporization of some of the raw fuel as well as the fairly low temperature of the mixture at time of ignition. After the cold start and during the warm-up period the HC and CO will gradually decrease as the engine parts get warmer. The period of time it takes an engine to "warm up" depends upon the design of the engine and inlet air system, the ambient temperature and the volatility of the gasoline, but "ball park" numbers are from two to five minutes.

In a throttle-controlled engine the intake manifold pressure increases as load increases. This increase in manifold pressure causes some of the fuel vapor in the vapor–air mixture in the manifold to condense out before reaching the cylinder causing the mixture actually reaching the cylinder to become momentarily lean. The opposite effect occurs, of course, when manifold pressure decreases. To compensate for this condensing effect with increased load the carburetor has a fuel pump that when the throttle is depressed sprays additional gasoline into the carburetor venturi. The quantity of fuel sprayed into the venturi has to satisfy both cold as well as warmed-up engine operating conditions, and is usually adjusted for cold conditions. Thus during acceleration the carburetor supplies a mixture which is unquestionably rich at the carburetor and most probably rich even at the cylinder. This increases both the HC and CO during this period, but should decrease NO.

Particulates can be seen emerging from the exhaust particularly during hard acceleration. The composition of the particulates in the exhaust is open to question. It may be carbon-rich particles from the rich mixture or may be combustion-chamber deposits broken loose by thermal shock due to the transient change in cylinder pressure and temperature and swept out the exhaust.

During the deceleration portion of the transient period, the raw gasoline that was condensed out in the manifold vaporizes and richens the mixture entering the cylinder. The effect of this momentary richness is an increase in HC and CO with a decrease in NO. As the manifold vacuum reaches 19–21 in. Hg, slow, partial or no flame propagation may occur.

B. Port Injection Spark Ignition

During cold state and warm-up the only difference is that a leaner fuel–air mixture can be supplied to the port-section in a shorter time since the port will warm up more rapidly than the entire manifold. Thus the total quantity of HC should be somewhat less than the carburetor equipped engine during the cold-start and warm-up period.

During acceleration not much fuel is condensed in forming a film since the effective manifold length is very short. Thus, the increase in HC and CO should be a minimum.

Likewise, during deceleration fuel from the film on the manifold should be a minimum and, in addition, since the fuel can be completely cut off, slow flame propagation at high manifold vacuums is not a problem.

C. Cylinder Injection Spark Ignition

This system should provide the least total amount of HC and CO during the cold start and warm-up transient period since the fuel–air mixture need not be enriched for starting during the warm-up transient period. Likewise during acceleration no extra fuel need be supplied except if a rich mixture is desired for power reasons. Also, since there is no film on the manifold and the fuel can be completely cut off, the emission of pollutants should be zero, or close to zero, during deceleration period.

D. Diesel Engine — Naturally Aspirated (NA)

With a cold intake manifold and engine the air that is inducted into the cylinder is cold at the beginning of the compression process. During the compression stroke the cold cylinder and the cold combustion chamber walls abstract a large amount of energy from the compressed air. Therefore, at the time fuel is injected into the cylinder the air temperature may be so low that ignition may not take place. However, with each succeeding compression stroke the engine wall will get warmer, although it may take a number of strokes before conditions are "right" for the engine to start. The number of strokes depends upon the compression ratio of the engine, engine design cranking rpm, and the cetane number of the fuel being used. Therefore, for the first few cycles raw fuel and/or white smoke may come out of the exhaust.

E. Diesel Engine — Turbocharged

The compression of a turbocharged engine is typically somewhat lower than that of a NA engine. For this reason it could take longer for the engine to start and the raw fuel coming out the exhaust pipe could be higher than that of an NA engine.

In a turbocharged engine the amount of air supplied to the combustion chamber is related to the energy available in the exhaust. During acceleration a portion of the exhaust energy must be used to accelerate the turbine and compressor wheels. Thus there is a time lag between the demand for a sudden load change and the availability of the air needed to provide that load. If the fuel supply is not carefully programmed severe smoke problems can be encountered. Typically the inlet maifold pressure is monitored and used to properly program the fuel injected. However, even if the fuel is properly programmed the cold walls may cause some momentary smoking problems. In general there is no problem during deceleration since the fuel is cut off.

F. Gas Turbines

The rotating speed attainable with an external starter is limited so that the pressure as well as the compressed-air temperature is low. For the first few seconds, whitish particulates as well as partially burned products may result from the colder air temperature available for combustion. The compressor blades, both rotor and stator, as well as the combustion chamber and nozzle are all at a low temperature. During acceleration the quantity of fuel has to be programmed to match the available air. During deceleration, because of the lower pressure and temperature of the air in the combustion chamber, the combustion process could deteriorate to the point where a haze can be seen from the turbine during this period.

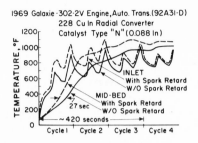

1969 Galaxie-302-2V Engine,Auto. Trans.(92A31-D)
228 Cu In Radial Converter
Catalyst Type "N"(0.088 In)

Fig. 25. System response under transient conditions.

G. External Combustion Engines*

Because of the thermal inertia of the heat exchange surfaces, some external combustion engines use off–on fuel controls; others use modulated fuel control. If an off–on control is used, a transient state is the method of operation under all conditions except the heaviest load.

Regardless of the type of fuel control used the combustion system is cold at start up. Increased heat transfer because of the cold surfaces may reduce NO, but is very likely to increase hydrocarbons due to quenching in the gas phase as well as to possible flame impingement on cold surfaces. The same situation exists to a lesser degree each time the burner comes on when using off–on control and when the burner is modulated. Pollution effects during acceleration and deceleration are noted only insofar as they affect the rate of firing. All evidence available now indicates that emissions are increased by several times when combustion occurs in other than the steady state.

V. TRANSIENT OPERATION — SYSTEM

"Hang-on" devices such as thermal reactors and catalytic mufflers are effective only when they are up to operating temperature. Thus their transient behavior must be considered as well as the transient behavior of the engine. These devices have thermal inertia with the magnitude of this thermal inertia being a design variable.

The interaction between the engine and a catalytic muffler is shown in Fig. 25 as well as the order of magnitude of time response. The temperature curves marked inlet are a measure of the thermal (but not chemical) energy output from the engine. Note that they are still rising slightly even after cycle one (approximately 2 min). The temperature of the mid-bed of the catalyst lags behind, of course, and is still rising slightly after cycle two (approximately 5 min). Note that the effectiveness of the catalyst is unimportant

---

*External-combustion engines do not use the products of combustion as a working fluid.

during cycle one from a practical standpoint — its temperature is too low to do any good. Thus preheating or extremely low thermal inertia hang-on devices will be necessary if emissions during the first cycle are to be eliminated.

VI. EVALUATION OF VEHICLE EMISSIONS

In contrast to the steady-state engine operation typical of much dynamometer testing, real motor vehicle operation involves a large number of transient conditions. As noted in the preceding discussion, transient operation may be of considerable importance for certain engine-pollution control systems; while for other systems transient effects may be of lesser importance. Clearly in the former case the method of evaluation of exhaust emissions must include transient operating modes. Traditionally, this problem has been handled by measuring emission rates as a vehicle is driven (on a chassis dynamometer) in conformance with a prescribed time–vehicle speed schedule or driving cycle. Dynamometer loading is programmed to simulate the speed–load relationship thought to be typical of the particular class of vehicle under test.

For the moment attention will be directed specifically to consideration of driving cycles. The attendant measurement of pollutant quantities will be discussed subsequently. However, it should be noted that in practice the two subjects are intimately related.

Ideally the driving cycle or cycles employed in vehicle emission tests should represent an average or composite of vehicle operation typical of the driving of a large segment of the driver population. Clearly the establishment of a single driving cycle representative of the wide variety of driving conditions encountered in various geographical areas is questionable. Not only does typical vehicle operation vary inter-regionally, or even intra-regionally, but in addition the relative importance of each of the several types of pollutants may vary geographically. As a consequence of the wide variety of vehicle operating conditions actually employed, it is possible that a truly realistic test cycle may have to encompass the entire speed–load range of the vehicle.

Formulation of a driving cycle for evaluation of exhaust emissions was first undertaken in California. In 1960, at the direction of the State Legislature, the California State Department of Public Health published a set of vehicle-emission standards based on hypothetical vehicle operation in eleven noncontiguous speed-based operating modes.[24] These operating modes were derived from a previous study sponsored by the Automobile Manufacturers Association.[25]

Because the eleven noncontiguous engine-operating modes selected did not adapt readily to a practical continuous driving cycle it was necessary for the California State Motor Vehicle Pollution Control Board to develop a

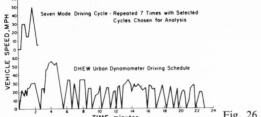

Fig. 26.   California seven-mode cycle.

practical cycle. The resulting cycle was the California seven-mode cycle.[26] This cycle, which is illustrated in Fig. 26, was subsequently adopted by the Federal Government for implementation of Federal regulations in 1968.[27] The Federal Government has recently proposed a revised driving schedule consisting of vehicle operation from a cold start over a 23-min cycle containing nonrepetitive idle, acceleration, cruise, and deceleration modes. This cycle, usually referred to as the DHEW Urban Dynamometer Driving Schedule, is proposed for implementation in 1972.

As previously mentioned, the question of exhaust sampling procedures really cannot be divorced from that of driving schedules. The major problem is the necessity for characterizing with a single numerical value the varying exhaust emissions rates emanating from the several vehicle operating modes. While the basic measurement obtained is always that of pollutant concentration, the desired characterization is the total quantity or mass of pollutant emitted.

For this reason it is necessary to derive effective mass-emission rates from exhaust pollutant concentration measurements.

The earliest sampling procedure adopted was that developed in conjunction with the California seven-mode cycle. During the course of the driving cycle, exhaust pollutant concentrations were continuously measured and recorded. The recorded pollutant concentrations for the several operating modes were then weighted and averaged in a prescribed manner yielding single cycle pollutant concentration values related to some extent to mass emissions.

More recently, proposed sampling techniques have been developed with the intent of obtaining accurate and direct measurements of total mass-emission rates. This has been accomplished generally by combining average overall cycle concentration values with total cycle exhaust mass flow measurements. Suitable concentration measurements can be obtained either by collecting, completely mixing, and subsequently analyzing the entire flow of exhaust products for a complete test cycle,[28] or by continuously collecting a fractional sample at a given rate throughout the cycle.

The method proposed for implementation by the Federal Government in 1972 is a variation of the latter technique in which the entire exhaust flow from the engine continuously is diluted with a high-volume air flow generated by a constant-speed positive-displacement pump. The total volume flow of exhaust gas plus dilution air can be accurately determined from the recorded pump speed. A continuous sample flow of diluted exhaust obtained from a sample probe located at a point just upstream of the pump is collected in a bag throughout the 23-min driving cycle. Subsequently, the sample is analyzed for pollutant concentrations which, combined with the measured total volume flow, yields mass emission rates. While this technique has the advantage of permitting a relatively straight forward determination of exhaust-mass flow, it suffers from the requirement of added analytical sensitivity due to exhaust dilution and also from the possibility of occurrence of chemical reaction within the sample during the driving cycle (23 minutes minimum).

In light of the present highly developed state of data-processing systems and the availability of continuous-response analytical instruments, modifications of the proposed Federal procedure might be considered.

VII. CONTROL TECHNIQUES

Some of the subsequent papers in this series will deal with current control techniques for specific types of engines. Thus at this time we shall only indicate broadly the control techniques available. In general, the control techniques can be classified as either before or after the fact, i.e., they can prevent formation or, once formed, destroy the emittants.

In reciprocating engines, hydrocarbons from the cold quench zone are almost inevitable and control techniques involve destruction after they are formed. Basically they involve destruction in the exhaust by further oxidation. It may be necessary to supply additional oxygen and in all cases exhaust energy (temperature) must be carefully conserved or catalysts used to achieve destruction in a reasonable size volume. Steady-flow combustion devices do not "scrape off" the quench zone and therefore hydrocarbons are usually the result of quenching in the gas phase. Careful combustion-chamber design usually yields low hydrocarbon levels although the destruction techniques mentioned above are theoretically applicable.

The formation of carbon monoxide is very directly related to fuel–air ratio. If formed it can be further oxidized either thermally or catalytically. It should be noted that more energy is released in CO destruction and, if done thermally, higher temperatures are required to achieve practical exhaust volumes (residence times).

The formation of oxides of nitrogen can be minimized by using a rich mixture with consequent increases in hydrocarbons and carbon monoxide (Fig. 18). Lower combustion temperatures, achieved by wall or diluent cooling (provided the diluent is not oxygen), will lower both the rate of formation and the equilibrium value, and is thus very effective. Catalytic destruction in the exhaust is possible although most current catalysts seem to require that CO also be present in the exhaust.

The formation of particulate matter is minimized by use of premixed combustion. However, exhaust-particulate traps for SI engines have been developed. It should also be noted that the catalyst often serves as a very effective particulate trap. While it is clearly an art, combustion-chamber design in heterogeneous combustion such as occurs in diesel and steady-flow combustion can be very effective in minimizing particulates.

REFERENCES

1. Myers, P.S., Automotive emissions—A study in environmental benefits versus technological costs, *SAE Trans.* Vol. 79, Paper 700182 (1970).
2. Barr, H.F., Automotive air pollution control, Paper 700248 presented at *SAE Automotive Engineering Congress*, Detroit, January 1970.
3. Barnes, G. J., Relation of lean combustion limits in diesel engines to exhaust odor intensity, *SAE Trans.* Vol. 77, Paper 680445 (1968).
4. Jackson, M.W., Wiese, W. M., and Wentworth, J.T., The influence of air–fuel ratio, spark timing and combustion chamber deposits on exhaust hydrocarbon emissions, published in TPS–6, *Vehicle Emissions — Part I*, Paper 486-A, Society of Automotive Engineers, Inc., New York.
5. National Academy of Engineering, *Effects of Chronic Exposure to Low Levels of Carbon Monoxide on Human Health, Behavior, and Performance*. Washington, D.C., 1969.
6. Myers, P.S., *Spark Ignition Engines Combustion and Lead*. Presented at Washington Academy of Sciences, January 21, 1971.
7. Newhall, H.K., Kinetics of engine generated nitrogen oxides and carbon monoxide, *Twelfth Symposium (International) on Combustion,* The Combustion Institute (1969), p. 603,
8. Huls, T.A., Myers, P.S., and Uyehara, O.A., Spark ignition engine operation and design for minimum exhaust emissions, published in PT–12, *Vehicle Emissions—Part II,* Paper 660405, Society of Automotive Engineers, Inc., New York, 1967.
9. Newhall, H.K., and Shahed, S.M., Kinetics of nitric oxide formation in high pressure flames, Paper 36 presented at *Thirteenth Symposium (International) on Combustion,* Salt Lake City, August 1970.
10. Shahed, S.M., and Newhall, H.K., Kinetics of nitric oxide formation on propane-air and hydrogen-air diluent flames, *Combustion and Flame,* **17,** 2 (1971).
11. Campau, R.M., and Neurman, J.C., Continuous mass spectrometric determination of nitric oxide in automotive exhaust, *SAE Trans.,* Vol. 75, Paper 660116 (1967).
12. Smith, D.S., *et al.*, Oxides of nitrogen from gas turbines, *J. Air Poll. Cont. Assoc.* **18,** 1 (1968).
13. Yumulu, V.S., and Carey, A.W., Exhaust emission characteristics of four-stroke, direct injection, compression ignition engines, Paper 680420 presented at *SAE Mid-Year Meeting,* Detroit, May 1968.
14. Landen, E.W., Nitrogen oxides and variables in precombustion chamber type diesel engines, Paper 714B presented at *SAE Summer Meeting,* Montreal, June 1963.
15. Daniel, W.A., Flame quenching at the walls of an internal combustion engine, *Sixth Symposium on Combustion,* Reinhold Publishing Co., 1957.
16. Daniel, W.A., Why engine variables affect exhaust hydrocarbon emission, *SAE Trans.,* Vol. 79, Paper 700108 (1970).
17. Stivender, D.L., Intake valve throttling (ITV) — A sonic throttling intake valve engine, *SAE Trans.,* Vol. 77, Paper 680339 (1968).

18. Bartholemew, E., Potentialities of vehicle emission reduction by design of induction systems, Paper 660190 presented at *SAE Automotive Engineering Congress,* Detroit, January 1966.
19. Street, J.C., and Thomas, A., Carbon formation in premixed flames, *Fuel,* **34,** 4–36 (1955).
20. Porter, G., Carbon formation in the combustion wave, *Fourth Symposium (International) on Combustion,* p. 248 (1952).
21. Parker, W.G., and Wolfhard, H.G., Carbon formation in flames, *J. Chem. Soc.* pp. 2038–49 (1950).
22. Flynn, P., An experimental determination of instantaneous potential radiant heat transfer within an operating diesel engine, Ph. D. Thesis, 1971.
23. Trumpy, D., Sorenson, S.C., and Myers, P.S., Discussion of Ref. 3.
24. Maga, J.A., and Hass, G.C., The development of motor vehicle exhaust emission standards in California, *Fifty-third Annual Meeting, APAC,* Cincinnati, 1960.
25. Teague, D.M., Los Angles traffic pattern survey, Paper 171 presented at *SAE National West Coast Meeting,* Los Angles, August 1957.
26. Hass, G.C., and Brubacher, M.L., A test procedure for motor vehicle exhaust emission, *J. Air Pol. Cont. Assoc.* **12,** 505 (1962).
27. Control of air pollution from new motor vehicles and new motor vehicle engines, *Federal Register* **33,** 2 (1968).
28. *Draft Regulation: Uniform Provisions Concerning the Approval of Vehicles Equipped with a Positive Ignition Engine With Regard to the Emissions of Gaseous Pollutants by the Engine.* Economic Council for Europe, April 11, 1969.

Chapter 2

The Chemistry of Spark-Ignition Engine Combustion and Emission Formation

J. B. Edwards

Department of Chemical Engineering
The University of Detroit
Detroit, Michigan and
Chrysler Corporation
Highland Park, Michigan

I. INTRODUCTION

The ability of the spark-ignition engine to release chemical energy via combustion reactions and transform this energy into mechanical energy is well known. The engine may be visualized as a combination of parallel but out of phase batch reactors. These are commonly called the combustion chambers. Four-, six-, or eight-batch reactors are arranged in parallel. The effluent from these batch reactors is combined in one or two pulsating-flow reactors, called the exhaust system.

Many different types of chemical reactions occur within these batch and flow reactors. Some result in the production of motive power. Other reactions are inconsequential with respect to the production of motive power but make major contributions to the chemical species found in the engine exhaust.

In this chapter the various chemical phenomena occurring within the engine will be examined in detail. In addition to examining each of these processes singularly, the engine will be viewed as a complex system involving parallel and sequential processes as well as feedback paths. This is necessary because changes in one process may alter the nature of other processes and complicate the assertainment of causal relationships.

Particular attention is paid to those chemical processes which result in the synthesis of molecules such as reactive hydrocarbons, oxygenated hydrocarbon derivatives, carbon monoxide, nitric oxide, and particulate matter. The design of the engine, the composition of the fuel injected, and the maintenance of the engine all may significantly influence the chemical phenomena which occur within the engine and ultimately the compounds exhausted into the atmosphere.

In a previous paper[1] the author discussed the basic principles of chemical phenomena occurring in a spark-ignition engine. This present discussion will update the previous one in two ways. First, the results of recent publications are included in this discussion. Recent investigations have led to a better understanding of many phenomena, particularly the generation of nitric oxide in spark-ignition engines. Secondly, the subject matter of the present discussion is broadened to include consideration of the relationship between chemical phenomena in spark-ignition engines and the production of chemical species of recent interest, such as aldehydes and particulate matter.

II. ENGINE PROCESSES

A. Inputs

The substances in the first four boxes shown in Fig. 1, the fuel, the oil, and the additives present in both, as well as the components of the air are rather obvious candidates for a variety of reactions which occur within the engine. Traditionally the materials of which the engine and exhaust system are constructed have not been considered to be consumable reactants; however, these materials are potential inputs for the production of particulates. Ac-

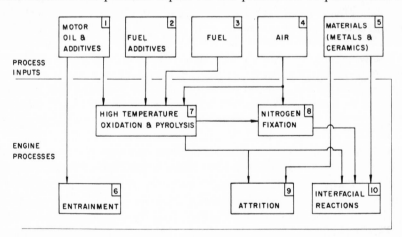

Fig. 1.   Inputs to and physical and chemical processes within a spark-ignition engine.

cordingly, the various metal and ceramic materials present in the engine have been included as a fifth possibility among the process inputs.

B. Physical and Chemical Processes within the Engine

The major physical and chemical processes occurring within the engine and their relationships to the inputs just discussed are also shown in Fig. 1. These engine processes include phenomena which occur in the combustion chamber as well as the exhaust system of the engine. The term exhaust system is used here to mean that volume beginning at the back of the exhaust valve and extending to the outlet of the tail pipe. It includes the manifold, muffler, and any devices such as catalytic or thermal reactors used to promote chemical reactions between these two spatial extremes.

Some of these processes are primarily physical in nature. For example, the entrainment of oil droplets in the combustion chamber. Another example is the attrition of surfaces to produce particulate matter.

The chemical phenomena can be subdivided into homogeneous processes and heterogeneous processes. The reaction in boxes 7 and 8 (see Fig. 1) are homogeneous in nature. The fuel molecules may undergo a variety of reactions including oxidation and pyrolysis. Theoretically it is possible to burn a homogeneous mixture of a fossil fuel and air without the production of particles. However, practically this ideal is seldom obtained. The rapid release of energy in the flame zone transfers sufficient quantities of energy to nearby fuel molecules to cause their fragmentation. If on a micro-scale sufficient oxygen is not available to insure oxidation of these fragments, which is often the case since not all fuel–air mixtures are truly homogeneous, reactions of a pyrolytic nature follow. These pyrolytic reactions can produce particles with high molecular weight and low hydrogen–carbon ratios. The oxidation of these particles is slow since it involves diffusion of gas species to and from the particle surface. As a result some of the particles, particularly the larger ones, may survive passage through the flame as well as oxidation in the post-flame gases.

The rapid release of energy in the flame zone also results in high gas temperatures. Other reactions such as the fixation of nitrogen to produce nitric oxide also occur. A variety of heterogeneous or interfacial reactions may occur between the working fluid of a spark-ignition engine and the many surfaces to which it is exposed during its passage through the engine. Surface catalysis is a timely example which will be discussed in some detail.

Another example of an interfacial-type reaction which was mentioned briefly above is the production of particulate matter. In this sense the spark-ignition engine can be considered to be a generator of the interfacial area. These particulates, or aerosols as they are sometimes called, which are generated as a combustion by-product are of considerable current interest.

C. Energy Consideration

Basically combustion is a chemical process wherein fuel molecules are oxidized to one or more products with the attendant release of energy. Figure 1 is a partial statement of the law of conservation of mass for a combustion process. Like mass, energy must also all be accounted for. A major part of the chemical energy is partitioned between the translational-, rotational-, vibrational-, and electronic-energy levels of the product molecules. This partitioning of energy is shown in Fig. 2. Ultimately the energy is either converted into mechanical work by the gases expanding against the cylinder face, or leaves the system as heat in the exhaust stream.

Part of the energy is expended to form bonds between atoms like nitrogen and oxygen. Though the nitric oxide thus formed is thermodynamically unstable in the lower temperature post-flame gases, this energy is not recovered due to the kinetic limitations on the decomposition of nitric oxide. Hence, hyperequilibrium quantities of nitric oxide are observed in the combustion products.

Finally, the simultaneous occurrence of large amounts of energy and clusters of atoms with low hydrogen–carbon ratios which are products of pyrolytic reactions, is precisely what is necessary for the formation of small particles. It would be surprising if a large number of small particles were formed by a process where only a small amount of energy was available, but is not particularly surprising that the high energy intensity associated with combustion processes can produce large numbers of fine particles. The energy

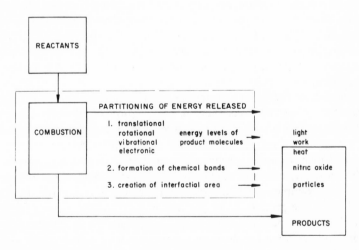

Fig. 2.   Partitioning of energy released by combustion in a spark-ignition engine.

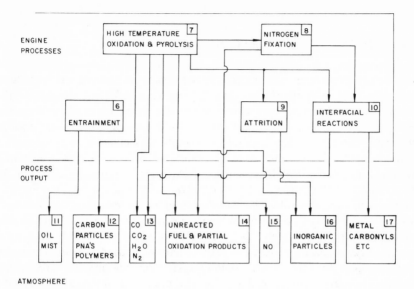

Fig. 3.   Relationship of outputs to the physical and chemical processes within a spark-ignition engine.

to create the necessary surface area is readily available. The fineness of these particles is a mixed blessing. On one hand their minuteness increases the possibility of subsequent complete oxidation in the post-flame gases. But on the other hand, if for some reason this does not occur, their small size will greatly complicate any attempt to remove them from the exhaust gas.

Moran,[2] in a recent study of particle emissions from spark-ignition engines, noted that, in general, 50% of the mass of exhaust particles were smaller than $0.1\,\mu$ (MMED).

D. Outputs

A summary of the various gaseous and particulate materials which are exhausted to the atmosphere along with their relationships to the engine processes discussed above is shown in Fig. 3. The principal gaseous products of the chemical phenomena discussed earlier are included in boxes 13–15. Particulate material ranges from minute polymeric particles with low hydrogen–carbon ratios, which are produced in the combustion process, to much larger particles. These larger particles may be produced by the flaking of deposits which were formed by the deposition of smaller particles in the combustion chamber or exhaust system. When Figs. 1 and 3 are combined, and if all inputs and outputs are accounted for, the result is a statement of conservation of mass for combustion in a spark-ignition engine.

III. ULTIMATE FATE OF ENGINE EMISSIONS

Before proceding with a detailed discussion of each of the important processes in a spark-ignition engine, it is worth considering the fate of the many compounds which are exhausted to the atmosphere. In most cases this fate is not limited to simple dispersion. Rather a polluted air mass is chemically and physically unstable and numerous changes can occur. For example, some of the primary gaseous pollutants can react to produce secondary pollutants which may be gaseous or particulate in nature (refer to Fig. 4). An example of this is the oxidation of nitric oxide emitted by a combustion source to nitrogen dioxide. This is followed by photolysis of nitrogen dioxide which in turn initiates a chain reaction converting the hydrocarbons present to secondary pollutants (refer to boxes 18–20 of Fig. 4). Certain hydrocarbon species tend to produce aerosols rather than gaseous products. There are many sources of atmospheric pollutants besides the internal-combustion engine. These are simply noted as other sources in Fig. 4 (boxes 22–25). The relative contribution of the internal-combustion engine varies from locale

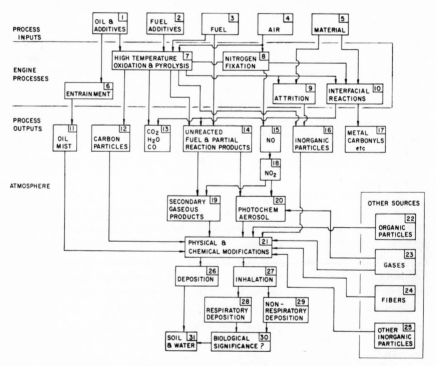

Fig. 4.   Atmospheric pollutants: their origin, modifications, and fate.

to locale and from time to time at a given location. All of these emissions, the ones from mobile as well as from stationary sources, undergo a variety of physical and chemical modifications in the atmosphere. Ultimately the emitted species or their reaction products are removed from the atmosphere. Some are deposited as dustfall while the remainder are removed by atmospheric cleansing processes such as precipitation. Still others may take a more circuitous route after being inhaled.

Figure 4 summarizes the relationship of spark-ignition engine processes to the inputs and outputs of the engine. It also illustrates in a general way why the various species emitted to the atmosphere may be of concern.

IV. THE ENGINE — A CHEMICAL REACTOR

Since the combustion chambers and exhaust system are being considered as reactors, the environment in which the chemical reactions occur will now be examined. In any analysis of chemical reactions or the kinetics of these reactions, it is important to consider the pressure, temperature, and time available for reaction.

At this point an important distinction must be made between measured and actual values of these properties. Most experimental measurements, particularly of temperature, give values that are averaged over a period of time or a volume of mass. In a spark-ignition engine where the reacting medium undergoes rapid and continual changes in pressure and temperature, large gradients may exist from point to point and from instant to instant at a given point. Absolute values often differ appreciably from average values. Chemical behavior depends on absolute rather than average conditions and prediction of chemical behavior based on averages may lead to erroneous conclusions.

V. THE COMBUSTION CHAMBER

Figure 5 illustrates the major regions of chemical reaction in a combustion chamber of a spark-ignited engine. Reactions in the unburned gases, U, ahead of the flame front are termed "precombustion" reactions and encompass a variety of gas-phase reactions, including low-temperature pyrolysis, slow oxidation, and cool flames. These reactions occur as a result of the increasing temperatures and pressures to which the mixture is subjected during compression. They usually begin prior to initiation of the flame, F, by the spark, S, and continue until the last of the unburned gas (end gas) is consumed by the flame. The term "unburned" is not synonymous with "unreacted," for the very existence of precombustion reactions implies that

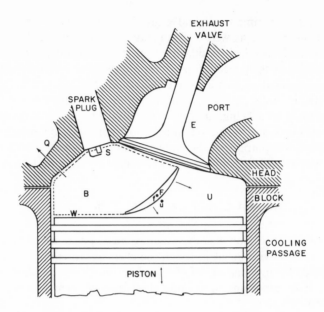

Fig. 5.   Major regions of combustion in a spark-ignition engine following spark ignition.   Unburned gases, U; flame, F; burned gases, B; quench layer, W; and exhaust reactions E.

the flame propagates through a mixture differing chemically from the ingested one. An abnormal condition termed "knocking combustion" results if the magnitude of the precombustion reactions is so great as to cause premature inflammation of the unburned gases. Relatively high temperatures are generated in the flame, F. The flame is that region where the fuel and oxygen undergo very rapid and nearly complete conversion to carbon monoxide, carbon dioxide, hydrogen, and water.

Reactions of the various species formed in the flame continue to occur in the burned gases, B, behind the flame. In this context the "burned" gases occupy the region through which the flame has passed, but "burned" does not imply complete combustion. Post-flame reactions in the burned gases include the recombination of atomic species, continued oxidation of carbon monoxide, and pyrolysis of any hydrocarbons present. The term "pyrolysis" refers to a variety of hydrocarbon reactions that occur in the absence of oxygen.

When the flame approaches a cooled surface, W, for example, the engine head, combustion-chamber walls, or piston face, it is extinguished. Thermal energy Q lost by the gases adjacent to the surface limits the temperature of these gases to a value below that required for continued propagation. This loss of thermal energy through the walls results in a peripheral layer of gas which is either unreacted or partially reacted. Partial oxidation occurs for the same reasons that precombustion reactions occur in the unburned gases. That is the temperatures, pressures, and available times are favorable. This peripheral layer of gas, known as the "quench layer" is important for it is the source of many of the unburned hydrocarbons and partially oxidized species which are exhausted from the combustion chamber. Spatially, the quench layer exists at the interface between the combustion gas and combustion-chamber walls. So, hydrodynamically it can be considered as an interfacial situation. Chemically though, the reactions which occur in this layer are primarily homogeneous. Any heterogeneous reactions are limited by:

—The short time available for reaction.
—The low mean-free paths of the molecules at elevated pressure.
—The limited surface area to which the quench gases are exposed.

The differences between the mechanisms and extents of reactions occurring in the bulk gases and the quench layer appears to be the result of differences in temperature rather than a transition from a homogeneous to a heterogeneous mechanism.

Mass transport from the low-temperature quench region to the burned gases results from turbulence and is followed by pyrolysis of the hydrocarbons in the oxygen-deficient burned gases. Conversely, exchange of hot gases from behind the flame with the boundary layer will promote partial oxidation reactions of the hydrocarbons remaining in this layer.

A. Reactor Conditions

1. Pressure

The working fluid of a spark-ignition (SI) engine undergoes continuous changes in pressure. Though pressure changes with respect to time during normal combustion, it is very nearly invariant with respect to spatial location within the combustion chamber. Any pressure gradient introduced by the combustion reactions is equalized at the velocity of sound in the gases, which is normally fast by comparison with the propagation velocity of the combustion wave. Typical combustion chamber pressures are shown in Fig. 6a. Units on the abscissa are crankangle degrees rather than seconds or milliseconds. The two are related through the rpm of the crankshaft. Values of pressure on the ordinate illustrate typical values throughout the cycle.

Absolute pressures in an engine vary with changes in intake manifold pressure, spark timing, equivalence ratio, and other operating conditions. The pressure trace will also vary from cycle to cycle within the same cylinder and from cylinder to cylinder in a multicylinder engine.[3-7] Variations in peak pressure from cycle to cycle affect not only power but also chemical

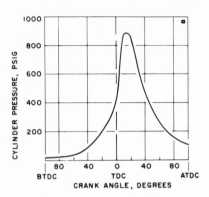
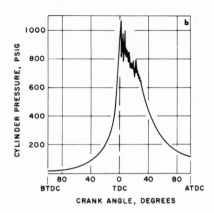

Fig. 6.   Pressure traces in combustion chamber. (a) Normal conbustion, (b) knocking combustion.

phenomena. For example, variations in peak pressure may significantly affect the quantity of nitric oxide produced. The initial increase in pressure prior to 40° BTDC is due primarily to compression of the gases by the piston. Thermal perturbations introduced by precombustion reactions are small in the early part of the cycle. The rapid rise in pressure culminating just past tdc is due primarily to combustion of the mixture and secondarily to compression or expansion of the gases due to changes in combustion-chamber volume resulting from piston motion. The pressure decreases rapidly after 15° ATDC as the gases expand on the power stroke.

Knocking combustion is said to occur when finite pressure waves (which travel at velocities exceeding that of sound in the working medium) are generated in the combustion chamber. These accompany extremely high rates of combustion. A pressure trace similar to that shown in Fig. 6b results. The spikes correspond to the initial passage and reflections of finite pressure waves at the sensor location. Knock is not usually desirable and results in excessive rates of heat transfer to combustion-chamber surfaces and possible mechanical damage to the engine.

2. Temperature

Unlike the situation for pressure, large thermal gradients with respect to spatial location do exist within the combustion chamber during normal combustion. The transport processes, convection, conduction, and radiation are incapable of equalizing the thermal gradients introduced by chemical reactions. Figure 7 illustrates the magnitude of thermal gradients that may exist within the chamber. Notation for the unburned, burned, flame, and wall regions is identical to that introduced in Fig. 5. This is an idealization. For example, a single temperature is ascribed to the wall whereas, in reality, variations around the circumference exist. Combustion-chamber deposits contribute to these variations via their effect on heat transfer.

At any given point on the wall, there will also be a small periodic fluctuation in temperature arising from the cyclical nature of combustion within the chamber. Larger difference will be found between average temperatures of the pistion face, cylinder wall, and engine head. Temperatures attained, particularly in the flame and post-flame regions, depend upon engine operating variables, such as fuel–air ratio, and upon combustion.

The flame front is depicted in Figs. 5 and 7 as a convex region. Photographs of actual combustion in an engine reveal rather irregularly shaped surfaces which do, however, traverse the chamber in a more or less systematic way.[8,9] It is sometimes desirable to assign a single temperature to the gas in the combustion chamber. Before the flame is initiated and after its passage is completed, a "pseudotemperature" of this nature may be calculated from an equation of state such as the ideal gas law and measured pressure. Experimental values of temperatures in the end gas and post-flame gas have been reported.[10-15]

During the period in which the flame is traversing the chamber, the moles of burned and unburned gases as well as their respective volumes are all unknown. If the rate of charge consumption can be estimated, then the properties of each region may be estimated by assuming that homogeneous properties exist within the burned and unburned gases. Calculation of peak temperature is usually based upon the assumption that constant-volume adiabatic combustion occurs. Even though this is a rough approximation

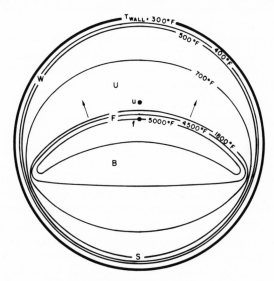

Fig. 7. Thermal gradients within combustion chamber.

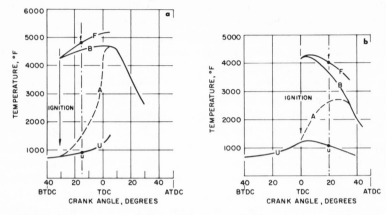

Fig. 8. Temperature variations within the combustion chamber. (a) Ignition
at 30° BTDC; (b) ignition at TDC.

it yields values for temperature which are more realistic for purposes of
chemical-kinetic and equilibrium calculations than are some of the other
alternatives.

Measurement of flame temperatures within the chamber is no simple
task. Large and very rapid variations (typically 10^5 °F/sec) occur. Relaxation
rates, particularly for vibrational energy levels, may be insufficiently rapid
for complete equilibration of molecular species under some conditions. The
observed temperature may then be dependent on the method of observation.[16]

A three-dimensional representation of the various temperatures encountered is given by Fig. 8. This simplified representation assumes no spatial
variations within each region at a given instant. More precise representation
of the temperature for each region would require five dimensions (x, y, z, t,
T) and is not justified for the purpose at hand. Curve U is the locus of temperatures for increments of mass lying ahead of the flame front. The lower case
u corresponds to that shown in Figs. 5 and 7, and represents the value of
temperature in the unburned gas at some particular point during the combustion process.

The locus of peak flame temperatures is shown as curve F. In general,
an increase in the temperature of the unburned gas prior to ignition (corresponding point on curve U) will result in an increase in peak flame temperature. However, the relationship between the two is not a simple linear one,
owing to molecular dissociation that occurs at elevated temperatures. An
increase in temperature increases dissociation and this in turn suppresses
the temperature rise. To further complicate matters, an increase in pressure
reduces dissociation and increases peak temperature.

Curve *B* represents the temperature of the burned gas. Its value may be less than the peak flame temperature as a result of thermal losses; or in some cases its value may be greater because of the compression accompanying combustion of the remainder of the charge. The latter is more likely with ignition substantially before tdc.

The broken curve *A* corresponds to a "pseudotemperature" obtained by averaging temperatures in the burned and unburned gases. This average temperature may be useful for rough heat transfer considerations. However, it is much less useful for chemical-kinetic calculations. Reaction rates are strongly temperature dependent. The extent of a reaction calculated from an average temperature may be very different than extent based on absolute point to point values. Herein lies one of the most formidable problems associated with the mathematical modeling of chemical phenomena in engines.

The temperatures in the flame, burned and unburned gases, as shown in Fig. 8a, increase monotonically with crankangle during the period of flame passage. This is not to imply universal behavior. In Fig. 8b, late ignition results in combustion during the power stroke. Temperatures for this latter case are significantly lower than for early ignition.

The temperature of the gas layer adjacent to the surface of the combustion chamber is important, for it is in this region that oxidation of the fuel is only partially completed. The temperature gradient near the wall is large as is shown in Fig. 9. Immediately adjacent to the wall is a zone where the temperature is sufficiently low that essentially no reaction of the fuel

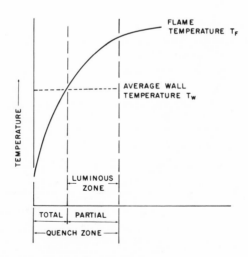

Fig. 9.   Temperature profile in the quench zone
(adapted from Ref. 17).

occurs. Further from the wall is the partial-quench zone. In this region the temperature is high enough for cool-flame and slow-oxidation reactions to occur but not sufficient for them to go to completion. This region is the source of many of the synthetic hydrocarbons observed in the exhaust. The total quench zone may extend into the deposits themselves to the extent that they are porous and can "absorb" gases as the pressure increases.

Finally there is the flame zone. Fuel and oxidant are rapidly and rather completely reacted in this zone. After the flame has passed, turbulence will transport hydrocarbons from the quench zone into the burned gases behind the flame.[18] The extent of mass transfer will depend upon the relative thickness of the thermal-quench layer and the hydrodynamic-boundary layer. This will provide reactants for pyrolytic reactions in the burned gases. These pyrolytic reactions also influence the nature of hydrocarbon species in the exhaust.

3. Residence Time

In addition to temperature and pressure which influence the rate of

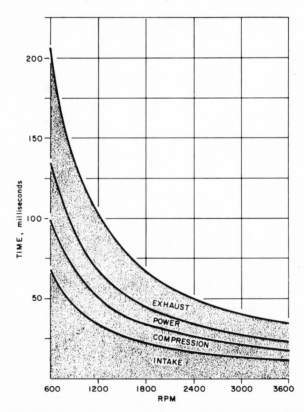

Fig. 10. Duration of engine events for a production 318 CID V–8 engine.

reaction, time is also important. In many cases, the time available limits the extent of conversion of a chemical reaction.

The time available for reactions to occur during the four events of a spark-ignition engine, that is intake, compression, power, and exhaust, is shown in Fig. 10. Basically the time available for reactions in the combustion chamber is a function of the rpm at which the engine is operated. This figure shows the maximum times which are available for reaction. In actuality the available time will be somewhat less. For example, if precombustion reactions are considered, appreciable reaction rates in the unburned gases are not obtained until late in the compression stroke when temperatures and pressures are higher than their suction values. Consequently the actual time available for precombustion reactions is somewhat less than that shown in Fig. 10 for the intake and compression events.

B. Low-Temperature Reactions

1. Studies of Motored and Fired Engines and of Rapid-Compression Machines

Reaction of hydrocarbons within the engine is not limited to combustion in flames. They can oxidize at relatively low temperatures in reactions that may or may not be accompanied by visible radiation. These are called "slow-" or "low-" temperature reactions, and if accompanied by a pale-blue luminosity characteristic of excited formaldehyde emission, they are referred to as "cool flames." Pyrolytic reactions of hydrocarbons can occur in the absence of oxygen. These reactions involve fission of bonds and/or abstraction of hydrogen with a resultant increase in the degree of unsaturation of the parent hydrocarbon. Low oxygen concentrations may even promote these pyrolytic reactions.

The existence and extent of slow oxidation and precombustion reactions within the combustion chamber are of importance for several reasons:

(1) Slow oxidation reactions occur in the quench layer not consumed by the flame, as illustrated in Figs. 5 and 7. They may continue as long as pressure and temperature are favorable. Partial oxidation reactions in the quench layer are probably responsible for many of the synthetic hydrocarbons found in the engine exhaust.

(2) When the flame does not consume the charge in the cylinder, as occurs during misfire, the total reaction products of slow oxidation and cool flames formed during the compression and expansion strokes are discharged into the exhaust system. A significant level of misfire may exist in a poorly maintained and/or adjusted engine. It may occur continuously in one or more cylinders or intermittently, and may develop only under certain conditions such as idle and deceleration where the residual fraction is large, rendering combustion more difficult.

Low rpm during idle and deceleration provide extended residence time for reactions within the combustion chamber. Intermittent misfire may be manifested as "eight cycling" in which only alternate charges in a given cylinder burn. This instability arises during marginal ignition conditions where the inert residual fraction renders incombustible the charge succeeding a combustible one.

Aside from increased fuel consumption and loss of power, an appreciable degree of misfire in an engine should not be overlooked with respect to its implications relative to air pollution. The partial oxidation products of most fuels have a greater propensity for atmospheric-photochemical reactions than do the parent hydrocarbons. Though the gas mixture accompanying a misfire is exhausted with little energy release, significant conversion of the fuel to form reactive intermediates (for example, olefins and carbonyls) may have occurred.

(3) To the extent that normal combustion is maintained, the partial oxidation products formed by the precombustion reactions will be consumed by the flame and will not appear in the combustion-chamber effluent. The low levels of hydrocarbons reported in the exhaust of knocking engines[19] indicate that partial oxidation products are consumed during knock also. This may not be true for compression ignition.[20]

Several investigators[9,21-23] have reported slow-oxidation and cool-flame reactions (CFR) in motored and fired engines. The term "motored engine" is used to describe one in which ignition does not occur due, for example, to the absence of a spark or ingestion of an air–fuel mixture exceeding the flammability limits. An external power source is used to turn over the engine. Observations have been made visually[9] and with a photo-multiplier tube.[22] In one case, gas-chromatographic analysis showed very similar reaction products for 2-methylpentane fuel reacting in low-pressure *in vitro* apparatus, a rapid-compression machine, and a motored engine, even though conditions (T, P, fuel–air ratio, and inerts) were very different.[21]

Reactions in a motored engine are often called "slow oxidation," and those in the unburned mixture ahead of the flame front in a fired engine are referred to as "precombustion reactions." Temperatures and pressures attained by the unburned gas in the fired engine exceed those in the motored engine.

The extent of the low-temperature reactions which occur may be reported in three ways. First, in terms of the percentage disappearance of fuel: second, by a description of the intermediates formed; and third, by estimating the percentage of available energy released. For example, 70% of the 2-methyl-pentane fuel disappeared before spontaneous ignition in a rapid-compression machine,[24] whereas 25% of the *iso*-octane fuel reacted in the end gas prior to flame arrival in a fired engine.[22]

In terms of intermediates formed, most of the 2-methylpentane was converted to olefins and oxygenates and the *iso*-octane was converted primarily to C_8 cyclic ethers, C_7 olefins, *di*-isobutylene, propylene, aldehydes, and ketones. In another study Fish *et al.*[25] report that the first-stage combustion of 2-methylpentane is very unselective. Sixty-five stable or moderately stable intermediates were observed. Johnson *et al.*[12] report that approximately 10% energy release occurred prior to flame-front arrival for 75–90 octane number primary reference fuels in a CFR-engine.

Two factors must be considered with respect to low-temperature oxidation studies of the type being described. First is the influence of residual gas, and second is the presence of additives that are intended to restrict the extent of these reactions (antiknocks).

A difference in composition of the reacting mixture for a fired and motored engine exists. It arises from differences in the composition of the residual gas (that which is not scavenged from the cylinder). In a motored engine large quantities of oxygen and partially-oxidized hydrocarbons are present in the residual fraction, whereas the residual fraction of a fired engine contains mostly inert gas. The effect of this residual fraction is not limited to charge dilution. Small concentrations of carbon monoxide (0.2–0.6%) present in the inert gases have been observed to accelerate the ignition of *iso*-octane/air mixtures in a rapid-compression machine.[26]

Additives are present in many commercial fuels. Organometallics such as tetraethyl lead (TEL) increase knock resistance. TEL has been shown to suppress CF activity and auto-ignition of *iso*-octane.[23] Salooja[27] reports that lead oxide, a product of TEL thermal decomposition at engine conditions, inhibits the precombustion reactions of paraffins and ethers but promotes combustion of carbonyls and hydroxyl-containing compounds. This indicates that TEL probably inhibits reactions that occur early in the reaction mechanism and prior to formation of intermediates, the combustion of which it promotes. It is also a slight promoter of combustion for refractory molecules like benzene,[28] possibly by contribution of its organic ethyl radicals to an otherwise unreactive system. Tetramethyl lead decomposes at higher temperatures and has been found effective for inhibiting auto-ignition of highly aromatic fuels.[29] This indicates that the lead-oxide particles formed during decomposition must be the right size at the exact time when precombustion reactions can be inhibited.

Trumpy *et al.*[30] have studied the kinetics of precombustion reactions for ethane in an SI engine. They were able to predict knock in agreement with experiment, based upon the assumption that it would accompany the simultaneous acceleration of temperature rise and radical and intermediate concentrations.

Ignition in a firing engine may result from three sources: the spark, a hot spot in the combustion chamber, or compression ignition. During normal combustion the spark ignites the mixture. A hot spot or compression ignition may result in either preignition or afterrunning (also called run-on). The influence of fuel structure on compression-induced after-running has been studied[31] as has the influence of the presence of small particles in the react-ants.[32] Rapid-compression-machine studies indicate that the intermediates formed by precombustion reactions are not completely destroyed when compression ignition occurs.[19]

2. In Vitro

Definitive experiments to establish the mechanism by which hydrocarbons react under the variable temperature and pressure conditions of an engine are extremely difficult to conceive and execute. It is desirable to establish the extent to which similitude of chemical behavior exists between processes within the engine and those in less complex laboratory studies.

Many early investigations of hydrocarbon oxidation were performed under conditions with an extreme excess of fuel that would not be ordinarily encountered in engine combustion. Sometimes pure oxygen was substituted for air. In addition many of these experiments were conducted at reduced pressures, often only a small fraction of an atmosphere. The choice of these conditions is understandable for *in vitro* (literally "in glassware") experiments performed without the aid of rapid and remote measuring techniques such as spectroscopic analysis.

A large excess of fuel constrains the temperature rise in two ways. First, the excess fuel is an energy sink. Second, the insufficiency of oxygen prevents complete release of enthalpy. Limiting the temperature rise increases the time required for the overall reaction through its effect on elementary reaction rates.

Reduced pressure also increases the time for reaction by increasing the elapsed time between collisions of molecules at a given temperature. However, at subatmospheric pressures a number of variables generally unimportant at higher pressures must be taken into consideration. As the time between molecular collisions increases, the probability of collision between chain carriers and the vessel walls also increases. Consequently the size of the vessel, the material of which it is constructed, and even the history of previous reactions that altered its surface structure become important experimental variables for low-pressure *in vitro* studies.

Much of chemistry believed to occur within the combustion chamber of an engine has been infered from data obtained with more controlled laboratory conditions.

a. Cool Flames. Cool flames (CF) have been a laboratory curiosity for many years. Their faint-blue luminosity, sometimes occurring periodically, has recently been attributed to emission from a very small fraction of the formaldehyde molecules present in the flame, which happen to be formed by infrequent radical–radical collisions.[16] Thus the visible radiation is more a side effect than a consequence of a major reaction path.

The conditions for occurrence of CF relative to other types of reactions are shown in Fig. 11. The boundaries depend upon composition, and a figure

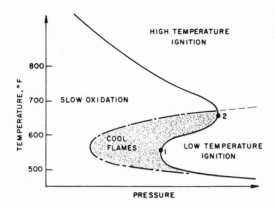

Fig. 11. Regions of hydrocarbon oxidation.

of this nature implies a constant initial composition. The solid curve passing through points 1 and 2 is the ignition limit. If a combination of temperature and pressure to the right of this line is imposed upon the system, ignition will result and be followed by a rapid rise in temperature, and possibly pressure. Oxidation can also occur to the left of the ignition limit in the regions marked "slow oxidation" and "cool flames." For initial conditions within the shaded region one or more CF will be observed.

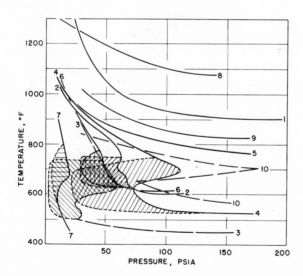

Fig. 12. Ignition and cool flame limits for oxidation of hydrocarbons in air: 1, methane[33]; 2, propane[34]; 3, n-heptane[33]; 4, i-octane[33]; 5, ethylene[35]; 6, propylene[35]; 7, propionaldehyde[34]; 8, benzene[33] 9, ethyl benzene[36]; 10, n-propyl benzene[36].

Not all hydrocarbons exhibit CF. This is evident from Fig. 12, which shows that ignition limits for methane, ethylene, and benzene indicate no CF region. They either oxidize slowly or ignite. It is evident from Fig. 12 that CF for a variety of substances oxidizing in air at near stoichiometric conditions occur over a similar temperature range. However, as molecular structure changes, there is a wide variation in the pressure limits. These pressures and temperatures are comparable to those prevailing in the unburned gases of an SI engine (Figs. 6 and 8). This, together with similar times noted earlier, increases the potential for CF reactions throughout the unburned gases in the bulk and quench layers of the combustion chamber.

b. Low-Temperature Reaction Mechanisms. The following discussion is intended to illustrate in a general way the types of low-temperature reactions that the fuel, RH ($R^\bullet$ is an alkyl radical and $H^\bullet$ a hydrogen atom), and oxygen may undergo. For greater detail, refer to Refs. 16, 37, and 38.

At low temperatures the reaction rate for direct attack of molecular oxygen upon the fuel is low because of the high activation energy (40–60 Kcal/mol) associated with this reaction.

$$RH + O_2 \longrightarrow R^\bullet + HO_2^\bullet \tag{1}$$

At best this reaction may initiate chains. Reaction of fuel molecules with a free radical such as the hydroxyl is more probable because the energy barrier is much lower (approximately 2 Kcal/mol).

$$RH + {}^\bullet OH \longrightarrow R^\bullet + H_2O \tag{2}$$

Once an alkyl radical has been formed, the addition of molecular oxygen can occur readily, for it too has an energy barrier of only a few Kcal/mol.

$$R^\bullet + O_2 \longrightarrow ROO^\bullet \tag{3}$$

The peroxy radical $ROO^\bullet$ may abstract a hydrogen from another molecule to form a peroxide, ROOH.

$$ROO^\bullet + RH \longrightarrow ROOH + R^\bullet \tag{4}$$

Alternatively the peroxy radical may undergo an internal isomerization in which a hydrogen atom is transferred from the alkyl group to the peroxy group, as in

$$ROO^\bullet \longrightarrow ROOH \tag{5}$$

In many cases a single path does not exist. Some fraction of the $ROO^\bullet$ radicals will undergo the reaction expressed by Eq. (4), while the remainder isomerize according to reaction (5).

The organic peroxide formed in reaction (4) is relatively unstable, and if temperature conditions are favorable, it will decompose homogeneously by fission of the oxygen–oxygen bond which has the lowest bond energy (42 Kcal/mol) in the molecule. Thus

$$ROOH \longrightarrow RO^\bullet + {}^\bullet OH \qquad (6)$$

This is a chain-branching reaction. The combination of Eqs. (4) and (6) constitute a degenerate chain with ${}^\bullet OH$ and ROOH as the alternate free radical and molecular intermediates.

The low activation energy for reaction (2) which occurs early in the sequence of steps that constitutes the reaction mechanism, is in part responsible for the large number of reaction products formed during the low-temperature oxidation of hydrocarbons. This is at least a partial explanation for the production of sixty-five stable intermediates during the low-temperature oxidation of 2-methylpentane as was reported earlier.[25] Any hydrogen atom of the alkyl group may be abstracted with relative ease. A variety of alkyl radicals, which differ only in the position of the unpaired electron, result. Subsequent reaction paths for these radicals may be quite different. For example, addition of oxygen at one location may be followed by an isomerization [reaction (5)] if the oxygen can bridge and form a five- or six-membered ring with a remote hydrogen. If a three- or four-membered ring is required, the high activation energy associated with bond distortion will favor the reaction expressed as Eq. (4). The pronounced lobes on the ignition curves of some hydrocarbons (for example, n-heptane in Fig. 12) have been attributed to differences in the ease of isomerization reactions with hydrogens located α, β, and γ from the peroxy group.[38]

The $RO^\bullet$ radical formed by [Eq. (6)] may undergo further decomposition to products including olefins, carbonyls (aldehydes and ketones), o-heterocyclics, and alkanes. These are all partial-oxidation products of the parent alkane. The distribution of products obtained depends strongly upon the molecular structure of the RO^- radical.

A well-known property of CF is their self inhibition, wherein the rate decreases and reaction ceases following the formation of some partial-oxidation products, even though additional oxygen is present. One explanation advanced to account for this behavior is that as temperature rises during the CF, the chain-branching step [Eq.(6)] is arrested because the thermally-sensitive peroxy radical $ROO^\bullet$ appearing earlier in the mechanism will no longer be formed. Another explanation is that partial oxidation of the parent fuel molecule forms intermediates that are more resistant to oxidation than the primary molecule itself. For example, referring to Fig. 12, propylene is observed to have a CF region requiring higher pressure conditions than propane or some higher molecular weight hydrocarbons from which it may be formed.

The phase relationship between reactants, intermediates, and products of complete oxidation is illustrated in Fig. 13. If a low-temperature oxidation is self-inhibiting, "products" typical of those at point a result. In terms of disappearance of fuel molecules, substantial reaction has occurred, whereas

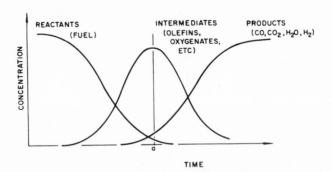

Fig. 13. Phase relationship between reactants, intermediates, and
products for oxidation of a hydrocarbon.

(in terms of appearance of products of complete oxidation, or in total energy
release) relatively little reaction has taken place.

Salloja[39] relates CF activity to methylene groups ($-CF_2-$) within the
reactant molecule. *Iso*-octane with a single methylene group is seen from
Fig. 12 to require greater minimun pressures to attain cool-flame oxidation
than *n*-heptane, which has five such groups. Methane with no methylene
groups exhibits no CF. This relationship appears contradictory to relative
carbon–hydrogen bond strengths (tertiary < secondary < primary). *Iso*-octane
with a tertiary carbon would appear more vulnerable to low-temperature
oxidation than *n*-octane, which has none. Resolution of this conflict lies in
understanding reactions that occur after the reaction designated as Eq. (6).
Initial attack at a tertiary carbon is indeed energetically easier; however,
reaction of the RO• thus formed produces a ketone that resists further
oxidation.

When initial attack is at a secondary carbon (methylene group) an
aldehyde may result from the RO⁻. The terminal hydrogen on the aldehyde
group is easily removed

$$RCHO + O^\bullet \longrightarrow RC^\bullet O + {}^\bullet OH \tag{7}$$

and oxidation may continue. The high reactivity of aldehydes is illustrated
by the relatively low pressures (shown in Fig. 12) required for a CF and for
ignition of propionaldehyde.

Thus far the nature of R• has been restricted to an alkyl (paraffinic)
group. Cases where it is either an aryl (aromatic) or allyl (olefinic) group are
also of interest. The strong influence of molecular structure on CF activity
is illustrated by comparison of the regions for the C_3 hydrocarbons, propane,
and propylene. Figure 12 indicates that the presence of a double bond shifts
the CF region toward higher pressures. This can be explained in terms of
bond and activation energies. Owing to inductive forces within the propylene

molecule, the C–H bond α to the double bond in propylene is weaker than any C–H bond in propane. Hydrogen abstraction at this point is easier when formation of a resonant stabilized allyl radical is involved.

$$CH_2 = CH - CH_3 + {}^{\bullet}OH \longrightarrow CH_2 = CH - CH_2{}^{\bullet} + H_2O \qquad (8)$$

The activation energy for this reaction is only 0.5 Kcal/mol[40] compared to 2.0 Kcal/mol for hydrogen abstraction from an alkane [see reaction (2)]. However oxygen addition to this stabilized radical is energetically more difficult than for the corresponding alkyl radical.[40] Consequently, in the low-temperature region the overall reactivity of propylene is less than propane. At higher temperatures where more energetic species are present, olefins are more reactive. This is evident from the intersection of the high-temperature ignition curves for these two compounds at 940°F, as shown in Fig. 12.

A similar relationship is found for C–H bond energies in benzene and toluene. Though removal of a hydrogen from the methyl group of toluene is easy, oxygen addition to the resultant resonance-stabilized aryl radical is difficult. The aromatics, benzene through ethylbenzene, exhibit no CF. The lower ignition limit for *n*-propyl benzene, Fig. 12, probably results from reaction at a point on the side chain removed from the aromatic ring.

Interactions may further modify CF activity for a mixture of hydrocarbons. The addition of an olefin to a paraffin has been found to inhibit slow combustion of the paraffin.[14] Inhibition probably results when chain carriers preferentially abstract the weakly bound hydrogen α to the double bond within the olefin and form a resonance-stabilized allyl radical instead of a more reactive alkyl radical.

The preceding discussion describes in a general way the types of reactions studied *in vitro* and also believed to be of significance in the early stages of hydrocarbon oxidation within the unburned gas and quench regions of the combustion chamber.

3. Discussion

Consider the variation in reaction time associated with hydrocarbon-oxidation reactions occurring under greatly different experimental conditions. Figure 14 shows the variation in the first induction period τ_1 for a wide range of reaction pressures. The first induction period, τ_1, plotted on the ordinate is the delay prior to the appearance of a cool flame. It is chosen here simply as a quantity representative of reaction time and is not the time required for completion of the reaction. Shaded areas highlight ranges of pressure and time that usually prevail for *in vitro* experiments, the engine combustion chamber, and the exhaust system. The upper and lower extent of these areas for the combustion chamber and exhaust are typical of the ranges normally encountered.

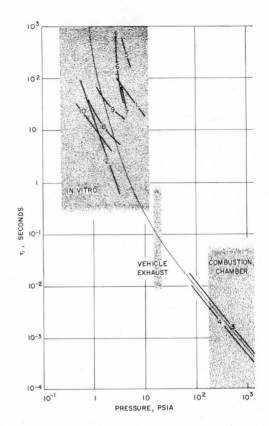

Fig. 14.   Characteristic reaction times for cool flames
over a wide range of pressures and temperatures.

The coincidence of these regions with experimentally observed values of induction periods for a variety of hydrocarbons, as shown in Fig. 14, is an indication that residence times and pressures normally encountered within both the combustion chamber and the exhaust are favorable for the occurrence of low-temperature hydrocarbon reactions (see Table I).

Just as similarities can be found, differences exist also. There is a fundamental difference in the temperature and pressure conditions prevailing during CF or slow oxidation in an engine and those for the *in vitro* experiments described earlier. For example, in glassware, the CF is largely responsible for the *P-T-t* path followed by the reacting gases; and, as noted earlier, self-inhibition is sometimes observed. In the combustion chamber of the engine, the *P-T-t* path is largely determined by factors other than the reactions themselves. Consequently, the extent of reaction attained may differ for the two. Whether this extent is increased or decreased for the fired engine relative

Table I. Reaction Times versus Pressures and Temperatures for Curves in Fig. 14

		Concentration (%)			Vessel		
Curve	Hydrocarbon	Hydrocarbon	Oxygen	Nitrogen	Type	Temp.,°F	Reference
1	2–Methyl pentane	66.6	33.3	0	*in vitro*	491	(21)
2	2–Methyl pentane	66.6	33.3	0	*in vitro*	547	(21)
3	2–Methyl pentane	0.0334<fuel/air<0.0526		79	Rapid compression machine	810	(21)
4	2–Methyl pentane	0.0334<fuel/air<0.0526		79	''	873	(21)
5	*n*-Butane	50	50	0	*in vitro*	554	(41)
6	*n*-Butane	50	50	0	*in vitro*	590	(41)
7	*n*-Pentane	50	50	0	*in vitro*	527	(41)
8	*n*-Pentane	50	50	0	*in vitro*	572	(41)
9	*n*-Hexane	15	85	0	*in vitro*	536	(41)
10	*n*-Hexane	15	85	0	*in vitro*	572	(41)

to either a motored one or an *in vitro* experiment will depend upon temperature and pressure conditions, available time, and composition of the reacting mixture.

The critical compression ratio (ratio above which knock occurs under fixed engine conditions) for several hydrocarbons is shown in Fig. 15. An additional similarity between laboratory *in vitro* experiments and combustion in an engine can be seen by comparing the information in this figure with the tendency of a hydrocarbon to knock in an engine. Methane and benzene, both of which do not exhibit cool flames, have relativly high knock resistances. Normal heptane which knocks easily has a cool-flame region that extends to low pressures, as can be seen from Fig. 12.

The higher the critical compression ratio is for a species the more knock resistant it will be when burned in a spark-ignition engine. Recently there has been considerable concern about the lead compounds which are present as anti-knock agents in gasoline and their derivitives which are emitted in the exhaust. Figures 12 and 15 illustrate some of the alternatives which are available to recover the octane lost by removal of lead from gasoline. One alternative is to increase the aromatic content of the gasoline. Note that aromatic species are resistant to cool-flame oxidation and have high critical compression ratios. It has been argued that this alternative is not particularly desirable because of the increased polynuclear aromatic content of the exhaust which is associated with the greater aromaticity of the inducted fuel. A second alternative which does not result in increased exhaust polynuclear aromatic content is the use of higher concentrations of branched chain hydro-

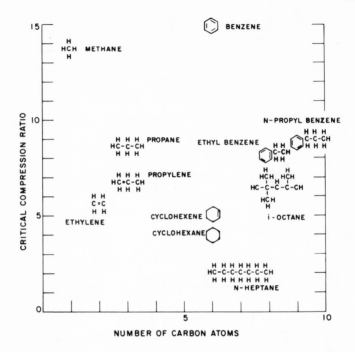

Fig. 15. Influence of molecular structure on critical compression ratio
in a spark-ignition engine (data from Refs. 42 and 43).

carbons in the fuel. These branch-chain molecules also have relativly high
critical compression ratios.

C. High-Temperature Reactions in an Engine

1. Flame Reactions

Mechanisms for oxidation of hydrocarbons at low and high temperatures
differ. The complex molecules and intermediates, which are stable at the
moderate temperatures that exist in CF and slow oxidations, dissociate at
higher temperatures.

The mechanism by which fuel molecules are converted to products in
a flame does not involve a series of peroxidic, olefinic, and aldehydic intermediates as is the case for low-temperature oxidations. Rather, dissociation
and other reactions with high activation energies occur readily. Conversion
to carbon dioxide, carbon monoxide, water, and hydrogen occurs more
directly. With equilibrium on the dissociative side the concentration of
radicals is high and radical–radical reactions become more probable than

at lower temperatures. Newhall[44] presents a summary of kinetic data for a number of bimolecular and termolecular reactions important at high temperatures.

As a consequence of the extensive fragmentation at high temperatures, detailed molecular structure of the fuel is not so critical as at low temperatures. At high temperatures there is little question of whether a particular fuel molecule will or will not react, nor for that matter is there question of the extent of reaction.

The physical relationship of fuel and oxidant also influences the nature of high-temperature combustion. When the two are uniformly distributed throughout the reacting mixture, the combustion reaction is referred to as a "premixed flame." Conversely, a diffusion flame exists when fuel and oxidant are separated and reaction occurs at their interface.

For the case of normal combustion of a premixed gas, a self-sustaining flame propagates through the mixture. The frontal nature of the reaction zone results from strong thermal and diffusion coupling.[45] Diffusion coupling involves the transport of active species from the high-temperature flame zone into the unburned gases, where they may accelerate precombustion reactions. Diffusion of active species is promoted by the large concentration and temperature gradients. The propagation rate of the flame is dependent on coupling between chemical reactions, thermal conduction, and diffusion of active species into the unburned gas. Control of peak flame temperature in a premixed flame can be accomplished by selection of the fuel–air ratio.

When the fuel is not completely vaporized but exists as small droplets a different situation exists. A diffusion flame is stabilized within an envelope some distance from the droplet surface. Within this envelope the fuel–air ratio is favorable for combustion. Combustion rates are strongly influenced by physical processes such as fuel-evaporation rate, diffusion of oxidant to the reaction zone, and diffusion of reaction products away from this zone. Reaction occurs over a range of equivalence ratios. Energy release in the flame zone is rapid, and quantities of thermal energy in excess of bond energies may be transferred to gaseous fuel molecules near the droplet surface. Pyrolytic decomposition results. Condensation of the fragments produces high molecular weight species that oxidize slowly in the flame and are a potential source of particulates in the exhaust gas.

Conflicting results have been reported with regard to the production of polycyclic aromatic hydrocarbons by some flames.[46,47] Acetylene, a precursor in the formation of these compounds, is also a product of pyrolytic reactions of larger hydrocarbon molecules.[48] Since some polycyclic aromatics have been reported to exhibit carcinogenic activity, it is important that a better understanding be obtained regarding the relative quantities of these substances produced by different fuels reacting in both premixed and diffusion flames.

2. Post-Flame Reactions

Reactions in the post-flame gases (region B, Fig. 5) include recombination and pyrolysis. Recombinations of reactions of simple species like $H^{\bullet}$, $N^{\bullet}$, $O^{\bullet}$, and $^{\bullet}CH_3$ involve a three-body collision. For example,

$$O^{\bullet} + O^{\bullet} + M \longrightarrow O_2 + M \qquad (9)$$

$$^{\bullet}CH_3 + {}^{\bullet}CH_3 + M \longrightarrow C_2H_6 + M \qquad (10)$$

M is any third body that acts as an energy sink. Without it the energy contained in the product molecules would be sufficient to cause them to dissociate again. These termination or recombination reactions have no activation energies; however, their overall rates are low, owing to two factors. First, the concentration of free radicals is generally low, reducing the probability of radical–radical collisions. Second, the statistical probability of a termolecular collision is low compared with the bimolecular ones associated with propagation reactions. Furthermore, decreasing pressures in the post-flame gases further reduce collision frequency. Kinetic calculations indicate that a large excess of free radicals and atomic species persist in the post-flame region.[44]

Pyrolytic reactions of hydrocarbons may also occur in the high-temperature post-flame gases where the oxygen concentration is low. Most of the hydrocarbons that appear in the post-flame region probably originate in the quench layer and are transported into the post-flame region by turbulence within the chamber. According to Ninomiya and Golovoy[48] who have studied the effects of varying fuel–air ratios on hydrocarbon emissions of an SI engine, relative quantities of hydrocarbons and oxygenates in the exhaust can be related to the combustion-chamber temperature, which is at maximum near stoichiometric, and the oxygen availability, which decreases with increasing fuel–air ratio.

For example, an increase in gas temperature increases the rate of unimolecular-decomposition reactions, which characteristically have high activation energies. An increase in oxygen concentration promotes oxidation of both reactants and products of their unimolecular reactions. Pyrolysis of *n*-pentane fuel has been studied in a motored engine by replacing the air with nitrogen.[49] Though temperatures and pressures were significantly lower than for the fired case, significant quantities of olefins were reported and their appearance was explained in terms of cracking reactions.

3. Nitric Oxide

High-temperature equilibrium predicts the occurrence of reasonably large quantities of nitric oxide at temperatures typical of those encountered in the flame of a spark-ignition engine. The higher oxides, such as nitrogen dioxide, are present in much lower quantities.

Early studies indicated that the quantities of nitric oxide observed in the exhaust of a spark-ignition engine correlated reasonably well with those which were predicted for constant-volume adiabatic combustion of hydrocarbon fuels. However there are three factors which make it difficult to distinguish whether or not this correlation is fortuitous or an indication that equilibrium is actually obtained at peak flame temperatures in the chamber. First, there is the extreme temperature sensitivity of nitric oxide equilibrium. Small variations of temperature result in relatively large variations in the concentration of nitric oxide. Second, is the realization that no single flame temperature prevails throughout the chamber. Third, there is a difficulty in measuring even a single temperature within the chamber, let alone the variations that occur from point to point. Not withstanding these uncertainties, in the early investigations it was usually assumed that the formation kinetics of nitric oxide were rapid, that equilibrium concentrations were obtained at temperatures corresponding to the peak flame temperature, and that the decomposition of nitric oxide was essentially frozen.

The "frozen equilibrium" just described was assumed to be due to two factors. First, the temperatures in post-flame gases fall rapidly during the expansion stroke. Second, the oxygen atom concentration is relatively low in the post-flame gases. The concentration of oxygen atoms enters into the Zeldovich mechanism for formation and decomposition of nitric oxide.

$$N^{\bullet} + O_2 \longrightarrow NO + O^{\bullet} \tag{11}$$

$$O^{\bullet} + N_2 \longrightarrow NO + N^{\bullet} \tag{12}$$

Instead of assuming no decomposition of nitric oxide in the post-flame gases, Newhall[44] performed a theoretical analysis of chemical kinetics in the internal-combustion engine expansion process. It was found that none of the elementary reactions directly involving nitric oxide were sufficiently rapid to effect an appreciable decomposition of nitric oxide once it was formed. The assumption that equilibrium concentrations of nitric oxide are attained in a spark-ignition engine was questioned by Ezyat et al.[50] Recent studies[51-53] have comfirmed that equilibrium is not obtained with respect to nitrogen oxide in the flame. The time available is too short and the formation kinetics are too slow. Nitric oxide formation continues into the post-flame gases.

Mathematical models which incorporate the formation kinetics of nitric oxide predict exhaust concentrations reasonably well for lean and stoichiometric fuel mixtures. Predictions for rich mixtures are approximately an order of magnitude too low. Possible reasons for this error may be; an incomplete elementary reaction scheme for nitric-oxide formation in rich mixtures, inaccurate rate constants, or inaccurate prediction of gas temperatures and cyclic variations in the combustion chamber.

In a recent study by Muzio et al.[54] it has been shown that a temperature difference on the order of 600 °C can be established in the burned gases within the combustion chamber. The first part of the mixture burned having the highest temperature. This difference in temperature can lead to a corresponding variation of nitric oxide formation within the combustion chamber. At a given engine operating condition this can easily lead to an order of magnitude variation in the concentration of nitric oxide from the first to the last part of the mixture which is burned.

Nitric oxide production in the combustion chamber can be reduced by operation at either extremely rich or lean conditions, by exhaust recirculation or water injection, and by increasing valve overlap. Recent investigations[55–58] have confirmed that these methods have a common mechanism on an atomic-molecular level. All reduce molal energy density of the combustion gases at peak flame conditions by distributing the energy released over a relatively larger mass of gas.

Lean operation, exhaust recirculation, water injection, and increased valve overlap all increase the inert fraction in the combustion chamber. The change in oxygen–nitrogen ratio[57] associated with an increase in inerts does not shift equilibrium concentrations sufficiently to account for reductions in nitric oxide. Rather, lower concentrations result from reduction of peak flame temperatures. The carbon dioxide and water present in recirculated gas dissociate at flame temperatures and increase the apparent heat capacity of the reacting mixture at precisely the time when this is most beneficial. Vaporization of the water injected as a liquid reduces the temperatures in the unburned gases, curve U in Fig. 5. This has a beneficial effect on knock but can hardly be expected to aid ignition at marginal conditions.

A further possibility for reducing nitric-oxide formation is control of the fuel composition. High hydrogen–carbon ratios in the fuel molecule result in low peak flame temperatures and correspondingly low concentrations of nitric oxide in the exhaust gases. Experimental studies confirm this relationship between hydrogen–carbon ratio for the fuel and nitric oxide levels observed in the exhaust.[59,60] A progressive change from paraffinicity to aromaticity is accompanied by an increase in nitric oxide emissions.

4. Carbon Monoxide

Hyperequilibrium quantities of carbon monoxide are observed in the exhaust gas of a spark-ignition engine. The elementary reaction,

$$CO + {}^{\bullet}OH \longrightarrow CO_2 + H^{\bullet} \tag{13}$$

thought to be responsible for carbon monoxide oxidation has been shown to be rapid enough to provide for equilibrium in the post-flame gases.[44] This discrepancy between kinetics and observed concentrations is related to the occurrence of other simultaneous reactions. Recombination of hydrogen is termolecular and does not occur rapidly enough for equilibration of this species in the post-flame gases. Hyperequilibrium concentrations of hydrogen shift the equilibrium for the reaction of Eq. (13) to the left, resulting in large concentrations of carbon monoxide in the exhaust.

Combustion studies of propane–air at atmospheric pressure indicate that minimum production of both nitric oxide and carbon monoxide can be obtained by quenching the post-flame gases fast enough to prevent the formation of large amounts of nitric oxide, but slowly enough to provide for equilibration of carbon monoxide.[61] Regulation of cooling rates by changes in rpm does not provide enough latitude to be an effective control technique.

VI. EXHAUST SYSTEM

A. Reactor Conditions

If conditions are favorable, reactions of species exhausted from the combustion chamber will continue in the exhaust ports, manifold, and even in the tailpipe as shown in Fig. 5. The effluent from the $(n-1)$st cycle continues to react during the intake, compression, combustion, and expansion events of the nth cycle in region E of Fig. 5. The nature of reactions that may occur in the exhaust system is influenced by the amounts of carbon monoxide, hydrogen, and partially oxidized hydrocarbons exhausted from the combustion chamber. The extent of any reactions will depend on temperature distribution within the exhaust and upon concentrations of oxidizing or reducing agents.

1. Temperature

The temperature distribution within the exhaust system depends upon the exhaust inlet temperature, thermal effects accompanying any chemical reactions in the exhaust gas, and upon thermal transfers or losses. The first of these factors, inlet temperature, depends upon engine design and operation. For example, decreased compression ratio, retarded-spark timing, and operation at near stoichiometric conditions all increase inlet temperatures.

Typical exhaust manifold temperatures for a production engine are shown in Fig. 16. The temperatures attained depend upon engine and manifold design. Recent efforts have been directed toward increasing these temperatures to promote oxidation reactions in the exhaust system. The temperatures shown are for steady-state engine operation. When the engine is operated in

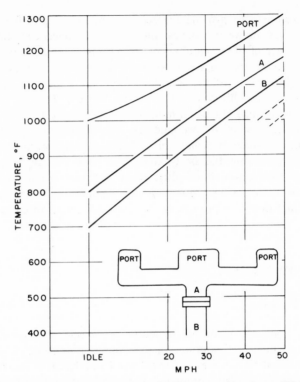

Fig. 16. Typical exhaust manifold temperatures for a 318 CID engine with CAS modifications; point *A* lies 16 inches on the average downstream from the valves, and *B* is 4 inches downstream from *A*.

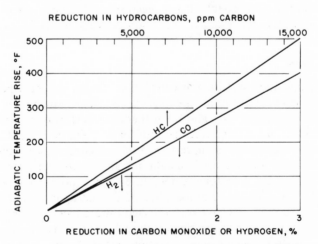

Fig. 17. Adiabatic temperature rises accompanying oxidation of combustible in the exhaust system.

transient fashion, the heat capacity of the exhaust system will introduce thermal lags. Consequently, gas temperatures will vary considerably from those shown. Internal thermal transfer of energy as well as losses to the ambient will depend upon gas-flow rates, the physical properties, and configuration of the exhaust system.

The oxidation of both carbon monoxide and hydrogen liberates considerable thermal energy. The adiabatic temperature rise accompanying oxidation of these combustibles is shown in Fig. 17. Under normal conditions the contribution due to oxidation of hydrocarbons is low compared with the other two. Though not usually reported, for it is not considered a pollutant, hydrogen is present and must be included in any calculations of thermal behavior.

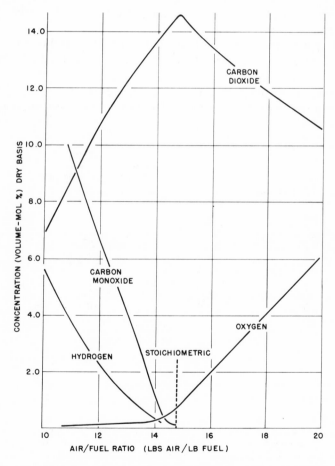

Fig. 18. Emissions of carbon dioxide, carbon monoxide, and hydrogen from a spark-ignition engine.

High-exhaust gas temperatures can be achieved by utilizing a rich air–fuel ratio in the combustion chamber to produce large amounts of combustibles. Figure 18 illustrates the concentrations of hydrogen and carbon monoxide which are obtained by varying the air–fuel ratio. Secondary air can be added and these combustibles can then be oxidized in the exhaust system.

2. Residence Time

The factors that determine residence time in the exhaust system are different from those in the combustion system where time is a function of rpm. In the exhaust system, residence time is essentially independent of rpm at constant load (neglecting flow-rate variations to compensate for friction differences), and is dependent on system volume, flow rate, and temperature of the gas. Exhaust-flow rates can be computed from air-flow rates by correction for molecular changes.

Average residence times for a standard manifold (62 in.³ volume) and for a typical exhaust manifold reactor (200 in.³ volume) have been calculated and are shown in Fig. 19.

Actually, a distribution in residence times exists, for some of the gas will follow a relatively short path from the reactor or manifold inlet to outlet

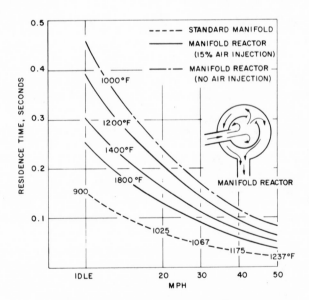

Fig. 19.   Steady-state residence times for 318 CID engine equipped with a standard manifold and with an exhaust-manifold reactor.

and have a residence time less than average. Other elements of the gas will flow into stagnant regions and have a residence time greater than the average. The broken line in Fig. 19 is for the standard manifold. Temperatures used in the calculation were calculated by assuming a temperature intermediate between values at the port and at point A in Fig. 16. Curves for the reactor with 15% air injection (based on engine intake air) operating at average temperatures of 1000, 1200, 1400, and 1800°F are also given in Fig. 19.

During unsteady-state operation of the engine, particularly for accelerations, the flow rate into the engine may be much greater than for steady-state. Other factors being constant, the residence time is inversely porportional to flow rate. This is an important consideration with exhaust systems, which must effect an appreciable reaction of some component that is produced in large quantities during acceleration because at the very time when a large quantity of the reactant is being produced, the time available for reaction is decreased.

B. Thermal Reactions in the Exhaust System

The nature of the reaction which hydrocarbons entering the exhaust system will undergo depends primarily upon the temperature and the availability of oxygen. If the concentration of oxygen in the exhaust system is zero, or near zero, and the temperature is sufficiently high, then the hydrocarbons will undergo pyrolytic reactions. This type of reaction may convert paraffinic substances to olefins and crack long-chain molecules to form shorter ones. Two recent studies[62,63] have confirmed that the reaction of certain individual hydrocarbons could result in the production of less desirable hydrocarbons from the viewpoint of smog-forming potential in the atmosphere.

When there is sufficient oxygen but the exhaust gas temperature is lower, the hydrocarbon species entering the exhaust system may be partially oxidized to produce intermediates such as the aldehydes. This type of situation can be illustrated by referring to Fig. 13. When the reaction temperature or the residence time in the exhaust system is too short for complete oxidation, the conversion may be frozen at point a. This results in a high conversion of reactant molecules to partially oxidized species which are subsequently exhausted to the atmosphere. The emission of aldehydes by spark-ignition engines is a subject of considerable current interest. Stahl[64] presents a summary of aldehyde emissions from mobile sources. High-temperature thermal oxidation reactions in the exhaust will reduce the most reactive species (refer to Fig. 12) proportionately more than unreactive species such as methane. Difficulties develop when the temperature and/or oxygen concentration is not sufficient for complete conversion in the time available.

C. Catalytic Reactions

Heterogeneous reactions may be promoted by the incorporation of a suitable catalyst and reactor in the exhaust system. Heterogeneous reactions of interest include the oxidation of carbon monoxide,

$$2CO + O_2 \longrightarrow 2CO_2 \qquad (14)$$

as well as the oxidation of hydrocarbons.

Since both nitric oxide and carbon monoxide are present in the exhaust, the heterogeneous reduction of nitric oxide by carbon monoxide is also possible:

$$2NO + 2CO \longrightarrow N_2 + 2CO_2 \qquad (15)$$

Under some conditions the reduction of nitric oxide may not be complete as shown above; instead, partial reduction to nitrous oxide may occur:

$$2NO + CO \longrightarrow N_2O + CO_2 \qquad (16)$$

Hydrogen, which is also present in exhaust gas, reduces nitric oxide:

$$2NO + 2H_2 \longrightarrow N_2 + 2H_2O \qquad (17)$$

In the case of hydrogen, partial reduction results in the formation of ammonia:

$$2NO + 5H_2 \longrightarrow 2NH_3 + 2H_2O \qquad (18)$$

All of the reactions shown in Eqs.(14)–(18) are thermodynamically feasible at temperatures which prevail in the exhaust of a spark-ignition engine.

It would be ideal if the oxidation of carbon monoxide and the complete reduction of nitric oxide could be achieved simultaneously in a single catalytic reactor. This arrangement is shown schematically in the upper half of Fig.

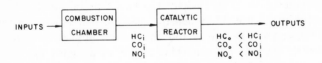

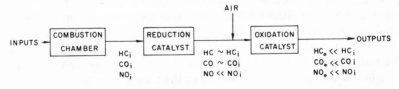

Fig. 20.   Single- and dual-bed catalytic exhaust reactors.

20. In practice, if the proper catalyst is chosen it is possible to decrease the concentration of hydrocarbons, carbon monoxide, and nitric oxide in a single reactor. However, this requires extremely close control of the input conditions to the combustion chamber, particularly of the air–fuel ratio. If there is too high a concentration of oxygen in the exhaust gases, reduction of nitric oxide will not occur. If there is too little oxygen then oxidation of the carbon monoxide and hydrocarbons is not stoichiometrically possible.

A more complex approach is to separate the catalytic oxidation and reduction process as shown in the lower half of Fig. 20. Close control of the air–fuel ratio is less important in this case. All that is necessary is that combustion in the engine be rich, for this will insure reducing conditions in the first reactor. Secondary air is added between the two reactors and catalytic oxidation of the excess carbon monoxide and hydrocarbons occurs in the second reactor. One of the difficulties with this dual reactor approach is that partial reduction of nitric oxide by hydrogen may occur in the first reactor. The ammonia produced by this reaction may then be reoxidized in the second reactor:

$$2NH_3 + {}^5/_2O_2 \longrightarrow 2NO + 3H_2O \tag{19}$$

This reoxidation reaction in the second reactor reduces the net decrease in nitric oxide realized by the combination of the two reactors.

VII. SYSTEMS APPROACH TO REACTIONS IN AN ENGINE

In this section the engine will be viewed as a "combustion system." Relationships among the various chemical phenomena described previously will be explored.

Figure 21 illustrates two simple types of relationships. A sequential process is shown in Fig. 21a. Products of the combustion-chamber reactions constitute the reactants for the exhaust system. A causal relationship exists in that a change in output stream from the combustion chamber will influence processes occurring in the exhaust system. This type of relationship is fundamental to engine-modification systems. For example, spark retard at idle can be utilized to promote reactions downstream. Another illustration is found in the combination of an engine- and exhaust-manifold reactor, a device designed to promote oxidation reactions in the exhaust. Rich operation of the engine provides the combustibles so important to sustenance of the elevated temperature in the reactor, which is necessary to obtain large reductions of both carbon monoxide and hydrocarbons.

Figure 21b illustrates a second important type of relationship that may exist within an engine. The simple causal relationship described above may no longer be applied, for now not only are there upstream reactions that

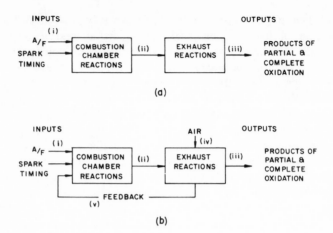

Fig. 21. Representation of the spark-ignition engine as a
combustion system.

influence those downstream, but also a feedback wherein changes downstream
affect the nature of the processes upstream. The nature of this feedback loop
may be thermal, as it is for the intake manifold stove where heat from the
exhaust is utilized to vaporize fuel in the intake. It may also arise from flow
reversal. For example, flow reversal during the valve overlap period may
transport some of the air injected into the exhaust system back into the
combustion chamber. Reactions in the chamber are altered because this
additional air decreases the fuel–air ratio relative to the case where only
exhaust products flow back into the chamber. Combustion at leaner conditions reduces the input of combustibles to the exhaust and alters reactions
therein. Because of the feedback that exists, the relative contributions of the
combustion chamber and the exhaust system cannot be evaluated by observation of the composition at point iii when the air is first injected into the
exhaust and then shut off.

To determine relative contributions of component parts of a complex
system such as Fig. 21b requires the facility to measure properties at the
intermediate location, point ii, without significantly perturbing the overall
process. Rapid fluctuations, spatial variation, and nonrepeatability from cycle
to cycle complicate measurements of temperature, pressure, and composition
within the combustion system. Timed sampling techniques,[14,65,66] expansion and dilution quenching,[16] and special sampling probes[48] have
all been utilized in efforts to obtain representative samples of gas in the
combustion chamber and exhaust.

Consider a system consisting of an engine coupled to an exhaust-manifold reactor with air injection at the ports. This system is capable of lowering
the emissions of all three pollutants. Rich combustion in the engine reduces

flame temperature and thereby the production of nitric oxide. Though relatively large amounts of carbon monoxide and hydrocarbons are produced in the combustion chamber, air injection into the exhaust ports upstream of the manifold reactor reduces these pollutants before they are exhausted to the atmosphere. If the air–fuel ratio is shifted toward the lean side of stoichiometric, nitric oxide production in the combustion chamber will increase and less carbon monoxide will be produced. This can be seen in Fig. 22 where the effect of changing the air–fuel ratio from 14 to 13.5 is shown by the shaded bands. Since less carbon monoxide enters the reactor, less thermal energy will be evolved and the elevated temperature necessary for oxidation of the carbon monoxide and hydrocarbons will not be maintained. Thus, the overall result of changing the value of one input to one component of the system can influence the performance of other components. In this case an excursion of air–fuel ratio in the combustion chamber may result in an increase in tailpipe emissions of all three pollutants. This discussion illustrates why maintenance of prescribed operating conditions of advanced emission-control systems is important.

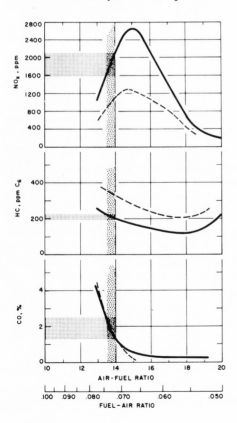

Fig. 22. Emission from a 318 CID engine: solid line – 2000 rpm, 38° BTDC, and 50 mph road load; broken line – 1200 rpm, 10° BTDC, and 30 mph road load.

VIII. SUMMARY

The spark-ignition engine may be thought of as a chemical reactor in which a variety of reactions may occur. In spite of our relatively incomplete knowledge and understanding of the many reactions which occur, some conclusions which appear valid, based on available information, are presented below.

1. A variety of chemical phenomena occur in a spark-ignition engine. Experimental studies conducted with both simulated and actual engines indicate that a variety of chemical reactions — including low-temperature oxidation, cool flames, pyrolysis, and high-temperature flames — are probable in the combustion chamber and exhaust system. Laboratory investigations of single component fuels reacting in steady state and batch reactors have elucidated reaction kinetics, mechanisms, intermediate species, and products. The influence of engine variables such as air–fuel ratio, spark advance, and load on the physical parameters of temperature, pressure, time, and composition, determine the relative importance of these various reactions.

2. Arrested chemical processes account for many undesirable species appearing in the exhaust gases. The majority of the hydrocarbons observed in the exhaust have their origin in an air–fuel mixture that is too rich, too lean, or too cool for complete oxidation to occur in the time allotted during passage through the engine. Carbon monoxide arises from the presence of too little oxygen for complete conversion to carbon dioxide, and from insufficiently rapid kinetics for equilibration of all atomic species during expansion of the post-flame gases. The hyperequilibrium concentration of nitric oxide found in the exhaust is a consequence of low oxygen concentrations and rapid decrease in temperature of the post-flame gases.

3. Chemical reactions within an engine do not occur independently. The nature and extent of chemical reactions at one location influence processes that occur at some other location through product-reactant and feedback relationships among the various components. This systems approach to chemical phenomena is a basic principle inherent in many current and future emissions-control systems, including engine modifications, air injection, and exhaust reactors.

4. Development of mathematical models for prediction of exhaust composition is hindered by the lack of absolute measurements of physical and chemical parameters within an engine. Chemical behavior on an atomic-molecular scale is responsive to local values of pressure, temperature, density, and composition. Most measured values of these quantities represent average properties over a much larger region of space or period of time. Average properties may differ significantly from point-to-point properties in an engine that is neither a steady-state nor a steady-flow reactor. Quantitative, numerical,

or analog computation of reaction rates and (ultimately) of exhaust composition requires a knowledge of absolute point-to-point values of physical parameters rather than of only average values.

5. Fuel composition influences exhaust composition. Significant variations in emission compositions have been noted for engine combustion of pure fuels or simple blends of hydrocarbons. It is possible to relate these differences to reactions within the engine. This is true for both nitric oxide concentrations and the distribution of hydrocarbons and oxygenated species observed in the exhaust.

6. Abnormal engine operation also influences exhaust composition. Chemical considerations predict that misfire will result in the formation of relatively high concentrations of photochemically reactive, partial oxidation products. This consideration, along with the sensitivity of nitric oxide concentration to variables such as spark advance and air–fuel ratio, plus the product–feedstream relationship of various components of advanced emission-control systems all emphasize the importance of proper engine maintenance.

ACKNOWLEDGMENTS

The author is indebted to Bryan H. Suits and Garnet K. Ross for aid in preparation of this manuscript.

REFERENCES

1. Edwards, J.B., and Teague, D.M., Unraveling the chemical phenomena occurring in spark ignition engines, Paper 700489 presented to the *SAE Mid-Year Meeting*, Detroit, May 1970.
2. Moran, J.B., Effect of fuel additives on the chemical and physical characteristics of particulate emissions in automotive exhaust, *National Technical Information Service*, Report PB–196–783, July 1970.
3. Patterson, D.J., Cylinder pressure variations, a fundamental combustion problem, SAE Paper 660129 presented at the *Automotive Engineering Congress*, Detroit, Mich., January 1966.
4. Peters, B.D., and Borman, G.L., Cyclic variations and average burning rates in a SI engine" SAE Paper 700064, presented at the *Automotive Engineering Congress*, Detroit, Mich., January 1970.
5. Hansel, J.G., A turbulent combustion model of cycle-to-cycle combustion variations in spark ignition engines, *Combustion Sci. Technol.* **2**, 223–225, 1970.
6. Barton, R.K., Lestz, S.S., and Meyer, W.E., An empirical model for correlating cycle-by-cycle cylinder gas motion and combustion variations of a spark ignition engine, Paper 710153 presented to the *SAE Automotive Engineering Congress*, Detroit, January 1971.
7. Hansel, J.G., Lean automotive engine operation — Hydrocarbon exhaust emissions and combustion characteristics, Paper 710164 presented to the *SAE Automotive Engineering Congress*, Detroit, January 1971.

8. Ball, G.A., Photographic studies of cool flames and knock in an engine, presented at the *Fifth Symp. (Internat.) Combustion,* Pittsburgh, Pa., August 30 to September 3, 1954. Reinhold Co., New York, 1955, pp. 366–372.

9. Wentworth, J.T., and Daniel, W.A., Flame photographs of light-load combustion point the way to reduction of hydrocarbons in exhausts, SAE Paper 425, presented at the *SAE Annual Meeting,* Detroit, Mich., January, 1965; *SAE Tech. Prog. Ser.* **6,** 121–136 (1964).

10. Chen, S.K., Beck, N.J., Uyehara, O.A., and Myers, P.S., Compression and end gas temperatures from iodine absorption spectra, *SAE Trans.* 503–526 (1954).

11. Wu, P.C.K., and Livengood, J.C., Data from sound velocity and absorption test compared, *SAE Trans.* **62,** 525–526 (1954).

12. Johnson, J.H., Myers, P.S. and Uyehara, O.A., End-gas temperatures, pressures, reaction rates and knock, *SAE Trans.* **74,** 748–768 (1966).

13. Myers, P.S. and Uyehara, O.A., Accuracy and representation of results obtained with an infrared pyrometer measuring compression temperature, *Institution of Mechanical Engineers, Proceedings* **180,** Part 3G, 95–103 (1965–6).

14. Bradow, R.L., and Alperstein, M., Analytical investigations of *iso*-octane and Di-isobutylene slow combustion in an Otto-cycle engine, *Combustion and Flame* **11,** 26–34 (1967).

15. Myers, P.S., and Uyehara, O.A., Fuel-engine research in universities, *SAE 1967 Horning Memorial Lecture* SP-340 presented at the Mid-Year Meeting, Detroit, Mich., May, 1968.

16. Minkoff, G.J., and Tipper, C.F.H., *Chemisty of Combustion Reactions,* Butterworths, London, 1962.

17. El-Mawala, A.G., and Mirsky, W., Hydrocarbons in the partial quench zone of flames: An approach to the study of the flame quenching process, SAE Paper 660112, presented at the *Automotive Engineering Congress,* Detroit, Mich., January, 1966.

18. Daniel, W.A., Why engine variables affect exhaust hydrocarbon emission, SAE Paper 700108, presented at the *Automotive Engineering Congress,* Detroit, Mich., January 1970.

19. Duke, L.C., Lestz, S.S., and Meyer, W.E., The relation between knock and exhaust emissions of a spark ignition engine, SAE Paper 70062 presented at the *Automotive Engineering Congress,* Detroit, Mich., January 1970.

20. Affleck, W.S., and Fish, A., Knock: Flame acceleration or spontaneous ignition, *Combustion and Flame* **12,** 244–252 (1968).

21. Afflect, W.S., and Fish, A., Two stage ignition under engine conditions parallels that at low pressure, presented at the *Eleventh Symposium (International) on Combustion,* Berkeley, Calif., August 14–20, 1966, The Combustion Institute, Pittsburgh, 1967, pp. 1015–1024.

22. Alperstein, M., and Bradow, R.L., Exhaust emissions related to engine combustion reactions, SAE paper 660781 presented at the *Fuels and Lubricants Meeting,* Houston, November, 1966.

23. Maynard, J.B., Legate, C.E., and Graiff, L.B., Pre-flame reaction products of *iso*-octane formed in a motored engine, *Combustion and Flame,* **11,** 155–165 (1967).

24. Kunc, J.F., and Roblee, H.E., Mechanism involved in vapor phase oxidation of hydrocarbons, *SAE Trans.* **69,** 458 (1961).

25. Fish, A., Read, I.A., Affleck, W.S., and Haskell, W.W., The controlling role of cool flames in two-stage ignition, *Combustion and Flame* **13,** 39–49 (Feb. 1969)

26. Martinengo, A., Melczer, J. and Schlimme, E., Analytical investigations of stable products during reaction of adiabatically compressed hydrocarbon–air mixtures, *Tenth Symposium (International) on Combustion, Cambridge, England, August 17–21, 1964.* The Combustion Institute, Pittsburgh, pp. 323–330 (1965)

27. Salooja, K.C., Effects of lead monoxide on the combustion behavior of oxygen derivatives of hydrocarbons, *Combustion and Flame* **11,** 247–254 (June 1967).

28. Salooja, K.C., Studies relating to the mechanism of anti-knock action of tetramethyl lead, *Combustion and Flame* **9**, 211–217 (September 1965).
29. Richardson, W.L., Ryason, P.R., Kautsky, G.J., and Barush, M.R., Organolead anti-knock agents — Their performance and mode of action, presented at the *Ninth Symposium (International) on Combustion,* Ithaca, N.Y., August 27–September 1, 1962. Academic Press, New York, 1963, pp. 1023–1032.
30. Trumpy, D.R., Uyehara, O.A., and Myers, P.S., The pre-knock kinetics of ethane in a spark ignition engine, Paper 690518 presented at *SAE Mid-Year Meeting,* Chicago, Ill., May 1969.
31. Affleck, W.S., Bright, P.E., and Fish, A., Run-on in gasoline engines: A chemical description of some effects of fuel composition, *Combustion and Flame* **12**, 307–317 (August 1966).
32. Haskell, W.W., Fuel ignition in a rapid compression machine: Sensitivity to flame ignition by particles, Paper 700059 presented to *SAE Automotive Engineering Congress,* Detroit, Jan. 1970.
33. MacCormac, M., and Townsen, D.T.A., The spontaneous ignition under pressure of typical knocking and non-knocking fuels: heptane, octane, *Diiso*-propyl ether, acetone and benzene, *J. Chem. Soc.* 238–245 (1938).
34. Kane, G.P., Chamberlain, E.A.C., and Townsen, D.T.A., The spontaneous ignition under pressure of the simpler aliphatic hydrocarbons, alcohols and aldehydes, *J. Chem. Soc.* 436–439 (1937).
35. Kane, G.P. and Townsen, D.T.A., The influence of pressure on the spontaneous ignition of inflammable gas-air mixtures: The simple olefins, *Proc. Royal Soc. (London)* **A160**, 174–187 (1937).
36. Burgoyne, J.H., Tang, T.L., and Newitt, D.M., The combustion of aromatic and alicyclic hydrocarbons, *Proc. Royal Soc. (London)* **A174**, 379–409 (1940).
37. Shtern, V. Ya., *The Gas-Phase Oxidation of Hydrocarbons* Macmillan, New York, 1964.
38. Jost, W., *Low Temperature Oxidation,* Gordon and Breach Science Publishers, New York, 1965.
39. Salooja, K.C., Studies of combustion processes leading to ignition of isomeric hexanes, *Combustion and Flame* **6**, December 275–285 (1962).
40. Burke, R., Dewael, F., and van Tiggellen, A., Kinetics of the propylene-oxygen flame reaction, *Combustion and Flame* **7** 83–87 (March 1963).
41. Malherbe, F.E., and Walsh, A.D., Experiments with cool flames, I — Induction periods, *Trans. Faraday Soc.* **46**, 824–835 (1950).
42. Lovell, W.G., and Cambell, J.M., Molecular structure of engine hydrocarbons and engine knock, presented at *Second Symposium (International) on Combustion 94th Meeting ACS,* Rochester, N.Y., September 9–10, 1937. Published by the Combustion Institute, Pittsburgh, Pa. (1965), pp. 343–353.
43. Lovell, W.G., Knocking characteristics of hydrocarbons, *Ind. Eng. Chem.* **40**, 2388–2438 (1948).
44. Newhall, H.K., Kinetics of engine generated nitrogen oxide and carbon monoxide, presented at the *Twelfth Symposium (International) on Combustion,* Poitiers, France, Jan. 14–20, 1968. Published by the Combustion Institute, Pittsburgh, Pa. (1969), pp. 603–613.
45. Fristrom, R.M. and Westenburg, A.A., *Flame Structure,* McGraw-Hill, New York, 1965.
46. Homann, K.H., and Wagner, H.G., Some new aspects of the mechanism of carbon formation in premized flames, presented at the *Eleventh Symposium (International) on Combustion,* Berkeley, Calif., August 14–20, 1966. Published by the Combustion Institute, Pittsburgh, Pa. (1967), pp. 371–379.
47. Tompkins, E.E., and Long, R., The flux of polycyclic aromatic hydrocarbons of insoluble burned material in premixed acetylene–oxygen flames, presented at the

Twelfth Symposium (International) on Combustion, Poitiers, France, July 14–20, 1968. Published by the Combustion Institute, Pittsburgh, Pa. (1969), pp. 625–634.

48. Ninomiya, J.S., and Golovoy, A., Effects of air-fuel ratio on composition of hydrocarbon exhaust from *Iso*-octane, diisobutylene, toluene, and toluene-*n*-heptane mixture, Paper 690504 presented at *SAE Mid-Year Meeting* (May 1969), Chicago, Ill.

49. Welling, C.E., Hall, G.C., and Stepanski, J.S., Concurrent pyrolytic and oxidative mechanisms in precombustion of hydrocarbons, *Trans. SAE* **69**, 448–457 (1961).

50. Eyzat, P., and Guibet, J.C., A new look at nitrogen oxide formation in internal combustion engines, Paper 680124 presented at the *SAE Automobile Engineering Congress,* Detroit, Mich., January 1968.

51. Heywood, J.B., Mathews, S.M., and Owen, B., Predictions of nitric oxide concentration in a spark ignition engine compared with exhaust measurements, Paper 710001 presented to *SAE Automotive Engineering Congress,* Detroit, January 1971.

52. Lavoie, G.A., Heywood, J.B., and Keck, J.C., Experimental and theoretical study of nitric oxide formation in internal combustion engines" *Combustion Sci. Tech.* **1**, 313–326 (1970).

53. Newhall, H.K., and Shahed, S.M., Kinetics of Nitric oxide formation in high pressure flames, Paper 36 presented to the *Thirteenth Symposium (International) on Combustion,* Salt Lake City (1970).

54. Muzio, L.J., Starkman, E.S., and Caretto, L.S., The effect of temperature variations in the engine combustion chamber on formation and emissions of nitrogen oxides, Paper 710158 presented to *SAE Automotive Engineering Congress,* Detroit, Jan. 1971.

55. Benson, J.D., and Stebar, R.F., Effects of charge dilution on nitric oxide emission from a single cylinder engine, Paper 710008 presented to *SAE Automotive Engineering Congress,* Detroit, Jan. 1971.

56. Quadar, A.A., Why intake charge dilution decreases nitric oxide emission from spark ignition engines, Paper 710009 presented to *SAE Automotive Engineering Congress,* Detroit, Jan 1971.

57. Ohigashi, S., Kuroda, H., Nakajima, Y., Hayashi, Y., and Sugihara, K., Heat capacity changes predict nitric oxide reduction by exhaust gas recirculation, Paper 710010 presented to *SAE Automotive Engineering Congress,* Detroit, Jan., 1971.

58. Kopa, R.D., and Kimura, H., Exhaust gas recirculation as a method of nitrogen oxide control in an internal combustion engine, *53rd Annual Meeting, Air Pollution Control Association,* May 1960.

59. Wimmer, D.B., and McReynolds, L.A., Nitrogen oxides and engine combustion, Paper 380E presented at *SAE Summer Meeting,* 1961.

60. Carr, R.C., Starkman, E.S., and Sawyer, R.F., The influence of fuel composition on emissions of carbon monoxide and oxides of nitrogen, Paper 700470 presented to *SAE Mid-Year Meeting,* Detroit, 1970.

61. Singer, J.M., Cook, E.B., Harris, M.E., Rowe, V.R., and Grummer, J., Flame characteristics causing air pollution: Production of nitrogen oxides and carbon monoxide, *U.S. Bureau of Mines, Report of Investigation* 6958 (1966).

62. Sigsworth, H.W., Myers, P.S., and Uyehara, O.A., The disappearance of ethylene, propylene, *n*-butene and 1-butene in spark ignition engine exhaust, Paper 700492 presented to *SAE Mid-Year Meeting,* Detroit, May 1970.

63. Sorenson, S.C., Myers, P.S., and Uyehara, O.A., The reaction of ethane in spark ignition engine exhaust gas, Paper 700471 presented to *SAE Mid-Year Meeting,* Detroit, 1970.

64. Stahl, Q.R., Preliminary air pollution survey of aldehydes — A literature review, *National Air Pollution Control Administration Publication* APTD 69–24, Oct. 1969.

65. Daniel, W.A., and Wentworth, J.T., Exhaust gas hydrocarbons — Genesis and exodus, SAE Paper 486B presented at *National Automobile Week Meeting,* Detroit, Mich., March 1962. Published *SAE Tech. Prog. Ser.,* Vol. 6, 1964, pp. 192–205.

66. Daniel, W.A., Engine variable effects on exhaust hydrocarbon composition (A single-cylinder study with propane as fuel), Paper 670124 presented at the *SAE Automotive Engineering Congress,* Detroit, Mich., January 1967.

Chapter 3

Mechanism of Hydrocarbon Formation in Combustion Processes

R. A. Matula

Environmental Studies Institute
Drexel University
Philadelphia, Pennsylvania

I. INTRODUCTION

Emissions from transportation systems that derive their energy directly from combustion processes include products of incomplete combustion, oxides of nitrogen, oxides of sulfur, and lead and other trace metals that are employed as combustion-improving additives. Emissions associated with the products of incomplete combustion can be further subdivided to include carbon monoxide, gaseous hydrocarbons, polynuclear aromatics, odor, and combustible particulates.

Table I, which was obtained from Ref. 1, presents an inventory of pollutant emissions in the United States for 1966. In this table, the emission sources are divided into a number of categories including transportation sources, which include continuous-combustion sources such as gas-turbine-powered aircraft and mobile gasoline- and diesel-powered cyclic combustion engines. Gaseous hydrocarbon emissions from these three classes of transportation systems were responsible for approximately 61 % of the total gaseous hydrocarbon emissions and transportation systems accounted for approximately 3.6 % of all polynuclear (PNA) emissions from all combustion sources in 1966. Projected annual emissions of gaseous hydrocarbons from both continuous-combustion sources and internal-combustion engines for the period 1970 to 1990 are graphically represented in Fig. 1. These data were also extracted from Ref. 1, and they indicate that unburned hydrocarbon emissions from transportation systems are presently, and will remain to be, significant fractions of the total contribution to these classes of pollutant emissions through 1990. Therefore, in light of the present and projected

77

Table I. National Inventory of Pollutant Emissions by Class of Source[a] for 1966, 10^6 tons/yr

Pollutants	Total, all sources	Total, all combustion sources	Total, all energy conversion combustion	Central-section power plants: Coal	Central-section power plants: Oil	Central-section power plants: Gas	Industrial processing	Industrial steam generation	Commercial & Residential Heating	Gas turbine: Stationary	Gas turbine: Aircraft	Mobile IC engines: Gasoline	Mobile IC engines: Diesel	Stationary IC engines: Gasoline	Stationary IC engines: Diesel	Stationary IC engines: Gas	Other non-ECC: Solid waste incineration	Other non-ECC: Miscellaneous combustion
1. Products of incomplete combustion																		
Combustible particulate[c]	—[f]	—	3.3	0.6	<0.1	<0.1	0.2	1.2	0.3	<0.1	<0.1	0.4	0.5	<0.1	0.1	n	—	—
CO	102.1	91.3	65.8	0.1	n^e	n	0.1	0.1	0.7	<0.1	0.3	62.3	0.2	1.4	0.1	0.5	7.6	17.9
Gaseous HC	31.5	27.9	17.8	<0.1	<0.1	n	n	0.1	0.2	<0.1	0.1	16.4	0.6	0.1	0.1	0.2	1.5	8.6
PNA[b]	12.	11.2	7.2	n	n	n	0.3	n	6.5	—	—	0.4	n	n	n	n	1.0	3.0
2. NO_x	18.9	18.6	16.4	2.7	0.3	0.5	0.4	2.0	0.8	0.1	0.1	6.3	0.9	0.1	0.2	2.0	0.5	1.7
3. Combustion-improving additives																		
Lead[d]	0.25	0.21	0.21	n	n	n	n	n	n	n	n	0.21	n	<0.01	n	n	n	n
4. Fuel contaminants																		
SO_x[d]	30.9	23.5	22.9	13.5	1.0	n	1.1	4.4	2.6	0.1	<0.1	0.2	<0.1	n	n	n	0.1	0.6
Ash (noncombustible particulate)[c]	—	—	6.8	4.9	<0.1	n	0.2	1.4	0.3	n	n	n	n	n	n	n	0.1	—
5. Total particulate	28.6	21.0	10.1	5.5	0.3	0.5	0.4	2.6	0.5	<0.1	<0.1	0.4	0.5	<0.1	0.1	n	1.0	9.9

[a] Sources of data: APCO/DAOED and DPCE plus APCO (NAPCA) reports.
[b] PNA shown in 10^3 tons/year.
[c] Combustible particulate apportioned from total data by Battelle judgment.
[d] Emission values for lead are for 1967.
[f] —, not available.
[e] n, emission considered negligible.

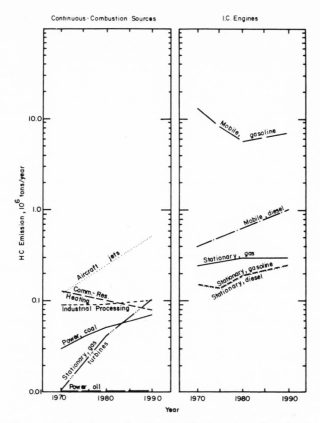

Fig. 1. Projected annual emissions of hydrocarbons from both
continuous- and cyclic-combustion sources.

importance of this class of emissions, the purpose of this chapter is to discuss
some of the important factors associated with the formation and emission of
hydrocarbons from transportation systems. Directly related phenomena such
as photochemical reactivity of exhaust gases, odor, and potential health
hazards associated with PNA emissions will also be considered.

Subsequent portions of this chapter are divided into three general sections.
In Section II, recent experimental and analytical results dealing with ignition
delays in low molecular weight hydrocarbon–oxidizer systems will be reviewed.
Section III deals with the production and emission of unburned hydrocarbons from engines. The effects of wall quenching of the flame on the production of unburned hydrocarbons, the sources of unburned hydrocarbons in
spark ignition, compression ignition, automotive and aircraft gas turbines,
and the effects of engine variables on hydrocarbon emissions from spark-
ignition engines are discussed in some detail. Finally, in Section IV, some of

the effects associated with hydrocarbon emissions on man and his environment, including photochemical smog, odor, and polynuclear aromatics, are discussed.

II. IGNITION DELAYS IN LOW MOLECULAR WEIGHT HYDROCARBON–OXIDIZER SYSTEMS

A. Introduction

The high-temperature oxidation mechanisms of hydrocarbon fuels play a role in determining the efficiency, design, and emissions of both cyclic and continuous-combustion devices, and hence a better understanding of these mechanisms may lead to the design of more efficient and lower-emission power plants. A relatively large number of investigators have studied the ignition and oxidation of hydrogen and carbon monoxide at high temperatures. However, despite their great technical importance, relatively little effort has been devoted to the high-temperature ignition of hydrocarbon–oxidizer mixtures. The ignition delays (induction times) of a number of low molecular weight hydrocarbons have been recently reported, and analytical studies have established the validity of induction times as a meaningful experimental parameter in terms of its application to the development of high-temperature oxidation mechanisms. Therefore these recent results are reviewed in subsequent paragraphs.

The induction time for a combustible mixture is often defined as the interval between the time that the mixture is heated to a specified temperature and the onset of the rapid exothermic combustion process. The onset of combustion is usually experimentally determined by measuring either the rapid increase of pressure or a specific chemical species that is characteristic of the combustion process. The rate of change of pressure is generally determined with the aid of piezoelectric pressure transducers and the increase in concentration of the important species are generally determined by either measuring the rapid change in absorption or emission of radiation by these species, or by mass spectrometrically following the depletion of a reactant or the appearance of a product during the course of the reaction.

Various experimental techniques, including closed-vessel studies,[2] flow methods,[3] and shock-tube experiments, have been utilized to measure induction times. However, since shock-tube techniques can be employed to rapidly and homogeneously heat the reactant mixture to readily controlled pressures and temperatures in the range of practical combustion devices, most recent induction time studies have been conducted in shock tubes. Both conventional shock tubes and single-pulse shock tubes, in which not only induction times but also stable quenched product distributions can be measured, have been used to determine induction times in hydrocarbon–oxidizer mixtures.

Induction times are usually found to be dependent on temperature, pressure, stoichiometry, and fuel type. The effect of these variables on the measured induction times are often correlated in terms of the equation

$$\tau \, [\text{Oxidizer}]^a \, [\text{Fuel}]^b \, [\text{Inert}]^c = A \, \exp \, (E/RT) \tag{1}$$

where

$\tau =$ Induction time, sec
[Oxidizer] = Concentration of oxidizer, moles/cc
[Fuel] = Concentration of fuel, moles/cc
[Inert] = Concentration of inert, moles/cc
$A =$ Apparent pre-exponential factor, sec-(moles/cc)$^{a+b+c}$
$E =$ Apparent overall activation energy, cal/mole
$R =$ Universal gas constant, cal/mole-°K
$T =$ Absolute temperature, °K
$a, b, c =$ Correlation parameters

Whenever possible, the induction-time data correlations discussed in subsequent paragraphs will be considered in terms of Eq. (1), and in all cases the experimental results discussed in subsequent paragraphs will be presented in terms of the above mentioned set of units.

B. Experimental Results

1. Methane

A number of investigators have studied the ignition of CH_4–O_2 mixtures in shock tubes. The early work on methane ignition was carried out by Skinner and Ruehrwein[4] in a single-pulse shock tube. In these studies, the induction times of rich CH_4–O_2 mixtures diluted in Ar were measured behind reflected shock waves. These studies were carried out in the temperature and pressure ranges 1200 to 1750°K and 3 to 10 atm, respectively. The induction time was defined as the time between heating of the gas by the reflected shock wave and the onset of a rapid rise in a photocell output that corresponded to the flash of light from the reacting mixture. Induction times for each stoichiometry at a given pressure were found to correlate with an empirical linear relationship between log τ and $1/T$, and these results yielded a set of apparent overall activation energies ranging from 38 to 63 Kcal/mole. The authors reported a noticeable effect of total pressure on the induction times. However, they were not able to correlate the measured induction times in terms of the concentration of either the fuel or the oxidizer.

Much of the subsequent work on methane ignition was undertaken in order to determine a general correlation between induction time, reactant concentration, and temperature. A number of investigators found that ignition times for relatively lean CH_4–O_2 mixtures could be correlated in terms of Eq. (1) if both b and c were set equal to zero and $a = 1.0$. The apparent activation energies for the ignition process were reported to be near 30 Kcal/mole by three of the investigators; Kistiakowsky and Richards,[5] Soloukhin,[6] and Higgin and Williams.[7] Asaba, et al.[8] and Miyama and Takeyama[9] reported lower apparent activation energies of 20.6 and 21.5 Kcal/mole, respectively.

Glass et al.[10] measured the induction times of CH_4–O_2 mixtures with the aid of a time-of-flight mass spectrometer that was coupled to the end of their shock tube. Their reactant mixtures were highly diluted in Ar, and their work covered the temperature and equivalence ratio ranges 2000 to 2300°K and 0.53 to 2.66, respectively. Their results were found to correlate in terms of Eq. (1) for $a = 1/2$, $b = 0$, and $c = 1/2$, and they reported an apparent activation energy of 56 Kcal/mole. For the sake of comparison, Higgin and Williams[7] attempted to correlate their data in this manner. They reported a poor correlation and a much lower apparent activation energy of approximately 30 Kcal/mole.

Both Asaba et al.[8] and Miyama and Takeyama[9] reported that relatively rich CH_4–O_2 induction times could be correlated in terms of Eq. (1) for $a = b = 1.0$ and $c = 0$. Although their data were scattered, Asaba et al.[8] reported an apparent activation energy of approximately 53 Kcal/mole. The results reported by Miyama and Takeyama[9] correspond to an apparent activation energy of 49.5 Kcal/mole. The induction times reported by Asaba et al.[8] are consistently shorter by approximately a factor of ten than the induction times reported by Miyama and Takeyama.[9]

Seery and Bowman[11] have recently reported a correlation between induction time, composition, and temperature for CH_4–O_2 mixtures diluted in Ar with equivalence ratios in the range 0.2 to 5.0. The experiments were carried out in the temperature and pressure ranges 1350 to 1900°K and 1.5 to 4 atm, respectively. The proposed correlation equation is given by

$$\tau_{CH_4} [O_2]^{1.6} [CH_4]^{-0.4} = 7.65 \times 10^{-8} \exp (51,400/RT) \tag{2}$$

In general the correlation given by Eq. (2) was reported to be in excellent agreement with results reported by Skinner and Ruehrwein[4] and Snyder et al.[12] and in fair agreement with the results of Miyama and Takeyama.[9] The experimental results of Williams and co-workers[7,13] can also be represented within a factor of approximately two to four by Eq. (2). Seery and Bowman[11] did not note the sharp decrease in the temperature dependence of the induction time below about 1500°K as reported by Asaba et al.[8]

In a subsequent study, Bowman[14] reported that the concentration dependence of induction times for the production of various products, including CO, CO_2, and H_2O, produced during the oxidation of CH_4 diluted in Ar could also be correlated in terms of relationships similar to Eq. (2). These studies were carried out for temperatures, pressures, and equivalance ratios in the ranges 1700 to 2500°K, 1.3 to 2.6 atm, and 0.25 to 4.0, respectively. These results further confirmed the fact that induction times for both fuel-rich and oxidizer-rich CH_4–O_2 mixtures can be correlated in terms of a single equation.

The correlation between induction times and reactant concentrations reported by Seery and Bowman[11] by the stoichiometric dependence of $[O_2]^{1.6}$ $[CH_4]^{-0.4}$ is in good agreement with the concentration dependence of the overall reaction rate deduced from other measurements. Kozlov[15] and Nemeth and Sawyer[16] found that the overall reaction rate for CH_4 oxidation could be represented in terms of $[O_2]^{1.5}$ $[CH_4]^{-0.5}$ in flow reactor experiments and Vandenabeele et al.[17] found an overall stoichiometric dependence of $[O_2]^{1.4}$ $[CH_4]^{-0.4}$ in flame studies.

Lifshitz et al.[18] have recently reported the induction times for CH_4–O_2 mixtures highly diluted in Ar. The experiments were carried out behind reflected shock waves in a single-pulse shock tube, and the measurements included temperatures, pressures, and equivalence ratios in the range 1500 to 2150°K, 2 to 10 atm, and 0.5 to 2.0, respectively. During the course of these experiments, the reactant mole fraction and test conditions were selected in a manner which allowed the values of a, b, and c in Eq. (1) to be directly evaluated from graphical plots of the experimental data. The proposed correlation equation is given by

$$\tau_{CH_4} [O_2]^{1.03} [CH_4]^{-0.33} = 3.62 \times 10^{-14} \exp (46{,}500/RT) \qquad (3)$$

Both the apparent activation energy and methane concentration dependence reported by Lifshitz et al.[18] are in good agreement with the results previously reported by Seery and Bowman.[11] However, the oxygen concentration power dependence of 1.03 is in disagreement with the 1.6 power dependence reported by Seery and Bowman.[11] The reasons responsible for these two significantly different reported values for the oxygen concentration dependence in the CH_4–O_2 system is not readily apparent. However, as discussed previously, oxygen concentration powers of both 1.0 and 1.6 can be supported by the results of other researchers.

Based on this review the most reliable data (in chronological order) on ignition delay measurements in CH_4–O_2–Ar mixtures are considered to have been reported by Skinner and Ruehrwein,[4] Higgin and Williams,[7] Seery and Bowman,[11] Bowman,[14] and Lifshitz et al.[18] Since a large quantity of experimental data are available relating induction times in CH_4–O_2 mixtures to temperature and reactant concentration, it is reasonable to seek a best-fit correlation for all of the data. Previous investigators have shown that the concentration of inert gas does effect the induction times in the CH_4–O_2 system, and hence it is proposed that the best-fit correlation equation for all of the data can be represented by Eq. (1) where the parameter c is set equal to zero. Equation (1) can be linearized in terms of the four unknown parameters

a, b, A, and E by taking the logarithm of both sides of the equation, and the numerical values of the four parameters that best fit a given set of experimental data which can be readily evaluated with the aid of standard multiple linear regression techniques.

Standard multiple linear regression techniques were utilized to determine the best-fit coefficients for the 99 data points reported by Skinner and Ruehrwein,[4] the 69 data points reported by Seery and Bowman,[11] and the 24 data points reported by Lifshitz et al.[18] The results of these computations are listed in Table II. The experimental data reported by Higgin and Williams[7] was not presented in a form that could be utilized to obtain best-fit correlation coefficients.

Table II. Correlation Parameters for Methane–Oxygen Induction Times Calculated by Multiple Linear Regression Techniques[a]

Data source	a	b	A	E
Skinner and Ruehrwein[4]	1.5	−0.012	2.04×10^{-18}	44,900
Seery and Bowman[11]	1.2	−0.29	5.26×10^{-16}	48,640
Lifshitz et al.[18]	1.13	−0.49	1.11×10^{-13}	45,770
Eq. (4)	1.37	−0.17	5.85×10^{-17}	45,940

[a] $\tau_{CH_4} [O_2]^a [CH_4]^b = A \exp (E/RT)$, where τ = seconds; [ ] = moles/cc; T = °K.

The results given in Table II indicate that the induction time data reported by Skinner and Ruehrwein[4] can be readily correlated in a manner similar to that suggested in Refs. 11 and 18, and that all three sets of data yield apparent activation energies near 46 Kcal/mole. The correlation parameters listed in Table II for the data in Ref. 18 are nearly equal to the parameters reported by the authors [see Eq. (3)]. The differences between the numerical values of the correlation parameters listed in Table II and Eq. (3) can be readily explained in terms of the fact that the authors based their parameters on more than 60 data points while the parameters in Table II were calculated from the 24 data points listed in Ref. 18.

The apparent activation energy listed in Table II for the best fit of the data reported by Seery and Bowman[11] is in good agreement with the results reported in Ref. 11, but the parameters a and b are both lower than those reported by the authors [see Eq. (2)]. It is suggested that the parameters listed in Table II yield a better representation of induction time data reported in Ref. 11 than the parameters given in Eq. (2). It should be noted that the numerical values of a and b listed in Table II for the data reported in Refs. 11 and 18 are in reasonably good agreement.

All of the data listed in Refs. 4, 11, and 18 have been employed to develop a proposed best-fit correlation equation for the prediction of induction times in CH_4–O_2 systems.

$$\tau_{CH_4} [O_2]^{1.37} [CH_4]^{-0.17} = 5.85 \times 10^{-17} \exp (45,940/RT) \qquad (4)$$

The proposed best-fit parameters for the calculation of CH_4–O_2 induction times, as given by Eq. (4), are listed in Table II, and Eq. (4) is shown graphically in Fig. 2. For the sake of comparison, a number of experimental data points were taken at random from Refs. 4, 7, 11, and 18, and these data are also plotted in Fig. 2. Inspection of Fig. 2, indicates that Eq. (4) can be effectively

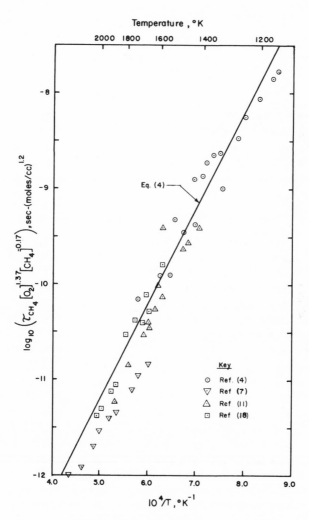

Fig. 2.   Correlation of measured induction times in CH_4–O_2
mixtures versus temperature.

employed to correlate a large number of experimentally determined induction times in CH_4–O_2 systems. The data points extracted from Ref. 7, represented by the inverted triangles in Fig. 2, are in reasonable agreement with Eq. (4). However, it should be noted that these data were obtained at very low pressures and that their data would be represented by an apparant activation energy near 30 Kcal/mole. As discussed previously, these data were not included in the evaluation of the best-fit coefficients for Eq. (4) and they are only shown in Fig. 2 to indicate that Eq. (4) can be effectively utilized to predict induction times in CH_4–O_2 systems over a wide range of temperatures and pressures. It is not suggested that Eq. (4) is unique in that either Eq. (2) or Eq. (3) could also be expected to reasonably correlate all of the data. However, Eq. (4) can be effectively employed to correlate the experimental data reported by a number of investigators.

2. Ethane

Only limited experimental data are available for ignition delay times of C_2H_6–O_2 mixtures. In a recent shock tube investigation of the comparative ignition delay times of the alkanes methane through pentane, Burcat et al.[20] reported that C_2H_6–O_2 mixtures had the shortest induction period of any of the fuels investigated. Bowman[14] studied the oxidation of C_2H_6 behind reflected shock waves using six C_2H_6–O_2–Ar mixtures and one C_2H_6–C_2H_4–O_2–Ar mixture with equivalence ratios in the range 0.25 to 4.0. The study was carried out in the temperature range 1300 to 2000°K and two initial reflected shock pressures of approximately 2.0 and 4.4 atm were employed. The reaction times for CO, CO_2, and H_2O were spectroscopically measured and correlated in terms of reactant concentrations and temperature. In this study, the characteristic reaction time was defined as the time interval between shock heating and the attainment of 90% of the equilibrium emission intensity from each of the three species. The product reaction times were reported to be inversely proportional to the total concentration raised to the 2.0 ± 0.2 power for both fuel-rich and oxidizer-rich mixtures. However, the dependence of product reaction times on initial reactant conditions was reported to be relatively complicated and different for fuel-rich and oxidizer-rich mixtures.

Cooke and Williams[13] studied the ignition of C_2H_6–O_2 mixtures diluted with 90 mole % Ar behind incident shock waves in the pressure range 0.25 to 0.40 atm. The experiments were carried out in the temperature and equivalence ratio ranges 1400 to 1800°K and 0.5 to 2.0, respectively. The experimentally determined induction times were not correlated in terms of reactant concentration, but plots of $\log_{10} \tau$ versus $1/T$ for each equivalence ratio yielded approximately linear relationships with an apparent activation energy of approximately 33 Kcal/mole.

The most extensive investigation of C_2H_6–O_2 ignition was recently reported by Burcat et al.[21] They reported induction times obtained in a single-pulse shock tube behind reflected shock waves in the temperature and pressure ranges 1235 to 1660°K and 2 to 8 atm, respectively. Highly diluted C_2H_6–O_2 mixtures in approximately 90 mole $\%$ Ar with equivalence ratios in the range 0.5 to 2.0 were studied in order to determine the dependence of induction time on reactant concentrations and temperatures. All of the experimental data were found to be correlated by

$$\tau_{C_2 H_6} [O_2]^{1.26} [C_2H_6]^{-0.46} = 2.35 \times 10^{-14} \exp (34,200/RT) \qquad (5)$$

Equation (5) is shown graphically in Fig. 3, and representative data points corresponding to equvalence ratios of 0.5, 1.0, and 2.0, as reported by

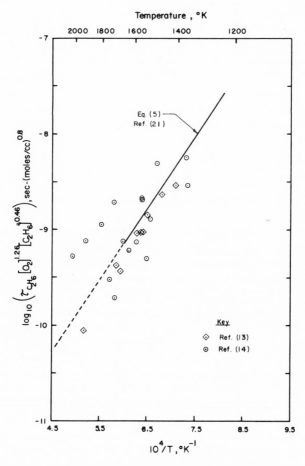

Fig. 3.   Correlation of measured induction times in C_2H_6–O_2 mixtures versus temperature.

Cooke and Williams[13] are also plotted on Fig. 3. Since Cooke and Williams reported that their data were obtained in the pressure range 0.25 to 0.40 atm, their results were correlated assuming that the pressure was 0.33 atm. The agreement between these two sets of data is good. The data reported by Bowman[14] are difficult to compare directly to the results given in Fig. 3 since he reported reaction times for the formation of products rather than induction times. However, for the sake of comparison the reaction times based on the production of H_2O is probably most representative of the induction time. Therefore, a few representative points from Bowman's results for both fuel-rich and oxidizer-rich mixtures have been correlated in terms of Eq. (5) and these data are also shown in Fig. (3).

In general, Eq. (5) correlates the data reported by both Cooke and Williams[13] and Bowman.[14] Some of the data reported by Bowman[14] correspond to longer induction times than predicted by the correlation equation. However, since the data reported by Bowman are based on induction times for the production of a reaction product, it is reasonable to expect that these results should yield slightly longer induction times than induction times based on the initiation of rapid exothermic reaction. It appears that the experimental data from three independent sources can be reasonably represented by Eq. (5).

3. Propane

The early high-temperature work on ignition delays in $C_3H_8-O_2$ mixtures was carried out in flow systems. Brokaw and Jackson[3] injected preheated fuel into a heated moving stream under conditions where the mixing time was short compared with the ignition delay and found that the ignition delay was only slightly dependent on the oxygen concentration, varied approximately inversely with the first power of the propane concentration and decreased as the temperature and pressure were increased. No correlation between ignition delay and either pressure or temperature was reported, except that it was noted that the temperature dependence could not be represented by an Arrhenius type expression with a constant activation energy. In a similar study, Miller[22] reported results that were in disagreement with those of Brokaw and Jackson.[3] For lean mixtures, they reported that the ignition times were independent of the fuel concentrations, varied inversely as the first power of the oxygen concentration and had an Arrhenius temperature dependence with an apparent activation energy of 32.5 Kcal/mole. Zimot and Trushin[23] reported results that were in reasonable agreement with Miller's[22] work. They reported that induction times were inversely dependent on the square of the pressure, could be correlated with the oxygen concentration raised to the minus 1.3 power, and had an apparent activation energy of 14 Kcal/mole.

Steinberg and Kaskan[24] appear to have reported the first shock-tube study of $C_3H_8-O_2$ ignition. They examined the ignition of stoichiometric

C_3H_8–O_2 mixtures over a small pressure range, and reported that the induction times were relatively independent of pressure and could be roughly correlated by an apparent activation energy of 19 Kcal/mole. Hawthorn and Nixon[25] reported ignition delays for C_3H_8–O_2–Ar mixtures for temperatures and equivalence ratios in the range 1100 to 1475 °K and 0.1 to 2.0, respectively. However, they did not attempt to correlate their data, and only discussed qualitative trends. Myers and Bartle[26] measured ignition delay times in highly dilute, shock heated mixtures of C_3H_8–O_2 in Ar. Their experiments covered initial temperatures, pressures, and equivalence ratios in the range 1000 to 1600 °K, 0.5 to 5.5 atm, and 0.1 to 1.5 respectively. Three ignition delay times based on; (1) the initial rise in OH emission; (2) the second rise in OH emission; and (3) the initial rise in CO_2 emission were measured. The two ignition delay times based on the second rise in OH emission and the onset of CO_2 emission were found to be identical over the entire temperature range. They attempted to correlate their experimental results in terms of a concentration dependence of the form $[O_2]^{1/2}[C_3H_8]^{1/2}$. Actual experimentally encountered influences of composition and total pressure were found which necessitated the use of complex correlating parameters involving a variable power law dependency on equivalence ratio to normalize their results. The complexity of their correlation tends to bury the influence of significant parameters in determining induction times and limits its general applicability. The authors also point out limitations associated with the measurement of ignition delay times behind incident shock waves.

Burcat *et al.*[27] have recently studied ignition of highly diluted C_3H_8–O_2 mixtures in Ar behind reflected shock waves in a single-pulse shock tube. The experiments included a temperature range from 1250 to 1600 °K, at pressures varying from 2 to 10 atm, and mixture equivalence ratios in the range 0.125 to 2.0. The following correlation equation was found to fit all of the experimental data for equivalence ratios in the range 0.5 to 2.0:

$$\tau_{C_3H_8}[O_2]^{1.22}[C_3H_8]^{-0.57} = 4.4 \times 10^{-14} \exp(42,200/RT) \qquad (6)$$

Several experiments were carried out below 1250 °K and, as previously reported by Myers and Bartle,[26] the apparent activation energy was found to significantly decrease for temperatures below 1250 °K. Two lean mixtures with equivalence ratios of 0.25 and 0.125 were also investigated. For these mixtures the concentration dependence of C_3H_8 on the measured induction times decreased from 0.57 to approximately 0.3.

Equation (6) is shown graphically in Fig. 4. Unfortunately, sufficient data are not included in Ref. 26 to allow Myers' and Bartle's data to be compared with the correlation given by Eq. (6). Hawthorn and Nixon[25] have reported 25 data points for induction times of C_3H_8–O_2 mixtures with equivalence ratios in the range 0.67 to 2.0. All of these data were obtained in highly diluted C_3H_8–O_2–Ar mixtures at a total pressure of approximately 1 atm.

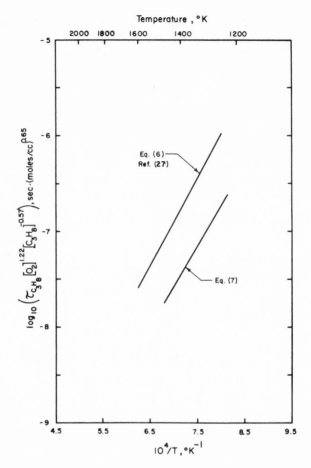

Fig. 4.   Correlation of measured induction times in C_3H_8–O_2
mixtures versus temperature.

These data covered the temperature range from approximately 1250 to
1475°K. When these 25 data points were correlated in terms of Eq. (6) and
plotted in Fig. 4, it was apparent that the data could be correlated by an
equation similar in form to Eq. (6). For the sake of comparison, the 25 data
points reported by Hawthorn and Nixon[25] were least-squares fit to an equation similar to Eq. (6) and the following result was obtained:

$$\tau_{C_3H_8} [O_2]^{1.22} [C_3H_8]^{-0.57} = 3.31 \times 10^{-14} \exp (38{,}551/RT) \qquad (7)$$

Equation (7) is also shown graphically in Fig. (4). Comparison of the results
shown in Fig. (4) indicates that the induction times reported by Burcat et
al.[27] are approximately four to six times longer than the induction times

reported by Hawthorn and Nixon.[25] It is not readily apparent that the experimental methods and conditions employed by the two groups of investigators should account for this large a difference in measured induction times for near stoichiometric C_3H_8–O_2 mixtures.

Since there is a measurable difference in the numerical values of the ignition times for C_3H_8–O_2 mixtures as reported by Burcat et al.[27] and Hawthorn and Nixon,[25] a satisfactory correlation equation for a wide range of equivalence ratios has not been established and the apparent activation energy is reported to undergo a significant change near 1250°K, more experimental work is required in order to develop a better understanding of ignition phenomena of C_3H_8–O_2 mixtures.

4. Higher Alkanes

Only limited induction-time data are available for C_4 and higher alkanes. Burcat et al.[20] have recently reported ignition delay times for highly diluted stoichiometric mixtures of C_1–C_5 alkanes in Ar. The C_4H_{10}–O_2 and C_5H_{12}–O_2 data covered the temperature and total pressure ranges 1240 to 1400°K, 9.2 to 10.6 atm and 1165 to 1400°K, 8.3 to 9.5 atm, respectively. The measured ignition delays for C_3H_8, C_4H_{10}, and C_5H_{12} were quite close to one another and the authors suggested that they might be well represented by a single curve of log τ versus $1/T$. The apparent activation energies for both C_4H_{10} and C_5H_{12} were reported to be near 40 Kcal/mole, and since only stoichiometric mixtures were studied, no attempt was made to determine the effect of reactant composition on the ignition delay times. Levinson[28] studied the high-temperature oxidation of n-heptane and reported ignition delay times for a number of C_7H_{16}–O_2 mixtures highly diluted in Ar in the temperature range 1180 to 1350°K. At 1250°K the observed ignition delays for lean and slightly rich mixtures were approximately proportional to the ratio of fuel to oxidizer concentrations for total pressures in the range 1.0 to 1.8 atm. In this pressure range, the delays were not influenced by the argon partial pressure. However, when the partial pressure of Ar was reduced below 1 atm, the ignition delays were reported to increase. It was also reported that the ignition delays for a mixture of olefins and CH_4 and C_2H_6 of approximately the same composition as might reasonably result from the cracking of heptane were nearly the same as for the corresponding mixture of uncracked C_7H_{16}.

Sufficient experimental data are not presently available to allow ignition delay times for the higher alkanes to be correlated in terms of Eq. (1), and a better understanding of ignition phenomena in this class of compounds requires additional experimental work.

5. Acetylene and Ethylene

A number of investigators have studied ignition delays of mixtures of C_2H_2–O_2 and C_2H_4–O_2 diluted in an inert gas, and it has been reported that over a wide range of temperatures the induction times for these two fuels are similar to the induction times for similar H_2-O_2 mixtures. Mullaney et al.[29] utilized the diverging nozzle technique to study induction times of $C_2H_4-O_2$ mixtures diluted in N_2. Seven data points were reported for an equivalence ratio of 2.0 in the temperature range 1000 to 1250 °K. Using the same technique they also measured the induction times for stoichiometric mixtures of H_2-O_2 diluted in Ar, N_2, and $C_2H_2-O_2$ mixtures with an equivalence ratio of 2.0 over the temperature range 1000 to 1600°K. For oxygen concentrations in the range 3.5×10^{-7} to 1.4×10^{-6} mole/cc, they reported that all of the induction-time data for the three fuels H_2, C_2H_2, and C_2H_4 could be correlated by an equation of the following form:

$$\tau[O_2] = 1 \times 10^{-12} \exp (14,650/RT) \tag{8}$$

White[30] measured induction times for very lean mixtures of H_2-O_2, $C_2H_2-O_2$, and $C_2H_4-O_2$ in the temperature range 1100 to 2200 °K. The induction times were measured by determining the first sign of exothermic reaction and the initiation of the exothermic reaction was determined by optical interferometry techniques. These data for the three fuels were reported to be correlated by

$$\tau_{H_2} [O_2]^{1/3} [H_2]^{2/3} = 3.63 \times 10^{-14} \exp (17,200/RT) \tag{9}$$

$$\tau_{C_2H_2} [O_2]^{1/3} [C_2H_2]^{2/3} = 1.55 \times 10^{-14} \exp (17,300/RT) \tag{10}$$

$$\tau_{C_2H_4} [O_2]^{1/3} [C_2H_4]^{2/3} = 1 \times 10^{-14} \exp (17,900/RT) \tag{11}$$

In the range of conditions considered, the induction times of the three fuels were similar and induction times, with respect to H_2, for C_2H_2 and C_2H_4 were about 0.5 and 0.4, respectively.

Kistiakowsky and Richards[5] measured the induction delay of $C_2H_2-O_2$ mixtures with equivalence ratios in the range 0.3 to 3.5. Their experiments were carried out in the temperature range 1400 to 2200°K, and they reported that their results could be correlated by

$$\tau_{C_2H_2} [O_2] = 2.69 \times 10^{-14} \exp (17,100/RT) \tag{12}$$

The general validity of their correlation for induction times in $C_2H_2-O_2$ systems was verified by several other experimental techniques. Gardiner[31] employed X-ray densitometry, Bradley and Kistiakowsky[32] utilized ionization techniques to measure induction times in $C_2H_2-O_2$ systems.

More recently, several investigators have simultaneously monitored several properties during the induction period and noted that the ignition delay may be dependent on the measurement technique. Stubbeman and Gardiner[35,36] reported that emission due to excited CH appears earlier than emission from ground state OH radical. Therefore, it was concluded that the ignition delays based on OH emission would be longer than induction times based on CH chemiluminescence.

Takeyama and Miyama[37,38] measured induction times in C_2H_2–O_2 systems obtained from simultaneous measurements of OH absorption, visible emission, and pressure in the temperature and pressure ranges 820 to 1630°K and 1.8 to 5.6 atm, respectively. The experiments employed mixtures with equivalence ratios in the range 1 to 3.76. In contrast to Stubbeman and Gardiner, they reported equal induction times based on all three measurements. Their induction times were correlated by an equation of the form

$$\tau_{C_2H_2}[O_2] = 6.92 \times 10^{-14} \exp(19,050/RT) \tag{13}$$

The apparent activation energy of 19.05 Kcal/mole is very similar to the value of 18.9 Kcal/mole that was obtained by the same authors in a previous H_2–O_2 study. However, Takeyama and Miyama[37,38] reported that the C_2H_2–O_2 induction periods were about one-half those of the corresponding H_2–O_2 mixtures.

Homer and Kistiakowsky[39] studied the oxidation and pyrolysis of C_2H_4 highly diluted in Ar behind reflected shock waves. Induction times were measured for C_2H_4–O_2–Ar mixtures with equivalence ratios of 0.5 and 1.5 in the temperature range 1500 to 2300°K. Induction times were defined by monitoring the infrared emission from CO and CO_2 and arbitrarily defined as the time interval between shock heating and the instant when the combined (CO + CO_2) concentration reached 10% of its final value. The experimentally determined induction times were correlated in terms of an expression similar to Eq. (1) with $a = 1/2$, $b = c = 0$ and an apparent activation energy of approximately 17 Kcal/mole.

Gay et al.[40] have measured the induction times of C_2H_4–O_2 mixtures highly diluted in Ar by observing the $CH(A^2\Delta \rightarrow X^2\pi)$ chemiluminescence and total chemi-ionization behind incident shock waves. Data were reported for temperature and equivalence ratios in the range 1475 to 2200°K and 0.3 to 4.0, respectively. None of the induction time measurements showed any significant dependence on the concentration of C_2H_4, and all of the induction time measurements were correlated by an equation of the form

$$\tau_{C_2H_4}[O_2] = 5.89 \times 10^{-15} \exp(23,110/RT) \tag{14}$$

The correlation equations for C_2H_2–O_2 mixtures, as reported by Kistiakowsky and Richards[5] and Takeyama and Miyama,[37,38] are graphically represented in Fig. 5. The single equation for H_2–O_2, C_2H_2–O_2, and C_2H_2–O_2 mixtures proposed by Mullaney et al.[29] and the correlation equations proposed by Gay et al.[40] for C_2H_4–O_2 mixtures are also graphically represented in Fig. 5 for the sake of comparison.

Direct comparison of the data is difficult due to the wide variety of experimental techniques and conditions that have been employed. However, a number of conclusions can be drawn from the results presented in the previous paragraphs. In the first place, the induction times for near stoichiometric mixtures of C_2H_4–O_2 and C_2H_2–O_2 can be correlated in terms of the

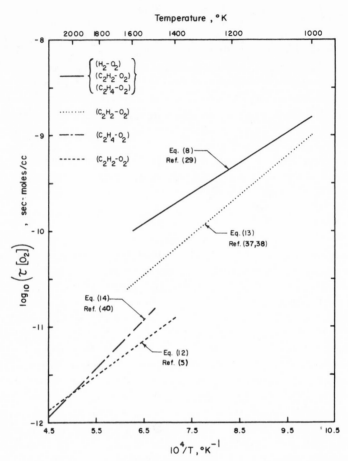

Fig. 5.   Correlation of measured induction times in $C_2H_2-O_2$ and $C_2H_4-O_2$
mixtures versus temperature.

oxygen concentration raised to the first power, and they are relatively independent of the concentration of fuel. For very lean mixtures of fuel and
oxidizer the concentration dependence of the induction times changes and
it is proportional to the oxidizer and fuel concentrations raised to the 0.33
and 0.67 power, respectively. It can also be concluded that, in the range of
temperatures and concentrations discussed in the preceeding paragraphs, the
induction times for mixtures of H_2-O_2, and $C_2H_2-O_2$ are within approximately
a factor of two of each other. Finally, extrapolation of induction time correlations outside of the experimentally reported temperature ranges can lead
to large errors.

C. The Prediction of Induction Times

1. Analytical Techniques

The induction time for a fuel–oxidizer mixture has been defined in Section II.A to be the interval between the time that the mixture is heated to a specified temperature and the onset of the rapid exothermic combustion process. As discussed previously, the onset of combustion is usually experimentally determined by measuring either the rapid increase of pressure or a specific chemical species that is characteristic of the combustion process. In order to analytically predict the induction time for a specific fuel–oxidizer mixture, it is necessary to compute the temperature, pressure, and chemical composition of the reacting mixture as a function of time and determine the time at which either the temperature or pressure begin to increase rapidly or the characteristic chemical species reaches a specified, critical concentration level. The analytical calculations require that; (1) the initial state of the reactant mixture is known, (2) a physical model for the combustion process is specified, (3) a chemical reaction mechanism be specified, and (4) the temperature dependence of all of the elementary rate constants in the chemical mechanism be known.

Shock tube ignition studies have been carried out behind both incident and reflected shock waves. The physical model for the induction process behind incident shock waves usually assumes that the combustion reactions take place in a one-dimensional, constant area, adiabatic flow field, and the induction process behind a reflected shock wave is usually modeled as an adiabatic, constant volume process.

Once the chemical mechanism and physical model for the combustion process have been specified, the variation of species concentrations and thermodynamic properties of the reacting mixture as a function of time can be calculated by numerically integrating the appropriate system of nonlinear chemical kinetic, conservation, and state equations. In most computational schemes, the reacting mixture is assumed to behave as a mixture of ideal gases, and the temperature dependence of the thermodynamic properties of each chemical species is specified in terms of a third- or fourth-order polynomial in temperature. The temperature dependence of the forward rate constant of each elementary step in the chemical mechanism is specified and generally the appropriate rate constants for all of the back reactions are determined by invoking the law of microscopic reversibility. This law states that the ratio of the forward and reverse rate constants for each of the elementary steps must be equal to the equilibrium constant, based on concentration, of the corresponding reaction. It should be noted that the computed time variation of the state of the mixture is significantly effected by the specified chemical mechanism and temperature dependence of the elementary rate constants.

2. Calculated Induction Times for CH_4–O_2 Mixtures

A number of investigators[7,11,13,14,19] have recently reported some success in analytically predicting induction times for CH_4–O_2 mixtures as a function of temperature, pressure, and composition of the reacting mixture. The most comprehensive computations have been carried out by Seery and Bowman,[11] Bowman,[14] and Skinner et al.[19] As discussed previously, an oxidation mechanism for CH_4 must be an integral part of any prediction scheme for induction times in these mixtures. A successful oxidation mechanism for CH_4 must address itself to the following important processes:

1. The initiation process
2. Oxidation of CH_3 radicals
3. Radical attack on CH_4
4. Oxidation of CO
5. Reactions important in H_2–O_2 system

The chain reaction can be initiated by any one of the following three processes:

$$CH_4 + M \rightarrow CH_3 + H + M \tag{15}$$

$$CH_4 + O_2 \rightarrow CH_3 + HO_2 \tag{16}$$

$$O_2 + M \rightarrow O + O + M \tag{17}$$

where M is any third body. In the temperature range 1000 to 2500°K, it can be readily shown that reaction (17) cannot contribute significantly to the initiation process, and hence either reaction (15) and/or reaction (16) must be responsible for initiation. Reaction (16) is the most important initiation step at low temperatures, but as the temperature is increased reaction (15) begins to contribute significantly and at still higher temperatures it dominates the initiation process. In the temperature range 1700 to 2500°K, both Higgin and Williams[7] and Seery and Bowman[11] have reported that reaction (15) is the only important initiation reaction. However, Skinner et al.[19] have reported that reaction (16) must be included in the reaction mechanism for temperatures below approximately 1750°K. It should be noted that all of the investigators report that the numerical values used for the initiation reactions significantly effect calculated induction times.

Various investigators have proposed alternate schemes for the oxidation of methyl radicals. Seery and Bowman[11] proposed that the oxidation of CH_3 could be considered to take place in the following manner:

$$CH_3 + O \rightarrow H_2CO + H \tag{18}$$

$$CH_3 + O_2 \rightarrow H_2CO + OH \tag{19}$$

$$H_2CO + OH \rightarrow HCO + H_2O \tag{20}$$

$$HCO + OH \rightarrow H_2O + CO \tag{21}$$

Since reliable rate constants were not available for these reactions, Seery and Bowman carried out a parametric study of the rate constants for reactions (18)–(21) in order to determine the relative importance of these reactions. The parametric study indicated that reaction (18) could be deleted from the mechanism with negligible effect on the induction times, and that reactions (19) and (20) could be combined to yield

$$CH_3 + O_2 \rightarrow HCO + H_2O \tag{22}$$

with only a slight effect on the calculated induction times. The parametric study also showed that predicted induction times were nearly independent of the values of k_{21} in the range 5×10^{12} to 5×10^{15} cc/mole-sec. It was also reported that the calculated induction times were only slightly dependent on k_{22} for $k_{22} > 10^{12}$ cc/mole-sec, but they were strongly dependent on k_{22} for $k_{22} < 10^{12}$ cc/mole-sec. In order to obtain good agreement between calculated and measured induction times Seery and Bowman were also forced to add reaction (23) to their reaction mechanism:

$$HCO + M \rightarrow H + CO + M \tag{23}$$

Skinner et al.[19] have postulated a very simplified scheme for the oxidation of CH_3:

$$CH_3 + O_2 \rightarrow H_2 + CO + OH \tag{24}$$

These authors do not propose that reaction (24) occurs in one step as written but that the two reactions

$$CH_3 + O_2 \rightarrow H_2CO + OH \tag{25}$$

$$H_2CO \rightarrow H_2 + CO \tag{26}$$

occur in rapid succession, and that the formaldehyde formed by reaction (25) rapidly decomposes to form the products H_2 and CO. An overall rate constant for reaction (24) is not available in the literature and Skinner et al.[19] assigned a value to the frequency factor and activation energy for the reaction of 4×10^{12} cc/mole-sec and 18 Kcal/mole, respectively.

The methane oxidation mechanism must include radical attack on CH_4,

$$CH_4 + O \rightarrow CH_3 + OH \tag{27}$$

$$CH_4 + H \rightarrow CH_3 + H_2 \tag{28}$$

$$CH_4 + OH \rightarrow CH_3 + H_2O \tag{29}$$

and reactions for the oxidation of CO and the high-temperature H_2–O_2 system:

$$CO + OH \rightarrow CO_2 + H \tag{30}$$

$$H + O_2 \rightarrow OH + O \tag{31}$$

$$O + H_2 \rightarrow H + OH \tag{32}$$

$$O + H_2O \rightarrow OH + OH \tag{33}$$

$$H + H_2O \rightarrow H_2 + OH \tag{34}$$

$$H + OH + M \rightarrow H_2O + M \tag{35}$$

Since Skinner et al.[19] included reaction (16) in their mechanism, they were forced to include the following reactions associated with the hydroperoxyl radical, HO_2:

$$HO_2 + Ar \rightarrow H + O_2 + Ar \tag{36}$$

$$HO_2 + CH_4 \rightarrow CH_3 + H_2O_2 \tag{37}$$

$$H_2O_2 + Ar \rightarrow OH + OH + Ar \tag{38}$$

$$HO_2 + H_2 \rightarrow H_2O_2 + H \tag{39}$$

$$H_2 + O_2 \rightarrow HO_2 + H \tag{40}$$

Skinner et al.[19] also included the following three-body recombination reactions:

$$CO + O + Ar \rightarrow CO_2 + Ar \tag{41}$$

$$O + O + Ar \rightarrow O_2 + Ar \tag{42}$$

$$H + H + Ar \rightarrow H_2 + Ar \tag{43}$$

$$H + H + H_2O \rightarrow H_2 + H_2O \tag{44}$$

Finally, Skinner and co-workers, included the recombination of methyl radicals in their proposed oxidation scheme

$$CH_3 + CH_3 \rightarrow C_2H_6 \tag{45}$$

Cooke and Williams[13] also proposed that the recombination of methyl radicals to ethane must be taken into account for relatively low temperatures or relatively rich mixtures where the concentration of methyl radicals becomes relatively large. Cooke and Williams[13] also include a large number of reactions dealing with the oxidation and decomposition of ethane in their proposed mechanism.

The proposed reaction mechanism and the temperature dependence of the forward rate constants for the various elementary rate constants as proposed by Seery and Bowman[11] and Skinner et al. (19) are presented in Tables III and IV, respectively. Seery and Bowman reported that numerical values for ten [reactions (15), (27)–(35)] of the thirteen steps in their mechanism were available in the literature. Of the remaining three reactions [reactions (21)–(23)], it was reported that k_{21} did not have an apparent effect on the calculated induction times, and hence two unknown rate constants k_{22} and k_{23} required determination before the mechanism could be integrated. The rate constant, k_{23}, was established from collision theory, and the value of k_{22} and k_{23} listed in Table III were evaluated by obtaining the best agreement between calculated and measured induction times for a group of experimental data points.

Skinner and co-workers[19] reported that numerical values, modified slightly within the estimated limits of experimental error to yield better agreement between experimental and calculated induction times, for nineteen [reactions (15), (27)–(34)–(36), (38), (39), and (41)–(45)] of the twenty-three steps in their proposed mechanism were available in the literature. The rate constants of the remaining four reactions [reactions (16), (24), (37), and (40)] have been estimated and the estimated values are listed in Table IV. The rate constant for reaction (16), one of the important initiation reactions, was estimated to have an activation energy slightly greater than the heat of reaction

Table III.   Reaction Mechanism for Methane Oxidation Proposed by Seery and Bowman[11]

No.[a]	Reaction	$K_f{}^b$[Seery and Bowman[11]]	$K_f{}^b$[Bowman[14]]
15	$CH_4+M \rightarrow CH_3+H+M$	$1.5 \times 10^{19} \exp(-50300/T)$	$1.0 \times 10^{18} \exp(-44500/T)$
21	$HCO+OH \rightarrow CO+H_2O$	$10^{12}-10^{15}$ (no effect on τ)	1×10^{14} (best fit)
22	$CH_3+O_2 \rightarrow HCO+H_2O$	2×10^{10} (best fit)	2×10^{10} (best fit)
23	$HCO+M \rightarrow H+CO+M$	$2 \times 10^{13} T^{1/2} \exp(14400/T)$ (best fit)	$2 \times 10^{12} T^{1/2} \exp(14400/T)$ (best fit)
27	$CH_4+O \rightarrow CH_3+OH$	$1.7 \times 10^{13} \exp(-4380/T)$	$2.0 \times 10^{13} \exp(-4640/T)$
28	$CH_4+H \rightarrow CH_3+H_2$	$6.3 \times 10^{13} \exp(6350/T)$	$6.9 \times 10^{13} \exp(-5950/T)$
29	$CH_4+OH \rightarrow CH_3+H_2O$	$2.8 \times 10^{13} \exp(-2500/T)$	$2.8 \times 10^{13} \exp(-2500/T)$
30	$CO+OH \rightarrow CO_2+H$	$3.1 \times 10^{11} \exp(-300/T)$	$5.6 \times 10^{11} \exp(-545/T)$
31	$H+O_2 \rightarrow OH+O$	$2.2 \times 10^{14} \exp(-8310/T)$	$2.2 \times 10^{14} \exp(-8340/T)$
32	$O+H_2 \rightarrow H+OH$	$4.0 \times 10^{14} \exp(-4730/T)$	$1.7 \times 10^{13} \exp(-4750/T)$
33	$O+H_2O \rightarrow OH+OH$	$8.4 \times 10^{14} \exp(-9120/T)$	$5.8 \times 10^{13} \exp(-9070/T)$
34	$H+H_2O \rightarrow H_2+OH$	$1.0 \times 10^{14} \exp(-10200/T)$	$8.4 \times 10^{13} \exp(-10100/T)$
35	$H+OH+M \rightarrow H_2O+M$	$2.0 \times 10^{-19} T^{-1}$	$1 \times 10^{19} T^{-1}$
46	$CH_3+O \rightarrow HCO+H_2$	Not used	1×10^{14} (best fit)

[a]See text.
[b]Units — cm, cal, °K, g-mole, sec.

Table IV.   Reaction Mechanism for Methane Oxidation
Proposed by Skinner et al. [19]

No.[a]	Reaction	$K_f{}^b$
15	$CH_4 + M \rightarrow CH_3 + H + M$	$4 \times 10^{17} \exp(-44{,}289/T)$
16	$CH_4 + O_2 \rightarrow CH_3 + HO_2$	$8 \times 10^{13} \exp(-28183/T)$ (see text)
24	$CH_3 + O_2 \rightarrow H_2 + CO + OH$	$4 \times 10^{12} \exp(-9059/T)$ (see text)
27	$CH_4 + O \rightarrow CH_3 + OH$	$2.1 \times 10^{13} \exp(-2290/T)$
28	$CH_4 + H \rightarrow CH_3 + H_2$	$6.3 \times 10^{13} \exp(-5989/T)$
29	$CH_4 + OH \rightarrow CH_3 + H_2O$	$2.1 \times 10^{13} \exp(-2446/T)$
30	$CO + OH \rightarrow CO_2 + H$	$5.6 \times 10^{11} \exp(-544/T)$
31	$H + O_2 \rightarrow OH + O$	$3.3 \times 10^{14} \exp(-8455/T)$
32	$O + H_2 \rightarrow H + OH$	$2.2 \times 10^{13} \exp(-4756/T)$
33	$O + H_2O \rightarrow OH + OH$	$5.8 \times 10^{13} \exp(-9059/T)$
34	$H + H_2O \rightarrow H_2 + OH$	$1.15 \times 10^{14} \exp(-10116/T)$
35'	$H + OH + H_2O \rightarrow H_2O + H_2O$	1.2×10^{17}
35''	$H + OH + Ar \rightarrow H_2O + Ar$	1.0×10^{16}
36	$HO_2 + Ar \rightarrow H + O_2 + Ar$	$1.8 \times 10^{15} \exp(-23100/T)$
37	$HO_2 + CH_4 \rightarrow CH_3 + H_2O_2$	$2.0 \times 10^{13} \exp(-9059/T)$ (see text)
38	$H_2O_2 + Ar \rightarrow OH + OH + Ar$	$3.2 \times 10^{16} \exp(-21643/T)$
39	$HO_2 + H_2 \rightarrow H_2O_2 + H$	$1.92 \times 10^{13} \mathrm{xpe}(-12079/T)$
40	$H_2 + O_2 \rightarrow HO_2 + H$	$3 \times 10^{13} \exp(-31706/T)$ (see text)
41	$CO + O + Ar \rightarrow CO_2 + Ar$	4×10^{13}
42	$O + O + Ar \rightarrow O_2 + Ar$	3×10^{13}
43	$H + H + Ar \rightarrow H_2 + Ar$	$2.5 \times 10^{14} \exp(-1384/T)$
44	$H + H + H_2O \rightarrow H_2 + H_2O$	$4.0 \times 10^{15} \exp(+1384/T)$
45	$CH_3 + CH_3 \rightarrow C_2H_6$	$1.4 + 10^{13} \exp(-201/T)$

[a]See text.
[b]Units — cm, cal, °K, g-mole, sec.

and a frequency factor typical for an atom-exchange reaction. The rate constant for reaction (24), probably the most controversial reaction in their mechanism, was assigned a frequency factor of 4×10^{12} cc/mole-sec and an activation energy of 18 Kcal/mole. Reaction (37) was assigned an activation energy slightly greater than the thermodynamic heat of reaction and a frequency factor typical for an atom-exchange reaction. It was also reported that the numerical value of k_{40} did not significantly effect the calculated induction times.

Seery and Bowman[11] employed the 13-step mechanism, all of the reactions listed in Table III except reaction (46), to predict induction times for CH_4–O_2–Ar mixtures in the temperature, pressure, and equivalence ratio ranges 1700 to 2000°K, 1.7 to 6.0 atm, and 0.2 to 5.0, respectively. Since their experimental results were obtained behind reflected shock waves, the reaction mechanism was integrated using a constant-volume, adiabatic process as the physical model for the process. The computed temperature and pressure profiles and the calculated concentration profiles of a CH_4–O_2–Ar mixture having an equivalence ratio of 0.5 and an initial temperature and

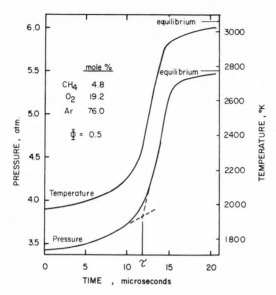

Fig. 6. Predicted temperature and pressure profiles for a reacting
CH_4–O_2–Ar system ($\phi = 0.5$; $T_0 = 2000\,°K$; $P_0 = 3.4$ atm).

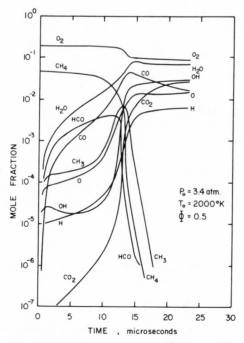

Fig. 7. Predicted composition profiles for a reacting CH_4–
O_2–Ar system ($\phi = 0.5$; $T_0 = 2000\,°K$; $P_0 = 3.4$ atm).

pressure of 2000°K and 3.4 atm, respectively, as reported in Ref. 11 are shown in Figs. 6 and 7.

The results shown in Fig. 6 indicate that the pressure and temperature increase very slowly during the initial stages of the reaction, and at some critical time begin to increase very rapidly toward their equilibrium values. The predicted induction time, τ, based on the rapid increase in pressure is shown in Fig. 6 to occur at approximately 12.5 μsec.

Inspection of Fig. 7 indicates that the fuel and oxidizer concentrations remain relatively constant during the initial stages of reaction, but that they begin to disappear rapidly at a critical time. The major reaction products (H_2O, CO_2, and CO) are shown to have significantly different time histories. The OH radical is shown to remain at a relatively constant value during the initial stages of the reaction and then at a critical time it begins an extremely rapid rate of increase in concentration. For the conditions shown in Fig. 7, this rapid increase in OH concentration occurs at approximately 12 μsec, and hence, in this case, induction times based on the rapid increase in either pressure or in OH concentration, that could be experimentally determined spectroscopically, would yield approximately the same values.

Seery and Bowman[11] reported that in the temperature range 1700 to 2000°K the computed induction times for CH_4–O_2–Ar mixtures, with equivalence ratios in the range 0.5 to 2.0, were within 30% of the experimentally determined values. However, for very lean ($\phi = 0.2$) and very rich ($\phi = 5.0$) mixtures, the calculated results were less satisfactory, and the computed induction times were reported to be within a factor of three of the experimental results.

In a subsequent paper, Bowman[14] followed the progress of high-temperature CH_4 oxidation by monitoring the infrared emission from CO_2, CO, and H_2O. Eight CH_4–O_2–Ar mixtures, with equivalence ratios in the range 0.25 to 4.0 were studied. The experiments included the temperature range 1700 to 2500°K, and most of the experiments were carried out behind reflected shock waves at a total pressure of approximately 2.6 atm. The experimental data were presented in terms of "reaction times" for CO_2, CO, and H_2O. The reaction times were defined to be the time between shock heating and the attainment of 90% of the equilibrium emission intensity for each species of interest.

The 13-step mechanism previously used by Seery and Bowman[11] to calculate induction times for the CH_4–O_2 system was used, with one addition, to predict the three experimentally determined reaction times as a function of experimental conditions. In the previous mechanism, the reaction

$$CH_3 + O \rightarrow HCO + H_2 \qquad (46)$$

was neglected in comparison to reaction (22) during the induction period when the concentration of atomic oxygen is relatively low. However, with the onset of rapid reaction, the atomic oxygen concentration increases significantly, and hence reaction (46) must be included in the mechanism that is employed to describe the overall reaction.

The temperature dependence of the rate constants utilized by Bowman[14] were updated from those employed by Seery and Bowman[11] to take into account more recent work which had appeared in the literature and these updated rate constants are also listed in Table III. As was the case in the earlier work the rate constants for reactions (21)–(23) had not been experimentally determined, and hence were unknown parameters in the mechanism. In addition, the rate constant for reaction (46) had not been experimentally evaluated in the temperature range of interest, and hence it became the fourth unknown parameter in the mechanism.

In order to assign the best values to these four unknown rate constants, a parametric study was carried out in order to assess the sensitivity of the calculated reaction times to the numerical values of the rate constants. The parametric study showed that variation of k_{21} in the range 1×10^{12} to 2×10^{14} cc/mole-sec had only a slight effect on the calculated reaction times. Variation of k_{22} in the range 4×10^9 to 1×10^{11} cc/mole-sec had a significant effect on the calculated reaction times, and it was noted that lower values of k_{22} resulted in a significant increase in the calculated reaction times over the entire temperature range. The pre-exponential factor in the expression for k_{23} was varied from 2×10^{10} to 2×10^{13} cc/mole-sec-$°K^{1/2}$, and it was found that variation in this coefficient slightly effected the temperature dependence of the calculated reaction times. Variation of k_{46} in the range 1×10^{11} to 1×10^{14} cc/mole-sec resulted in a significant change in the calculated reaction times. The calculated reaction times increased as k_{46} decreased over the entire temperature range. Low numerical values for k_{46} resulted in calculated CO reaction times that were much longer than the calculated H_2O reaction times, which was contrary to the experimental observations.

Based on the parametric study, a set of rate constants for reactions (21)–(23) and (46) were determined that best fit the data for a series of experiments, and these results that were used by Bowman[14] for his subsequent calculations are listed in Table III. It is interesting to note that the best-fit value for k_{21} is in the range of numerical values of k_{21} which were previously shown not to effect the computed induction times and that the same k_{22} was employed in both the induction time and reaction time studies. The best pre-exponential coefficient for k_{23} used in the reaction time studies is ten times smaller than its best value in the induction time computations. The 14-step mechanism was employed to predict reaction times for both lean and rich CH_4–O_2–Ar mixtures, in the temperature and pressure range 1700 to 2500 °K and 2.0 to 4.4 atm, respectively. For lean mixtures, the H_2O reaction times were in good agreement with the experimental results over the entire range of conditions, and the predicted CO_2 reaction times were in reasonable agreement with the limited experimental data. However, it was noted that the predicted CO_2 reaction times for the leanest mixtures were significantly larger than would be expected. Both the experimental and analytical results indicated that the CO_2 reaction times should be much greater than the H_2O reaction times for lean mixtures. Unfortunately, due to experimental difficulties CO reaction times could not be measured for lean mixtures.

For rich mixtures, the calculated and measured reaction times for both CO and H_2O were in good agreement, but the predicted values tended to have a smaller apparent activation energy than the experimental results. The CO_2 reaction times were in reasonably good agreement with the experimental results. However, the predicted CO_2 reaction times were larger than the observed times and errors of 200 to 500% were encountered between the calculated and measured values near 2500°K. The difference between the two values at higher temperatures is primarily due to the fact that the apparent activation energy is lower for the calculated results than for the experimental results. For fuel-rich mixtures, it was found that the reaction time for CO_2 was greater than the reaction time for H_2O, and that the reaction time for H_2O was longer than the reaction time for CO. However, in contrast to lean mixtures, the differences in the magnitudes of the reaction times for the three products were reported to be small for rich mixtures.

The 14-step mechanism employed by Bowman[14] for high-temperature CH_4 oxidation predicts reaction times for CO_2, H_2O, and CO that are generally in good agreement with the experimental results. In all cases, the temperature dependence of the predicted reaction times is smaller than that observed experimentally, and the predicted CO reaction times are slightly larger than the predicted H_2O reaction times, which is contrary to the experimental results. In general, the differences between the calculated and measured reaction times are greatest for rich mixtures.

Skinner et al.[19] have employed the mechanism listed in Table IV to compute induction times for a few H_2–O_2–Ar, H_2–CH_4–O_2–Ar, and CH_4–O_2–Ar mixtures, and these results have been compared to the corresponding experimentally measured induction times. The 13 reactions, given by reactions (31)–(34), (35)–(36), (38)–(40), and (42)–(44), were used to estimate induction times in the H_2–O_2–Ar systems, and all 23 steps in Table 4 were employed for the other computations. Since the experimentally measured induction times were obtained behind reflected shock waves, the reaction mechanism was integrated assuming that the combustion process was modeled as a constant-volume adiabatic process.

The observed and calculated induction times for six H_2–O_2–Ar mixtures in the temperature range 1050 to 2500°K, with equivalence ratios of 1.0 and 2.0, were reported to be in agreement within approximately 11%. At 1000°K, where the induction time for H_2–O_2–Ar mixtures is changing extremely rapidly with temperature, the computed and experimental induction times for one mixture were reported to agree within approximately 65%. The inhibition of H_2-O_2 explosions by small quantities of CH_4 in the temperature range 1050 to 1150°K was also reported to be reproduced within approximately 70% by the complete 23-step mechanism.

The 23-step mechanism was also employed to compute induction times for a number of CH_4-O_2-Ar mixtures for temperatures, pressures, and equivalence ratios in the ranges 1600 to 2000°K, 2.5 to 10 atm, and 0.5 to 2.0 respectively. In this range of conditions, all of the computed and experimentally observed induction times agreed to within approximately 40%. However, a number of the results were within 10%. Unfortunately, the mechanism was not tested for very lean or very rich mixtures. Based on the limited number of computations presented by the authors, it is very difficult to ascertain any general trends associated with either temperature of equivalence ratio.

The mechanisms proposed by Williams and co-workers[7,13] for the oxidation of CH_4 reduces to a form very similar to that proposed by Seery and Bowman[11] for lean mixtures. However, Cooke and Williams[13] were first to suggest that the oxidation of rich CH_4-O_2 mixtures is much more complicated than the oxidation of lean CH_4-O_2 mixtures. For rich mixtures, they suggested that the recombination of methyl radicals via reaction (45) is an important step in the reaction mechanism. It was further suggested by these authors that the formation and subsequent reactions of ethane present a significant complication, and hence they proposed that the analysis of ignition of rich CH_4-O_2 mixtures must also involve an analysis of ethane oxidation. The authors suggest that this complication requires the addition of 17 steps dealing with the oxidation and pyrolysis of ethane, ethylene, and acetylene to the reaction mechanism.

Based on their computations at approximately 1875°K, Cooke and Williams[13] reported that the recombination of methyl radicals to ethane and its subsequent reaction does not influence the computed ignition delays for a lean CH_4-O_2 mixture with an equivalence ratio of 0.5. However, in the same temperature range, methyl radical recombination was shown to play a role in the conbustion process for a rich mixture with an equivalence ratio of 2.0.

Based on the information discussed in the preceeding paragraphs, it is obvious that significant progress has been made in the development of analytical models that can be utilized to predict induction times of CH_4-O_2 mixtures. It has been shown that induction times for a variety of CH_4-O_2 mixtures with equivalence ratios between 0.5 and 2.0 can be predicted to within 30% of the experimentally measured values in the temperature range 1600 to 2000°K. It has also been shown that the difference between predicted and experimental induction times tends to increase for either very lean or very rich mixtures.

At the first glance the two mechanisms proposed by Bowman and co-workers[11,14] and Skinner and co-workers (19) appear to be significantly different from one another. Upon close inspection the two mechanisms differ only in three important areas; (1) relative importance of reaction (16) as a potential initiation step, (2) oxidation path for methyl radicals, and (3) the relative importance of methyl radical recombination.

All of the investigators have either directly or indirectly stated that the calculated induction times are very sensitive to the rate of initiation, and hence reliable induction time computations require reliable rate constant data for the two initiation rate constants, k_{15} and k_{16}. The numerical values of the initiation rate constant k_{15} as reported by Seery and Bowman,[11] modified by Bowman[14] in light of new experimental data, and modified still further by Skinner et al.[19] based on additional experimental information are significantly different in the temperature range of interest (see Tables III and IV). The numerical value of the other potential initiation rate constant, k_{16}, is not well known, and hence the relative importance of reaction (16) as an initiation reaction is difficult to assess. In order to determine the relative importance of the initiation rate on the calculated induction times, it is suggested that a parametric study should be carried out to determine the effects of these rate constants on the two mechanisms for a range of experimental conditions.

It is difficult to ascertain which of the two proposed paths for methyl radical oxidation is the most reasonable. However, it is important to note that the two oxidation paths proposed have been used in conjunction with the rest of the mechanism to predict experimentally observed quantities. Since chemical mechanisms are in general not unique, more work will be required to evaluate which of the proposed CH_3-oxidation schemes has the most general applicability. Finally, the relative importance of the methyl radical recombination reaction to form ethane, reaction (45), and the potential additional reactions which C_2H_6 may participate in must be considered in more detail if induction times are to be predicted with a high degree of confidence for CH_4–O_2 systems over a wide range of temperatures and equivalence ratios.

3. Calculated Induction Times for Other Hydrocarbon Systems

Analytical predictions of induction times for hydrocarbons other than CH_4 are essentially nonexistent. Bowman[14] employed a greatly simplified mechanism for ethane oxidation and compared calculated and measured reaction times for CO_2, CO, and H_2O for rich ($\phi = 2.0$ and 4.0) C_2H_6–O_2 mixtures diluted in Ar in the temperature range 1450 to 2000 °K. The greatly simplified mechanism, which assumed that ethane only reacts via the decomposition reaction

$$C_2H_6 \rightarrow CH_3 + CH_3 \qquad (47)$$

and the methyl radicals only react via the reactions listed in Table 3, was used to predict reaction times for CO_2, CO, and H_2O. The predicted temperature variation of the reaction times was considerably lower than the experimentally measured ones. However, over the range of variables considered, the computed and measured reaction times were never different by a factor of more than eight to ten, and over most of the range they were much closer than that. Clearly, this simplified mechanism can not be utilized to study the details of C_2H_6 combustion.

As discussed previously, Cooke and Williams[13] have proposed a 34–step mechanism that can be utilized to calculate induction times in either C_2H_6–O_2 or rich CH_4–O_2 systems. However, a comparison of the predicted and measured induction times for C_2H_6–O_2 systems over a range of conditions has not been published and hence no firm conclusions regarding the ability of this mechanism to predict induction times can be drawn at the present time.

III. PRODUCTION AND EMISSION OF UNBURNED HYDROCARBONS

A. Introduction

It was previously shown in Section I of this chapter that gaseous hydrocarbon emissions from mobile spark-ignition and diesel-powered cyclic-combustion engines and gas turbine powered aircraft represented approximately 61 % of the total gaseous hydrocarbons emitted in 1966. The projected annual emissions of unburned hydrocarbons from transportation sources, see Fig. 1, indicates that these devices will continue to be significant contributors to the total gaseous hydrocarbon emissions in the period 1975–1990. Therefore, it is important to develop an understanding of the processes that control the production and emission of gaseous hydrocarbons in present transportation system power plants and to subsequently utilize these results to improve engine design in order to minimize emissions of this class of pollutants. It is beyond the scope of the present chapter to discuss the application of control devices such as thermal or catalytic reactors as a potential means of controlling gaseous hydrocarbon emissions.

The following discussion will be divided into three main areas; (1) discussion of flame quenching phenomena, (2) consideration of the processes in spark ignition, compression ignition, and gas turbine combustors that are responsible for the production of unburned hydrocarbons, and (3) discussion of techniques that have been developed to predict the effect of engine variables on hydrocarbon emissions from spark-ignition engines.

B. Flame Quenching Phenomena

1. Background Information

Since it has been well established that quenching of flames by the walls of the combustion chamber contributes significantly to unburned hydrocarbons in spark-ignition engines, some experimental and theoretical results concerning this phenomenon will be presented before proceeding to the second two topics. Numerous researchers have shown that a laminar flame propagating through a combustible mixture, well within its flammability limits, may be extinguished if it is forced to propagate through a small constriction. It has been postulated that the walls comprising the constriction quench the flame by acting as a sink for energy and/or active chain carriers. Most research on flame quenching has been confined to studies employing slot burners, flame tubes, or parallel plate systems. In the case of the slot burner or the parallel-plate system, the quenching distance is defined as the minimum distance between the flat plates through which a flame will propagate. The quenching distance is usually found to be a function of pressure, temperature, and reactant composition. The quenching diameter is defined as the minimum diameter tube through which a flame will flash back if the flow is suddenly interrupted. The quenching diameter has been shown to be larger than the quenching distance, and it is usually found to be approximately one and one-third times the quenching distance. When a flame is quenched by a single wall, as would be the case in the combustion chamber of a spark-ignition engine, the distance of closest approach of the flame to the wall is smaller than the quenching distance, and for convenience, this distance is often called the "dead space." In general, the dead space has been assumed to range from 0.33 to 1.0 of the parallel-plate quenching distance.[41-43] Recent experimental evidence[44] has indicated that the dead space, for propane–air flames propagating normal to the end wall of a right circular cylinder, is approximately equal to 40% of the corresponding parallel-plate quenching distance.

2. Experimental Results

Friedman and Johnston[45] measured the laminar quenching distance for propane–air mixtures using rectangular slot burners, and reported the effect of pressure, temperature, and equivalence ratio on the quenching distance. They reported quenching distances for pressures in the range 0.08 to 2.8 atm and air to fuel ratios from 11 to 24. They also studied the effect of temperature from 80 to 595 °F on the quenching distance. Based on the results of these experiments, they proposed that the variation of quenching distance with pressure and temperature could be correlated by an equation of the form

$$q_d = (q_d)_r \left[\frac{P_r}{P} \right]^\alpha \left[\frac{T_r}{T} \right]^\beta \tag{48}$$

where q_d = quenching distance at (P,T); $(q_d)_r$ = reference value of quenching distance at (P_r, T_r) that is dependent on the stoichiometric ratio of the propane–air mixture; P and T = pressure and temperature of the compressed quench zone mixture when flame extinction occurs.

The numerical values reported by these investigators for $(q_d)_r$, α, and β are listed in Table V for $T_r = 300\,°\text{K}$ and $P_r = 1$ atm, respectively. Friedman and Johnston[45] also reported that the quenching distance of copper, mica, glass, and platinum surfaces were the same, and hence they concluded that the quenching effect was independent of the surface material.

Table V. **Correlation Parameters for Predicting Quenching Distances in C_3H_8–Air Systems as a Function of Temperature, Pressure, and Equivalence Ratio**

Equivalence ratio	$(q_d)_r$ (in.)[a]	α	β
0.7	0.170	0.80	1.5
1.0	0.083	0.90	0.60
1.3	0.084	0.96	0.49

[a]See Eq. (48).

Green and Agnew[46] measured the quenching distance of propane–air mixtures for pressures in the range encountered in actual internal-combustion engines. The quenching distances for five propane–air mixtures covering the spectrum from very rich ($A/F = 11.5$) to very lean ($A/F = 22.0$) were determined for pressures in the range 3 to 90 atm. The experiments were carried out in closed spherical bombs maintained at ambient temperature. Two parallel flat plates with a variable slot opening were inserted into the cavity through the wall of the vessel, and a fast response surface thermocouple, that was employed to determine if the flame succeeded in passing between the plates, was inserted on the inner wall of the quenching plates. The quenching distance as a function of pressure was determined by simultaneously recording the pressure and temperature at the plates as a function of time after the flame was centrally ignited. The spacing between the plates was decreased until the flame was extinguished by the plates, and this critical distance was defined as the quenching distance. The experimentally determined quenching distances for five C_3H_8–O_2 mixtures as a function of pressure are shown graphically in Fig. 8. It is apparent from these data lean mixtures have significantly larger quenching distances than do stoichiometric or rich mixtures

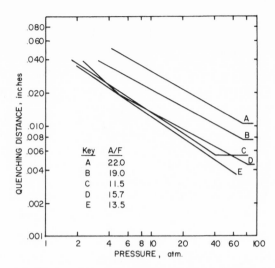

Fig. 8.   Quenching distances for C_3H_8–air mixtures as
a function of pressure.

at any given pressure. It should also be noted that the quenching distances of all of the mixtures except mixture E become pressure independent at high pressures. Experimental data could not be obtained for the slightly rich mixture E at pressures higher than 60 atm. For this mixture, that has the highest burning velocity of any of the mixtures tested, autoignition occurred at pressures above 60 atm, and therefore, it was impossible to obtain data at higher pressures

Green and Agnew's[46] results for the two extreme air–fuel (A/F) ratios studied are compared with the extrapolated low-pressure results of Friedman and Johnston[45] in Fig. 9. The high-pressure data yield a smaller pressure dependence than the low-pressure data, and extrapolation of Friedman and Johnston's data to pressures in the range of those encountered in engines yield quenching distances in the range 1/2 to 1/3 of the measured value. Therefore, it should be noted that extrapolation of the low-pressure quenching data to conditions encountered in engines by the use of Eq. (48) may result in significant errors. At pressures where data are available from both studies, the data reported by Green and Agnew are slightly lower than the data reported by Friedman and Johnston. This may be due to the fact that the former measurements were carried out at constant volume while the latter experiments were carried out at constant pressure.

Gottenberg et al.[47] studied flame quenching in a constant volume analog of a rapid-compression machine. The experiments were carried out in a constant-volume combustion chamber that was charged and exhausted through automatically activated poppet valves. Provisions were also made to vary the surface to volume ratio and the wall temperature of the com-

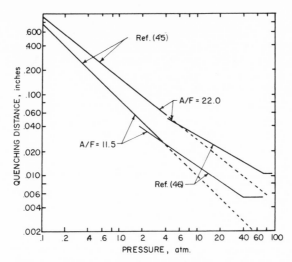

Fig. 9. Comparison of high-pressure C_3H_8–air quenching
distances with extrapolated low-pressure results.

bustion chamber. The unburned hydrocarbons were determined as a function of surface to volume ratio, wall temperature, stoichiometry, and turbulence for *iso*-octane–air mixtures. The turbulence level in the combustion chamber was varied by changing the time interval between inlet valve closure and the ignition of the spark. However, the turbulence level was nonuniform, decaying, and unmeasured. They reported a direct linear relationship between total exhausted hydrocarbons and surface to volume ratio, a direct linear relationship between a representative measured quench distance and the quantity of unreacted hydrocarbons in the combustion products for zero turbulence combustion and that the quantity of unburned hydrocarbons was dependent on the turbulence level.

Up to the present time, all experimental measurements of quenching distance have been carried out with laminar flames. The results reported by Gottenberg et al.[47] indicate that turbulent flames could be expected to have different quenching distances than corresponding laminar flames. Hence, utilization of laminar quenching distances for calculation of the effect of quenching in the production of unburned hydrocarbons in the highly turbulent flows encountered in the combustion chamber of engines may lead to significant errors. Therefore, it is suggested that the effects of turbulence on quenching distances should be studied both experimentally and analytically

3. Prediction of Quenching Distances

Kurkov and Mirsky[48] developed a simplified analytical model to predict the quantity of unburned hydrocarbons that are produced by a laminar flame that is extinguished as it propagates normally toward a cold wall.

The model assumes that the specific heat of all gaseous components is equal and independent of temperature, all binary diffusion coefficients are equal, the thermal conductivity of the mixture is directly proportional to the temperature, and that the reaction mechanism can be represented by a simple first-order step of the type $A \rightarrow B$. The model was employed to estimate the quench-zone thickness and the quantity of unreacted fuel left near the wall after the flame was extinguished. The authors reported that the calculated results were only weakly dependent on the activation energy of the reaction. However, they were strongly dependent on the numerical value of the Lewis number. When the Lewis number was set equal to one, mass diffusion became important enough to significantly reduce the quantity of unburned fuel near the wall by allowing it to diffuse out of the quench layer. The predicted values of the quench-zone thickness calculated at typical engine operating conditions were reported to be several times lower than the measured values reported by Daniel.[41]

Green[49] considered three different models for the prediction of quenching distances for high-pressure propane–air flames. The first model was based on the heat losses due to flame stretching. Based on this model, the flame should extinguish when a critical Karlovitz number is reached. Since the Karlovitz number was shown to be a function of gas propeties, mixture burning velocity, and the quenching distance, the quenching distance can only be calculated as a function of pressure if the critical Karlovitz number and the variation of the flame velocity with pressure are known. Green showed that the Karlovitz number at flame quenching as calculated from measured quenching distances and experimentally determined flame velocities, ranged from 0.30 to 0.45 for stoichiometric propane–air mixtures in the pressure range 2 to 60 atm. Based on these results, he suggested that the critical Karlovitz number was 0.38. As would be expected, the experimental and theoretical quenching distances for stoichiometric propane–air mixtures were in good agreement over the entire pressure range when the critical Karlovitz number of 0.38 was utilized. However, the prediction technique could not be compared to experimental data at other stoichiometries since high-pressure flame velocity data were not available.

Green[49] employed a second model based on thin-flame theory to predict quenching distances. This model consists of a thin flame existing between two parallel plates. The reactants are assumed to flow in a one-dimensional coordinate system, and a single-step mechanism, $A + B \rightarrow C$, is assumed. For a specified heat-transfer mechanism, it can be shown that quenching occurs at a critical heat loss parameter. Subsequently, the quenching distance can be determined in terms of gas properties, the critical heat loss parameter, and the pressure dependence of the flame velocity. The predicted quenching distance based on thin-flame theory, for stoichiometric propane–air flames in the pressure range 2 to 60 atm were in good agreement with the experimental results. Unlike the flame stretch results, experimentally determined quenching distances were not required to obtain the correlation. However, the pressure dependence of the flame velocity must be known to calculate quenching distances from this model. Since high-pressure flame speeds are not presently available for a range of gas mixtures, the general applicability of this model is limited.

Finally, Green attempted to develop a model, based on a flame extinction limit theory of Berlad and Yang[50] for the prediction of quenching distance that did not require an a priori knowledge of the pressure dependence of flame velocity. When applying this model, the temperature profile through the flame is assumed to be represented by a Gaussian error function and energy losses to the wall by both conduction and radiation are taken into account. The flame is assumed to be dominated by a single reaction, $A + B \rightarrow C$, and effects due to diffusion are neglected. The quenching distance as a function of pressure was calculated by employing a double iteration scheme until a stable solution was found. The advantage of this technique is that the solution does not require an a priori knowledge of either high-pressure flame velocity or experimentally determined quenching distances. The theoretical prediction of quenching distance of stoichiometric propane–air flames were carried out for pressures in the range 2 to 60 atm, and unfortunately in all cases, the theoretical value was smaller than the experimentally determined quenching distance. The slope of the theoretical curve was also found to be greater than the experimental curve, and the theoretical results were smaller than the experimental results by a factor of approximately two and five at 10 and 60 atm, respectively.

In general, it can be concluded that some progress has been made on the theoretical prediction of quenching distances. However, experimental results are considered to be presently more reliable than the theoretical predictions, and should be used in computations that require a numerical value for the quenching distance whenever data are available.

C. Sources of Unburned Hydrocarbons in Engines

1. Spark-Ignition Engines

Hydrocarbon emissions from spark-ignition engines vary significantly with the mode of operation. Hence, it is common practice to use composite emission levels, averaged over the Federal 7-mode driving cycle, to evaluate emissions from spark-ignition automotive engines. Hydrocarbon emissions from new automotive engines have been subjected to Federal regulation since 1968. Typical hydrocarbon emissions on the 7-mode Federal driving cycle for pre-1968 uncontrolled vehicles, vehicles which met the 1968 standards, and vehicles which met the 1970 standards have been reported in Ref. 1 to be

11.5, 3.3, and 2.2 g/mile, respectively. For a typical 4000-lb vehicle operating on the 7-mode cycle, these average mass emissions correspond to approximate average concentrations of 900, 275, and 180 ppm, respectively. Average hydrocarbon-emission standards, based on the Federal cycle, are projected to be reduced to approximately 0.4 g/mile by 1975.

Due to the important technical applications of the spark-ignition engine, and the pressure to reduce pollutant emissions from all engines, considerable effort has been expended in order to determine the source of unburned hydrocarbons produced in this type of engine. Daniel[41] proposed that the unburned hydrocarbons that are exhausted during the cruise and acceleration modes are due to the quenching of flames by the walls of the combustion chamber and piston. He measured the thickness of the dark zone between the flame and the combustion-chamber wall in a single-cylinder engine employing propane as a fuel that was fitted with a quartz head. The dark zone, or the dead space, was measured by taking stroboscopic pictures of successive cycles through the quartz cylinder head, and he showed that the quantity of propane trapped in the dead space was sufficient to account for the unburned hydrocarbons emitted from the engine. In order to make the calculations, he assumed that all of the propane contained in a volume equal to the area of the combustion-chamber surface exposed to the flame front times the thickness of the dark zone did not burn. He also reported that the thickness of the dark zone as a function of temperature and pressure compared favorably with the values calculated using the correlations obtained by Friedman and Johnston.[45] Subsequent work by Gottenberg et al.,[47] Shinn and Olson,[51] Daniel and Wentworth,[52] and Daniel[53] has supported Daniel's initial hypothesis and begun to shed some light on the quantity and distribution of hydrocarbons emitted from spark-ignition engines as a function of engine operating conditions. Wentworth[54] showed that a significant source of unburned hydrocarbons was due to the quenching in the crevices and the region between the piston crown and the first compression ring. Wentworth and Daniel[55] have also shown that some of the mixture in the combustion chamber may not be flammable due to dilution by residual gases at closed throttle conditions encountered during deceleration. This phenomena may also be a source of unburned hydrocarbons during this mode of engine operation.

Tabaczynski et al.[56] proposed that there are four separate quench regions in the cylinder of a spark-ignition engine whose aerodynamic behavior is expected to be different. As shown in Fig. 10a, these four quench layers may be expected to be exhausted from the cylinder at different times during the exhaust stroke. Regions 1 and 2 shown in Fig. 10 are the head and side-wall quench layers, respectively. Region 3 represents the piston face quench layer and region 4 corresponds to the quench volume between the cylinder wall, piston crown, and first compression ring. It has been proposed that the head

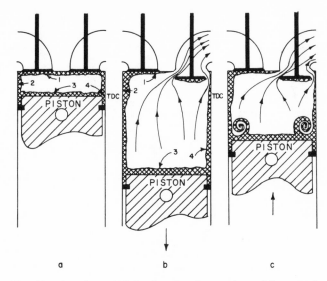

Fig. 10.   Aerodynamic behavior of various regions of the quench
zone during the exhaust process.

quench layer and part of the side-wall quench layer nearest the exhaust
valve will leave the cylinder when the exhaust valve opens (see Fig. 10b). Due
to the low flow velocities near the piston face, the piston face quench layer
will probably not leave the cylinder at any time during the stroke. During the
expansion stroke, the hydrocarbons from the crevice between the piston
crown and the first compression ring are laid along the cylinder wall. As
the piston begins its upward stroke, it has been shown that a vortex is
formed,[57] see Fig. 10c, which "scrapes up" the hydrocarbons along the wall
and forces them to be exhausted near the end of the exhaust stroke. The
authors[56] postulated that this mechanism could explain the high concentration
of hydrocarbons that were observed by Daniel and Wentworth[52] at the end
of the exhaust stroke, and the results reported by Wentworth[59] that showed
that reduction of the volume between the piston crown and the first compression ring substantially reduced average hydrocarbon emission levels.

In order to study the relative contributions of the vortex mechanism
which occurs late in the exhaust process and the quench layers that are
exhausted early in the exhaust process, a series of experiments were carried
out in a single-cylinder engine in which the mass flow rate of hydrocarbons
emitted from the engine as a function of crank angle were measured. The
variation of hydrocarbon mass flow rate versus crank angle was determined
by simultaneously measuring the instantaneous gas mass flow rate leaving
the cylinder and determining the time resolved hydrocarbon concentration
in the exhaust manifold. The instantaneous exhaust gas temperature as a

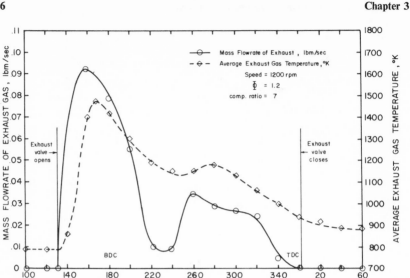

Fig. 11.   Total mass flow rate and temperature of the exhaust gases emitted from a single-cylinder spark-ignition engine as a function of crank angle.

function of crank angle for an engine speed of 1200 rpm, as reported in Ref. 56, are shown in Fig. 11. The exhaust gas mass flow rate is characterized by a sharp rise in flow rate as the exhaust valve opens followed by a continual decrease until the upward motion of the piston begins to influence the exhaust process. All temperature measurements were characterized by the initial rise as the exhaust valve opened. The initial flow is most probably cool, because it was mixed with the residual gas that was present in the exhaust manifold prior to valve opening. In all cases, the gas temperature was reported to reach a maximum at a crank angle of approximately 170° degree after which it began to decrease due to the expansion process.

The hydrocarbon concentration as a function of crank angle, that was measured at a sampling station located 5 in. downstream of the exhaust valve, is shown in Fig. 12. These results were obtained for an air purge of 140 liters/min which was introduced into the exhaust manifold just downstream of the exhaust valve. The air purge was employed in order to sweep out the stagnant mixture remaining in the manifold after valve closure between cycles. The total hydrocarbon mass flow rate as a function of crank angle, which was obtained by multiplying the total mass flow rate (see Fig. 11) times the hydrocarbon concentration, are shown as the solid line in Fig. 12. Figure 12 was also obtained from Ref. 56. These results show that as the exhaust valves open, the hydrocarbon mass flow rate increases rapidly, due to both an increasing exhaust gas mass flow rate and an increasing hydrocarbon concentration. After the initial surge, the mass flow rate decreases rapidly until approximately at a 300° crank angle when the unburned hydrocarbon mass flow rate again increases due to the vortex phenomena which was described earlier. Experiments in which the air purge rate was changed and the air purge was replaced with a corresponding nitrogen purge rate indicated that air injection into the exhaust manifold does reduce the total hydrocarbon emission. However, it was also shown that only the hydrocarbons that leave the cylinder early in the exhaust process when the temperature is high are oxidized to any significant extent in the exhaust manifold.

The results of these experiments further justify the concepts that wall quenching phenomena both in the open regions and crevices in the combustion chamber are the principal source of unburned hydrocarbons in spark-ignition engines. The results also indicate the transient character of the exhaust process which is an important factor that must be taken into consideration when estimating the quantity of unburned hydrocarbons that may be oxidized in the exhaust manifold.

2. Compression-Ignition Engines

The concentration of unburned hydrocarbons in diesel exhaust varies from a few parts per million to approximately 1000 ppm depending on engine type, speed, and load.[58–65] Typically hydrocarbon emissions range from 10 to 300 ppm at idle and 100 to 1000 ppm at full load. The wide range of emission levels reported from diesel engines is primarily due to the fact

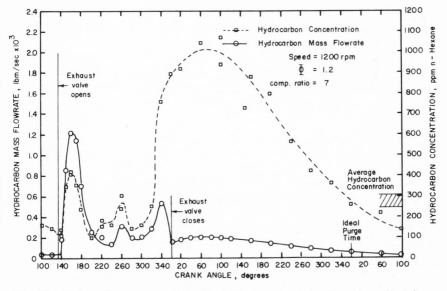

Fig. 12. Hydrocarbon concentration and total hydrocarbon mass flow rate emitted from a single-cylinder spark-ignition engine as a function of crank angle.

that a large number of different types of diesel engines are commercially available; 2-stroke cycle versus 4-stroke cycle, naturally aspirated versus turbo charged, and direct injection versus precombustion-chamber ignition.

As in the case of spark-ignition engines, the direct measurement of the concentration of exhaust hydrocarbons does not yield an adequate measure of the contribution of an individual engine to the pollution problem. Therefore, the California Air Resources Board and the Engine Manufacturers Association, with assistance from SAE and the Coordinating Research Council, have developed a mode test cycle for measuring emissions from diesel engines[65,66] This test procedure will be employed for the 1973 and 1975 California Diesel Emissions Standards. It is anticipated that this new diesel emission cycle, which is used to obtain grams of pollutant per bhp/hr for a specific driving cycle, will be very helpful in determining the realtive contribution of various types and sizes of diesel engines to the pollution problem.

In contrast to the relatively homogeneous combustion processes that takes place in a spark-ignition engine, the heterogeneous combustion processes taking place in a diesel combustion chamber are extremely complicated and not well understood. This is primarily due to the fact that the fuel-injection, vaporization, turbulent mixing, ignition, and combustion processes taking place in a diesel combustion chamber are not only extremely complex and difficult to model experimentally, but have not been studied in as great a detail as combustion phenomena in spark ignition engines. Another complicating factor is that the various engine design features contribute significantly in determining the combustion characteristics of the individual classes of diesel engines.

However, a number of important characteristics of diesel combustion are common to all types of diesel engines. In all diesel engines, the combustion process is initiated by autoignition in a heterogeneous mixture in which the stoichiometry has significant spatial variations. Secondly, physical phenomena, including droplet atomization and vaporization, turbulent mixing, mass transport, and thermal radiation, play a significant role in the combustion process. Finally, excess air is available during the combustion process, the power stroke, and as the combustion products are passing through the exhaust system. Therefore, excess oxygen is readily available to further oxidize hydrocarbons that have not been burned during the primary combustion process.

Even though the complex physical and chemical phenomena leading to the production of hydrocarbons are not well understood, a number of factors effecting hydrocarbon emissions from this class of engines can be discussed. In the first place, the hydrocarbon portion of diesel exhaust includes both original fuel components and products that result from thermal cracking during the combustion process. The relative contributions of the original fuel

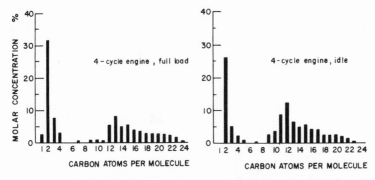

Fig. 13. Distribution of exhaust gaseous hydrocarbons as a function of number of carbon atoms in the molecule for a 4-stroke cycle diesel engine.

components and the products due to thermal cracking may vary significantly as a function of operating conditions. The percentage distribution of various gaseous hydrocarbons as a function of the number of carbon atoms in the molecule in the exhaust of a 4-stroke cycle diesel engine operating at idle and full load, as reported by Hurn,[63] is shown in Fig. 13. Experiments with engines operated on a single pure hydrocarbon fuel[67,68,69] have shown that the diesel combustion process synthesizes almost the complete spectrum of hydrocarbons. It was also shown that the exhaust of such an engine contains molecules with higher molecular weight than the fuel, as well as lighter molecules and molecules with significantly different structures than the fuel molecules; that is, aromatic compounds were found in the exhaust of diesel engines operating with pure paraffins as fuels.

The fuel-injection system has been found to play an extremely important role in determining hydrocarbon emissions from diesel engines.[54,65] In order to minimize hydrocarbon emissions, it is important that injection system provide a sharp start and end of injection and distribute fuel properly in the combustion chamber without impinging the spray on the piston or cylinder head, which could lead to enhanced hydrocarbon emissions due to wall quenching. Ford and co-workers[64] have shown that hydrocarbon emissions from a 2-stroke cycle diesel engine can be significantly reduced by careful design of the injector valve which minimizes the volume which contains uncontrolled fuel below the injector valve. This effect is graphically represented in Fig. 14 where total hydrocarbon emissions as ppm carbon are shown as a function of engine load for three different injectors. All of the tests were identical except for the fact that three different sets of injectors with different volumes of uncontrolled fuel below the injector valve were employed. Inspection of the results in Fig. 14 clearly indicates that total hydrocarbon emissions in all operating modes can be significantly reduced by proper injector design.

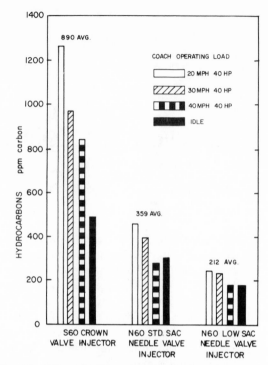

Fig. 14.   Effect of injector design on hydrocarbon exhaust
emissions from a 2-stroke cycle diesel engine.

Ford *et al.*[64] also showed that fuels with higher than 10% point distillation temperatures tended to produce lower hydrocarbon emissions under
the same operating conditions than fuels with lower distillation temperatures.
Based on these results and the results discussed earlier, which related total
hydrocarbon emission to the volumes of uncontrolled fuel below injector
valves, the authors[64] proposed that the uncontrolled fuel contained below
the injector valve is "pushed out" of the injector tip due to expansion and
increasing vapor pressure at higher temperatures. These results indicate that
the physical properties of the fuel play an important role in determining
hydrocarbon emissions from this class of engines.

Experimental results consistently show that hydrocarbon emissions
from diesel engines are a function of engine speed, load, and A/F ratio.
Hurn[63] has shown that normalized hydrocarbon emissions at idle exceed
normalized emissions at full load for a 2-stroke cycle air-scavanged engine,
a 4-stroke cycle nominally aspirated engine, and a 4-stroke cycle turbocharged
engine by factors of approximately three, three, and ten, respectively. Although
there are not definitive data to explain these results, it is reasonable to postulate

that these large differences are due primarily to poor injection performance at off design conditions and the very lean F/A ratios in the neighborhood of 0.01 that are encountered at idle.

Hurn[63] also reports that the quantity of unburned hydrocarbons in diesel exhaust is not significantly influenced by the chemical structure of the fuel. A series of experiments in which both 2-stroke cycle and 4-stroke cycle engines were operated on ten different fuels having significantly different compositional character, while meeting the specification of the engines, were tested. The experimental data indicated that the unburned hydrocarbons were not significantly influenced when operating the engines on ten different fuels.

Based on the experimental evidence which is presently available, one may intuitively conclude that physical properties of the fuel such as viscosity, surface tension, density, and volatility which effect the injection, vaporization, and mixing processes should have a more important effect on determining the level of unburned hydrocarbons in diesel exhaust than the chemical composition of the fuel. It has been shown that injector design is also an important variable in determing the production of unburned hydrocarbons which further substantiates the fact that the physical processes significantly effect the combustion process in diesel engines.

Some progress has been recently reported[70,71] on the development of theoretical models which can be utilized to determine the relative importance of various physical processes in the diesel combustion process. Conclusive results have not been obtained at the present time, but it is anticipated that these studies hold some promise for the development of a better understanding of the mechanisms for pollutant formation in diesel engines.

3. Automotive and Aircraft Gas Turbines

As in the case of diesel engines, the combustion process in gas turbine combustors is heterogeneous in nature and little is known about the details of the complex physical and chemical phenomena that control the production of pollutants in these systems. However, it has been established that both automotive[72-75] and aircraft[76-78] gas turbines emit relatively low quantities of unburned hydrocarbons. In fact, at the present time automotive gas turbines[72,73] can meet proposed 1980 standards for unburned hydrocarbons. Unfortunately, these combustion devices can not meet the proposed NO_x emission standards for 1980, and hence extensive research and development is presently in progress with the goal of reducing NO_x emissions from this class of potential automotive engines. In general, combustor design modifications that reduce NO_x emissions tend to increase unburned hydrocarbon and CO emissions, and therefore any combustor modifications that reduce NO_x emissions must be carefully checked to insure that the corresponding hydrocarbon emissions do not increase to unacceptable levels. It has also been shown that unburned hydrocarbons emitted by aircraft gas turbines are significantly higher in the idle and taxi modes than in the takeoff, cruise, and approach modes of operation. Therefore, hydrocarbon emissions from aircraft gas turbines in these low-load "off design," operating modes may cause local problems near airports and combustor design will have to be modified to minimize these local problems.

The primary function of a gas turbine combustor is to raise the temperature of the gases leaving the exit plane of the combustor to the peak temperatures which can be withstood by the turbine blades. In modern engines, peak turbine inlet temperatures are limited to approximately 1300 °K, and hence the overall equivalence ratio of the combustor is in the range of 0.25. Since hydrocarbon–air mixtures with equivalence ratios of 0.25 are well below the lean flammability limits for hydrocarbon–air mixtures, the overall combustion process is carried out in two zones. The combustor is divided into a primary zone, where sufficient air is mixed with the fuel to essentially complete the combustion process and a secondary or dilution zone where the rest of the air is mixed with the combustion products.

Basically, the combustor, shown schematically in Fig. 15, consists of a liner in which the highly turbulent mixing and combustion processes take place and an annulus through which air flows until it is injected into the liner. Details of the combustion process are not well understood, but the following qualitative description of the combustion process is generally accepted. The fuel enters the primary zone in the form of a spray that rapidly breaks up into small fuel droplets. Since the gases are at a much higher temperature than the boiling point of the fuel, the droplets begin to vaporize as they are formed. A very fuel-rich region exists near the injector where the droplet evaporation is taking place but where there is little combustion due to the lack of air. Either air or air plus hot combustion products flows by the fuel-rich zone entraining and mixing with the fuel vapor. Most of the combustion takes place as the flow travels down stream in the outer region of the mixing zone. In the secondary zone, dilution air is mixed with the combustion products and additional oxidation reactions take place. More details concerning

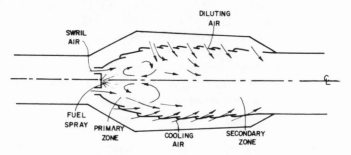

Fig. 15.   Schematic diagram of a gas-turbine combustor.

the models which have been proposed to model gas-turbine combustion can be found in Refs. 72 and 79–87.

Even though the details of the phenomena which control pollutant formation in gas-turbine combustors are not well understood the low hydrocarbon emission levels associated with these devices can be understood in a qualitative manner. Low hydrocarbon emissions from gas-turbine combustors are primarily due to the following factors; (1) essentially no surface quenching occurrs in gas-turbine combustors, (2) continuous combustion take place at a relatively high temperature and in an overall lean fuel–air mixture, and (3) adequate residence time is available in the secondary zone of the combustor to complete the oxidation of the unburned hydrocarbons that leave the primary combustion zone.

D. Effect of Engine Variables on Hydrocarbon Emissions from Spark-Ignition Engines

It has been shown in Section II. C. 1 that the production of unburned hydrocarbons in spark-ignition engines is dependent on a number of factors. At closed throttle condition encountered during deceleration some of the fuel–air mixture in the combustion chamber may not be flammable due to dilution by residual gases.[55] At other operating conditions flame quenching in crevices[54] and near the walls of the combustion chamber[41,51,52] are primarily responsible for the production of unburned hydrocarbons. However, it has also been shown that the quantity and distribution of unburned hydrocarbons finally exhausted from the engine is effected by other processes including; (1) oxidation and pyrolysis of some of the quench gases after flame passage,[53] (2) preferential exhausting of the burned products from the combustion chamber[51,52] and additional oxidation and/or pyrolysis in the exhaust system.[51,53,56,88]

Based on a number of previous studies, Daniel[43] has proposed a semi-empirical model that describes the effects of several engine variables including air–fuel ratio, air-flow rate, compression ratio, engine speed, and ignition timing on total hydrocarbons emitted from a single-cylinder spark-ignition engine operating with propane as a fuel. The unburned hydrocarbons are assumed to be produced by wall quenching and in crevices which are separated from the main combustion chamber by sufficiently narrow passages which do not allow the flame to propagate into the crevice. The quantity of fuel that fails to burn when the flame propagates through the combustion chamber would not be emitted from the engine exhaust system if subsequent processes did not take place. However, gas motion inside the cylinder mixes some of the unburned hydrocarbons with hot combustion products and additional oxidation and pyrolysis may occur before the exhaust valve opens. Not all of the gases in the cylinder are exhausted during the exhaust stroke, and hence some of the unburned hydrocarbons remain in the cylinder to be reprocessed during the next cycle. Finally, the unburned gases leaving the cyliner are mixed during the exhaust process and may continue to react in the exhaust system.

Based on the considerations outlined above, a number of possible relationships may be proposed to predict the quantity of hydrocarbons emitted from the exhaust system per cycle. However, Daniel found that a relationship of the following form provided the best empirical fit to the experimental data.

$$N_e = (N_q X_q + N_c X_c)\, Y_f Y_e \qquad \text{moles/cycle} \qquad (49)$$

where

N_e = moles of hydrocarbons emitted from the exhaust system per cycle

N_q = moles of unburned hydrocarbons per cycle due to flame quenching in the open part of the combustion chamber

N_c = moles of unburned hydrocarbon per cycle due to flame quenching in crevices

X_q, X_c = fractions of unburned hydrocarbons formed by quenching in open part of chamber and crevices that do not oxidize in the cylinder before the exhaust valve opens

Y_f = fraction of unburned hydrocarbons in the combustion chamber at the time the exhaust valve opens that are exhausted into the exhaust system

Y_e = fraction of unburned hydrocarbons entering the exhaust system that are emitted from the exhaust system

Sufficient data were not available to allow the absolute value of hydrocarbon emissions per cycle, N_e, to be directly calculated. However, the absolute quantity of hydrocarbons formed due to flame quenching in the open part of the combustion chamber, N_q, was computed and relative effects of engine operating variables on each of the other factors in the exhaust hydrocarbon emission model were estimated. Details of the trial and error procedures employed to obtain an empirical fit to Eq. (49) are outlined in Daniels's paper.[43] In general, reasonable initial values were assigned to the unknown parameters and each parameter was varied in an iterative manner to minimize the standard error of estimate between the unburned hydrocarbon emissions predicted by Eq. (49) and the measured unburned hydrocarbons emitted by the single-cylinder engine operating with propane as a fuel. The effect of air–fuel ratio, compression ratio, ignition timing, air-flow rate and engine speed on each of the parameters in Eq. (49), and the total exhaust hydrocarbon emissions are reported in Ref. 43. The measured and computed exhaust hydrocarbon concentrations as a function of various

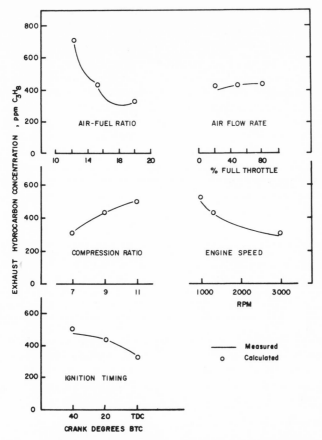

Fig. 16.   Effect of engine operating variables on exhaust hydrocarbon
emissions from a single-cylinder spark-ignition engine.

engine operating conditions are shown in Fig. 16. The results in Fig. 16 indicate that the air–fuel ratio has the greatest effect on total hydrocarbon emissions of any of the operating parameters that were studied.

Based on the results of these studies, Daniel[43] suggested that the quenching of the flame near the open walls of the chamber and flame quenching in chamber crevices were about equally important in determining the exhaust hydrocarbon concentration. He also reports that, on the average, only about one-third of the hydrocarbons which the flame failed to burn were emitted to the atmosphere. Approximately one-third of the unburned hydrocarbons were subsequently oxidized in the combustion chamber after flame passage; about one-half of the hydrocarbons in the chamber when the exhaust valve opened entered the exhaust system; and about one-tenth of the hydrocarbons entering the exhaust system were oxidized before being admitted to the atmosphere.

The results of this study cannot be considered complete. However, it does provide some insight into the relative importance of certain engine variables on exhaust hydrocarbon emissions from spark-ignition engines, and it provides a reasonable starting point for the development of models that can be employed to predict the effect that variations of engine operating parameters will have on emissions of unburned hydrocarbons.

IV. SOME EFFECTS OF HYDROCARBON EMISSIONS ON MAN AND HIS ENVIRONMENT

A. Introduction

A compound or group of compounds is generally classified as an air pollutant if it produces either a short- or long-range deleterious effect on man or his environment. More specifically, hydrocarbon emissions would be classified as air pollutants if they have either a direct or indirect negative effect on the health of man, animals, or plants, dilute the quality of the environment, or directly or indirectly cause economic loss. It has been suggested that various forms of hydrocarbon emissions meet some or all of the above mentioned criteria. Among other things, hydrocarbon emissions from transportation systems participate in the formation of photochemical smog, have been identified as contributors to odor, and may directly or indirectly effect the health of individuals. Subsequent sections of this chapter will deal briefly with each of these three topics.

B. Hydrocarbons and Photochemical Smog

1. Background Material

Although the mechanism is not completely established, photochemical smog results from the atmospheric reaction between certain hydrocarbons and the oxides of nitrogen in the presence of sunlight. The photochemical reactions produce oxidants, primarily ozone, that have undesirable effects on health, vegetation, materials, and the general welfare.[89] The most prominent manifestations of photochemical smog include eye irritation, potential effects on the respiratory system, especially in cases where the respiratory system is marginal for other reasons, reduced visibility, and plant damage.

Laboratory simulations of photochemical smog formation have shown that the rate of formation and maximum concentration of oxidants is strongly dependent on a number of experimental variables including the concentration and structure of the hydrocarbons participating in the reaction, the concentrations of nitric oxide and nitrogen dioxide, the spectral distribution and intensity of the light, and the temperature of the reactant mixture. A discussion of the chemical mechanism responsible for this process is beyond the scope of the present discussion, but the interested reader is referred to Refs. 89–92 for a discussion of the mechanism of photochemical smog formation.

2. Reactivity Scales

It has been clearly established that various hydrocarbons differ markedly in their photochemical behavior. Therefore, the photochemical reactivity of emissions from transportation systems is not only dependent on the total quantity of hydrocarbons emitted but it is also strongly dependent on the concentration of the individual hydrocarbons comprising the exhaust emissions.

Two general approaches have been employed to determine the reactivity of exhaust emissions; (1) calculations based on a summation of the reactivities of individual species in the exhaust appropriately weighted by their concentrations in the exhaust and (2) direct measurements based on experimental observations of photochemical phenomena in smog chambers. Jackson[93] and Altshuller[94] have proposed reactivity scales for individual hydrocarbons that have been widely used to calculate the reactivity of exhaust emissions. In general, specific reactivity scales have shown that saturated hydrocarbons have very low levels of reactivity, aromatics are intermediate, and olefinic material, especially internally bonded diolefins, are very reactive. When applying this technique, the total reactivity of the exhaust is determined by summing the product of the concentration of each individual hydrocarbon multiplied by its individual reactivity over all hydrocarbons in the exhaust. In addition to a knowledge of the reactivities of individual hydrocarbons, application of this method for evaluating exhaust reactivity requires a complete chemical identification of the concentrations of individual hydrocarbons in the exhaust.

McReynolds et al.[95] and Dimitriades et al.[96] have used direct-measurement techniques to determine the reactivity of exhaust gases. When applying the direct method, samples of exhaust gases from a vehicle are collected, introduced into a smog chamber, photoirradiated for a specific period of time, and measurements of the reactivity are obtained by various methods at the end of the irradiation period. Additional information may be attained if the composition of the exhaust gases are also determined.

3. Photochemical Reactivity of Exhaust Emissions

Jackson[93] has employed the summation technique to determine the effects of some engine variables on the reactivity of the exhaust emitted from a spark-ignition engine. The results of this study indicated that the reactivity of the exhaust as a function of engine operating parameters cannot be directly correlated with the total concentration of hydrocarbons in the exhaust. The experimental results indicated that retarding the spark timing at rich A/F ratios increased the reactivity of the exhaust while at the same time decreasing the total hydrocarbon concentration in the exhaust as measured with the aid of a standard nondispersive infrared analyzer system. He also reported that an engine modification based on a combination of lean A/F ratio and retarded-spark timing, reduced the reactivity of the exhaust considerably less than the corresponding decrease in total hydrocarbon emissions. Similarly, leaning the A/F ratio did not reduce the reactivity of the exhaust nearly as much as it reduced the total hydrocarbon concentration. In a subsequent study, Jackson and Everett[97] showed that changes in the reactivity of evaporative emissions as a function of fuel composition could not be directly correlated with the total quantity of evaporative losses.

Dimitriades et al.[96] have employed the direct-measurement technique to assess the effect of three fuel factors; (1) the effect of fuel volatility, (2) the effect of fuel front-end compositon, and (3) comparative evaluation of emissions from leaded and nonleaded fuels on the photochemical reactivity of hydrocarbon emissions. Their work considered emissions from both the fuel system and the exhaust. In their experiments the reactivity of the emissions were characterized by the rate of NO_2 formation in their smog chamber. These authors reported that the photochemical reactivity of both the evaporative and exhaust emissions were significantly reduced by either reducing fuel volatility or by replacing the light olefinic fractions of the fuel with saturated compounds. Variations in fuel composition that were required in order to provide acceptable fuel octane numbers in the absence of lead additives were also reported to generally increase the photochemical reactivity of the emissions.

Dimitriades and co-workers[96] also compared the photochemical reactivity index, as measured by their direct smog chamber technique, R_{ch}, and as calculated from the summation method, R_s, based on the measured composition of the emissions and the reactivity scale proposed by Jackson[93] for a simulated national average city trip plus evaporative losses during the subsequent one hour soak period. Sixteen vehicles were employed in the study and the results presented in Fig. 17 were obtained by averaging the individual results from tests at ambient temperatures of 70 and 95 °F. Inspection of the results presented in Fig. 17 indicate that the direct measurements predict a more rapid rate of rise in photochemical reactivity than do the summation measurements. The results presented in Fig. 17 also indicate the photochemical reactivity of the exhaust emissions from cars that were equipped with exhaust emission controls were slightly lower than for vehicles without the controls.

It is not obvious why the results predicted by the direct and summation methods are not in better agreement with one another. It may be possible

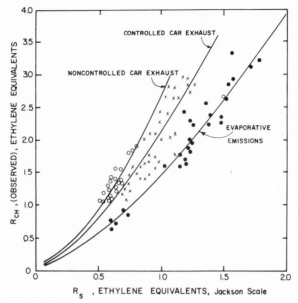

Fig. 17. Comparison of photochemical reactivity of hydrocarbon
emissions as obtained from the direct and summation methods.

that the direct-summation techniques do not give adequate weight to highly
reactive components or that synergistic effects may add to the reactivity of
complex exhaust mixtures. On the other hand, the specific reactivity of the
summation scale has been determined based on measurements of eye irritation levels while the reactivities in the direct method have been evaluated
based on the rate of NO_2 formation. There is some question as to whether
the rate of NO_2 formation is directly proportional to the eye irritation level,
and hence the differences in the methods employed to measure reactivity may
account for the variations in the results. In any event, it is important to note
that there are at least two independent methods that can be utilized to measure
the photochemical reactivity of hydrocarbon emissions.

Since the photochemical reactivity of vehicular emissions are not directly
proportional to total hydrocarbon mass emissions, consideration should be
given to not only controlling total hydrocarbon emissions but also to limiting
the photochemical reactivity of the exhaust emissions. Unfortunately, measurements of total hydrocarbon emissions by either nondispersive infrared
or hydrogen flame ionization detectors, which have been written into present
standards, are misleading since they are not only used to measure total hydrocarbons but also because of the relative insensitivity of these measuring
devices to highly reactive species. The complexities associated with a complete exhaust analysis from a standards point of view are recognized. However,

the true assessment of the photochemical reactivity of the exhaust can only be presently evaluated in this manner, and legislated standards that do not take this important factor into account may be considered to be inadequate.

C. Hydrocarbon Emissions and Odor

1. Background Material

Since they are readily noticeable to the public, odorous emissions have been the subject of considerable adverse public reaction. Some odor is characteristic of the exhaust of spark-ignition engines, but odor emissions from this class of engines are not presently of much concern. It has been reported that properly operated automotive gas turbines are relatively odor free.[72] Aircraft gas turbines have been reported to emit substantial odorous products during the idle and taxi modes of operation, and it is safe to assume that these undesirable emissions will come under legislative scrutiny in the near future. Since almost everyone has been subjected to the odorous emissions from either a diesel-powered truck or bus, diesel engines have been the subject of considerable adverse public opinion. Therefore, most research directed towards developing a better understanding of odorous emissions has centered around the diesel engine. The information discussed in the following paragraphs is especially pertinent to the diesel odor problem, however it is anticipated that the results may be readily extended to other odor sources.

2. Odor-Measurement Techniques

Effective chemical or physical methods for the measurement of odor have not been developed, and therefore the human nose plays a significant role in all odor studies. Unfortunately, this introduces certain subjective influences in any experimental odor tests. When the nose is subjected to an odor, the physiological response to the odor can be classed by either intensity or intensity and quality. The Turk[98] kit, contains a number of different "standard odors" that are classified as; (1) burnt/smoky, (2) oily, (3) pungent/acid, and (4) aldehydic/aromatic. It has been generally accepted as a standard for rating the intensity and quality of an unknown odorous sample.

Odor-detection methods that have been developed to date may be placed in two general categories. The first category includes methods that only rate the overall odor intensity, while the second group is employed to classify odors by quality and intensity. The threshold-dilution technique[67,99,100] and the natural dilution technique[101] that are described below fall into the first category.

The general method[67,99,100] of determining threshold-dilution ratios consists of presenting raw diesel exhaust synthetically diluted with variable quantities of odor-free air to a panel of "sniffers." A series of diluted samples, both above and below the threshold-dilution ratio, are randomly presented to the panel, and the individual panel members are asked to determine whether or not any odor is detectable. The odor intensity is assumed to be proportional to the dilution ratio at which the odor is just detectable to the panel. The main advantage of this technique is that a panel does not have to be trained to differentiate between various odor qualities. The obvious disadvantages of this method are that threshold-dilution ratios yield no information concerning either the quality of the odor or the relative intensities of the various odor qualities.

A method for the evaluation of exhaust odor intensity using natural dilution has been reported by Colucci and Barnes[101] This method was developed in order to determine if full-size diesel-powered vehicles could meet the motor vehicle exhaust odor and irritation standards[102] set by the state of California. During the course of these tests, a panel is seated at varying distances from a full-scale vehicle. Both the vehicle and the panel are located inside a large municipal hangar, in order to minimize any effects due to winds, and the panelists are asked to determine whether or not they can detect any odor from the vehicle. Their responses are utilized to evaluate threshold-response distances.

Variations of the direct method[99,103,104] have been used to rate the quality and intensity of diesel odor, and hence they fall into the second category of odor-detection methods. When applying this method, the exhaust from the diesel engine is usually diluted with odor-free air at the engine exhaust pipe, and the resulting mixture of gases which consists of raw diesel exhaust mixed with odor-free air in ratios ranging from 1 to 200 flows dynamically through a presentation system to the panelists through either a "sniff-tube" or "sniff-box" arrangement. The panelists who have been previously trained to evaluate both quality and intensity as determined by the Turk kit[103,104] or some other criteria[99] are asked to record their response to the test gases as a function of dilution ratio and experimental parameters. An important advantage of the direct technique over the threshold-dilution technique is that the direct technique allows one to rate both the quality and intensity of the exhaust odor. The disadvantages of this technique are that relatively elaborate and expensive facilities must be designed and constructed and large numbers of panelists must be selected, trained, and administered.

3. Odor Relevant Compounds

Until recently very little was known about the compound or compounds that contribute to the odorous and acrid qualities of diesel exhaust. In 1957, Rounds and Pearsall[99] attempted to correlate odor with the aldehydes in diesel exhaust gas, and for a period of time, this class of compounds was commonly cited as the principal contributor to the odor problem. However, Vogh[105] showed that aldehydes were not significant contributors to the overall odor problem. Work by Reckner and co-workers[103,106] contributed to the general knowledge of the composition of diesel exhaust, but little was learned of the chemical nature or the mechanisms responsible for the production of odorous species. Vogh[105] also showed that neither SO_2 nor particulates contribute significantly to diesel odor emissions.

Barnes[67] reported good correlation between exhaust odor threshold-dilution ratios and the lean flammability limits of fuel–oxidizer mixtures. Based on these results, he suggested that the products of partial oxidation that are produced in regions that are too lean to burn at the start of combustion contribute significantly to diesel odor. Hurn[63] suggested that Barnes' hypothesis is plausible for the group of odors that are normally classified as acrid, pungent odors. However, he proposed that the odors that are commonly classified as nauseous may be associated with cracked-fuel components and high molecular weight oxygenated compounds that are produced by low-grade autoignition processes.

Recent work[107] at both the Illinois Institute of Technology Research Institute (IITRI) and Arthur D. Little, Inc. (ADL) have contributed significantly to the development of a better understanding of the chemical nature of the odorous constituents in diesel exhaust. The experimental method developed at IITRI is based on the collection of the organic compounds emitted by a commercial diesel engine (operating on commercial fuel and at a fixed set of operating conditions) on a substrate that does not absorb significant quantities of water or other major exhaust gas components. The collected organic fraction is subsequently released by heating the substrate and passing a dry stream of helium through it. The eluted compounds are subsequently trapped in a stainless steel needle, cooled by liquid nitrogen, and subsequently transferred to a dual-column gas chromatographic system where the compounds are separated and detected with the aid of flame-ionization detectors. The individual gas chromatographic fractions are routed to a sniff port where an observer characterizes the odors based on the Turk[98] classification method. Only about 100 of the over 1000 fractions that have been separated with the aid of gas chromatography were reported to be odor relevant. Once a fraction was determined to be odor relevant, it was trapped in a stainless steel needle cooled in liquid nitrogen for subsequent mass spectrometric identification. The mass-to-charge ratio of the molecular ion and the fragmentation pattern are useful in the identification of odor relevant compounds.

Based on the IITRI work, several classes of compounds have been eliminated as important odorants. Lower series alcohols, simple alkyl benzenes, and alkanes were found as significant exhaust products, but generally at concentrations below their relatively high odor thresholds. The presence of n-alkonoic acids was found, but they did not contribute significantly to the odor.

Various high molecular weight cyclic and aromatic hydrocarbons, including naphthalenes, indans, tetralins, and cyclo paraffins, some with olefinic and/or paraffinic side chains, were reported as major contributors to the burnt odor note of the exhaust. Various nonaromatic hydrocarbons with more than one double or triple bond were also reported to contribute to the

Table VI.   Compounds Contributing to Diesel Odor That Have Been Identified by IITRI

Compound	Elemental composition	Odor note
Toluene	C_7H_8	Nonodorous
Ethylbenzene	C_8H_{10}	Pungent
p-Xylene	C_8H_{10}	Pungent
m-Xylene	C_8H_{10}	Pungent
C_4-substituted benzene	–	Pungent-fuel
Trimethylbenzene	C_9H_{12}	Pungent
p-Ethylstyrene	$C_{10}H_{12}$	Unpleasant-burnt
C_5-substituted benzene	—	Pungent-burnt
Allyltoluene	$C_{10}H_{12}$	Burnt-pungent
C_6-substituted benzene	-	Burnt-pungent
Dimethylcumene	$C_{11}H_{17}$	Strong burnt-pungent
Naphthalene	$C_{10}H_8$	Naphthalene
Methylnaphthalene	$C_{11}H_{10}$	Naphthalene
Methylindan	$C_{10}H_{12}$	Burnt-fuel
Dimethylindan	$C_{11}H_{14}$	Burnt-fuel
Methyltetralin	$C_{11}H_{14}$	Burnt-rubber
Decane	$C_{10}H_{22}$	Nonodorous
Possibly 1,7-octadiyne	C_8H_{10}	Pungent-burnt
Substituted cyclohexane and cyclohexene	—	Pungent-burnt
Cyclic olefin or alkyne	—	Unpleasant-burnt
Trimethylthiophene	$C_7H_{10}S$	Strong foul
C_2-substituted benzothiophene	—	Burnt
Ethanal	C_2H_4O	Sweet
n-Propanal	C_3H_6O	Sweet
n-Butanal	C_4H_8O	Unpleasant-aldehyde
n-Pentanal	$C_5H_{10}O$	Unpleasant-aldehyde
n-Hexanal	$C_6H_{12}O$	Aldehyde
n-Heptanal	$C_7H_{14}O$	Citrus aldehyde
n-Octanal	$C_8H_{16}O$	Aldehyde
Benzaldehyde	C_7H_6O	Cherry-pungent
C_2-substituted benzaldehyde	—	Pleasant-sweet
Ethylbenzaldehyde	$C_9H_{10}O$	Sweet-floral
Tolualdehyde	C_8H_8O	Pleasant-floral
Acetone	C_3H_6O	Pleasant
Possibly acetoin and methylallylketone	$C_4H_8O_2$	Strong foul
Acetophenone	C_8H_8O	Pleasant-floral
Methylacetophenone	$C_9H_{10}O$	Pleasant-sweet

burnt odor note. Furan aldehydes, aromatic benzene, and paraffinic aldehydes from ethanal to *n*-octanal were found to be important odor contributors and have individual odors that varied from pleasant to pungent. Some heterocyclic sulfur compounds, thiophene and benzothiophene derivatives, were also reported to be odor contributors. The odor relevant compounds and their specific odor notes, at concentrations characteristic of their levels in the exhaust, isolated by IITRI[107] are listed in Table VI.

The experimental method developed by ADL is based on the collection of the fraction of diesel exhaust condensable at 0 °C from a large volume of diesel exhaust, 20,000 to 60,000 liters, that is emitted by a commercial engine operating on a specific fuel.

Odor characterization was carried out by a panel located in a specially constructed odor room. The panel was asked to characterize the odorous fractions of the condensate and to relate the odorous compounds to the original exhaust odor notes. The odor panel characterized the odor into two distinct notes; (1) oily-kerosene and (2) smoky-burnt, and it was found that the two character notes contributed approximately equal odor contributions to the overall exhaust odor.

The liquid condensate, that was collected at 0 °C, separated into an aqueous phase and an oily phase upon standing. The organic fraction of the condensate was solvent extracted with pentane and chloroform. Silica gel liquid chromatography was used to separate the odorous components of the extract from the nonodorous components.

These two important odor-relevant extracts have a characteristic oily-kerosene and a smoky-burnt odor, respectively. The two extracts were subsequently separated with the aid of dual-column gas chromatography and a high-resolution mass spectrometer was employed to help identify the chromatographic peaks. Odor panelists were employed to select and characterize odor relevant peaks as they were eluted from the chromatograph.

ADL has reported more progress on determining the classes of compounds that are responsible for the oily-kerosene than for the smoky-burnt odor note. This is primarily due to the fact that the compounds responsible for the oily-kerosene odor were similar to fuel fractions, and hence they were more readily separated and identified than the more polar compounds found in the smoky-burnt fraction. The compounds responsible for the oily-kerosene fraction were also found at concentrations at least a factor of ten higher than the compounds in the smoky-burnt fraction, and this fact also contributed to the relative ease of identification.

Table VII summarizes the odor relevant compounds and specific odor notes of the various compounds, reported by ADL, that contribute to the oily-kerosene odor note. ADL reports that all of the compounds listed in Table VII were present in the fuel. It was also reported that the more easily oxidized compounds, i.e., alkyl benzenes, were present in lower relative

Table VII. Compounds Contributing to the Oily-Kerosene Odor Note of Diesel Exhaust That Have Been Identified by ADL

Compound	Elemental composition	Odor note
Indan and tetralin compounds		
Methylindan	$C_{10}H_{12}$	Irritation
Tetralin	$C_{10}H_{12}$	Rubbery sulfide
Dimethylindan	$C_{11}H_{14}$	Kerosene[a]
Methyltetralin	$C_{11}H_{14}$	Naphthenate[a]
Methyltetralin	$C_{11}H_{14}$	Naphthenate[a]
Dimethyltetralin	$C_{12}H_{16}$	Kerosene[a]
—	$C_{12}H_{16}$	Indole-like
Dimethyltetralin	$C_{12}H_{16}$	—
Dimethyltetralin	$C_{12}H_{16}$	Kerosene[a]
Trimethylindan	$C_{12}H_{16}$	Kerosene[a]
Trimethylindan	$C_{12}H_{16}$	Irritation
Trimethylindan	$C_{12}H_{16}$	—
Trimethylindan	$C_{12}H_{16}$	Kerosene[a]
Alkyltetralin	$C_{12}H_{16}$	Solvent (kerosene)[a]
Alkyltetralin	$C_{12}H_{16}$	Kerosene[a]
Alkyltetralin	$C_{12}H_{16}$	Light Kerosene[a]
Trimethyltetralin	$C_{13}H_{18}$	Naphthenate[a]
Alkyltetralin	$C_{13}H_{18}$	Irritation
Alkyltetralin	$C_{13}H_{18}$	Kerosene[a]
Alkyltetralin	$C_{13}H_{18}$	Pungent, acrid
Alkyltetralin	$C_{13}H_{18}$	Light kerosene
Indene, acenaphthene, and benzothiophene compounds		
Alkylindene	$C_{12}H_{14}$	Heavy oil
Alkylindene	$C_{13}H_{16}$	Heavy oil
Dimethylbenzothiophene	$C_{10}H_{10}S$	None
Dimethylbenzothiophene	$C_{10}H_{10}S$	None
Acenaphthene	$C_{13}H_{12}$	None
Naphthalene compounds		
Monomethyl	$C_{11}H_{10}$	Feel (mothballs)
Monomethyl	$C_{11}H_{10}$	Irritation
Dimethyl	$C_{12}H_{12}$	None
Dimethyl	$C_{12}H_{12}$	None
Dimethyl	$C_{12}H_{12}$	None
Alkyl-benzene compounds		
—	$C_{10}H_{14}$	Rubbery
—	$C_{10}H_{14}$	Musty oily[b]
—	$C_{10}H_{14}$	Tarry
—	$C_{11}H_{16}$	Strong
—	$C_{11}H_{16}$	—
—	$C_{11}H_{16}$	Oily metallic[b]
—	$C_{11}H_{16}$	Oily[b]
—	$C_{11}H_{16}$	—

Table VII   Continued

Compound	Elemental composition	Odor note
—	$C_{12}H_{18}$	Wax[b]
—	$C_{12}H_{18}$	Irritation
—	$C_{12}H_{18}$	Rubbery (not sulfur)
—	$C_{12}H_{18}$	—

[a]Kerosene-related odor notes.   [b]Oily-related odor notes.

concentrations in the exhaust than in the fuel, and that the lower molecular weight compounds in each structural class were relatively less abundant in the exhaust than in the fuel. The total exhaust gas concentration of the compounds related to the oily-kerosene note was reported to be approximately 3 ppm and individual odor-relevant compounds ranged from approximately 6 to 60 ppb.

Significantly less quantitative results were reported for the smoky-burnt odor note. The tentative assignments given to odor relevant compounds in the smoky-burnt fraction as reported by ADL are listed in Table VIII. Based on the present state of the art two general trends were reported. Oxidized derivatives of the alkyl benzenes, the most easily oxidized of the oily-kerosene compounds, were found to be most prevalent in the smoky-burnt fraction. It was also reported that the oxidized derivatives of the indans and tetralins appeared to be related to the smoky-burnt odor note.

The work carried out by IITRI and ADL has contributed to our knowledge of the compounds that contribute to the odorous nature of diesel exhaust and has clearly demonstrated the complexity of the analysis problem. It

Table VIII.   Tentative Identification of Compounds in Diesel Exhaust Relevant to the Smoky-Burnt Odor Note as Reported by ADL

Structure class[a]	C range	Odor note
Alkenone	$C_5–C_{11}$	Oxidized oily
Furan	$C_6–C_{10}$	Irritation, pungency
Dieneone	$C_9–C_{12}$	Sour, oxidized oily
Furfural	$C_6–C_7$	Burnt, oily
Methoxybenzene[b]	$C_8–C_9$	Smoky, pungency
Phenol[b]	$C_7–C_{12}$	Burnt, smoky, particle size, pungency
Benzaldehyde[b]	$C_7–C_{10}$	Burnt, smoky, metallic, pungency
Benzofuran	$C_8–C_9$	Particle size, smoky
Indanone[b]	$C_9–C_{13}$	Metallic, smoky, sour
Indeneone	$C_9–C_{10}$	Linseed-oily, sour
Naphthol[b]	$C_{10}–C_{14}$	Smoky, burnt
Naphthaldehyde	C_{11}	Smoky

[a]Includes hydroxy and methoxy derivatives.   [b]Most abundant classes.

should be noted that both the intensity and character of diesel odor are dependent on engine design and operating conditions. Therefore, even though it was shown that the chosen operating conditions produced an exhaust odor that was typical of diesel exhaust, additional experimental justification is required before the present results can be extrapolated to other engine designs and operating conditons.

The diesel odor research program at Drexel University (DU) has been implemented in order to attempt to elucidate the chemical and physical processes responsible for odorant production and to suggest potential methods for diesel odor abatement. In order to simplify the problem as much as possible without loosing the important phenomena, the DU program has employed a single-cylinder Waukesha CFR cetane rating diesel engine operating on a pure, single-component hydrocarbon fuel as a diesel exhaust source. It has been found that a diesel engine operating on a pure hydrocarbon fuel such as n-heptane or n-hexadecane produces an exhaust whose odorous intensity is comparable, but not exactly the same, as that from an engine operating on commercial fuel. Gas chromatographic analysis of the exhaust gases emitted by an engine operating on a single component fuel indicates that the number of exhaust compounds are greatly reduced. Therefore, diesel odor research, employing single-component fuels in a CFR diesel engine, meets a number of important criteria including production of realistic diesel odor samples, greatly simplified qualitative and quantitative procedures, and the ability to systematically study the effect on varying engine operating parameters, such as injection timing, compression ratio, duel cetane rating, speeds, and load, on both odor production and exhaust gas composition.

Aaronson and Matula[68] utilized combined gas chromatographic-mass spectrometric techniques to analyze the aqueous condensate collected at 0 °C from a lightly loaded CFR diesel engine operating on a pure n-hexadecane fuel at the compression ratio of 14 to 1. A number of species including saturated and unsaturated aldehydes, alcohols, and acids, both homo and heterocyclic compounds, and aromatic compounds were qualitatively identified. Since commercial lubricating oil contains significant quantities of cyclic compounds, the pyrolysis and/or partial oxidation of any oil blow-by was investigated as the possible source of the aromatic compounds that were found in the exhaust. Additional experiments indicated that cyclic compounds were formed even when the commercial oil was replaced with an oil that did not contain any cyclic compounds. Based on the results of both their engine and spray-burner studies, the authors concluded that a large number of aromatic hydrocarbons are synthesized during the heterogeneous combustion of even simple, pure hydrocarbon fuels.

4. Odor Control Techniques

Numerous studies have also been undertaken with the goal of reducing odor emissions from commercial diesel engines. The odor intensities determined by three different measurement techniques using human panels, of two General Motors, 2-stroke cycle engines (type 6V–71E and 6V–71N) have been reported by Merrion.[60] The author reported that the most significant differences between the two engines tested were the type of injector and the combustion chamber shape and volume. It should also be noted that the compression ratio of the type of 6V–71E and 6V–71N engines were 17 to 1 and 18.7 to 1, respectively. The type 6V–71E engine was found to produce exhaust gases with the highest odor intensity. These results were subsequently confirmed by Springer and Hare.[104] Unfortunately, these test results were not obtained with only one independent variable, and hence the relative effects of compression ratio, injector design, and combustion chamber shape and volume on odor production could not be independently evaluated.

Springer and co-workers[104] have tested a number of possible methods for the control of diesel odor. The addition of dilution air at various points throughout the exhaust system of a 2-stroke cycle engine produced no reduction in odor intensity. Engine derating and fuel injection alterations were not found to be satisfactory methods of reducing diesel odor. The installation of a Jacobs engine brake on a single bus did not significantly reduce odor emissions. A number of odor suppressant fuel additives were also tested, and it was found that none of the additives was effective as a consistent odor modifier. On the positive side, a special catalytic muffler was found to slightly reduce odor levels and the odor intensity emitted by a diesel bus was found to be decreased if the standard "S" injectors were replaced by the newer type "N" injectors.

Ford et al.[64] have shown that not only total hydrocarbon emissions but also odor emissions from 2-stroke cycle diesel engines can be significantly

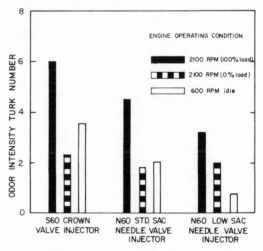

Fig. 18.   Effect of injector design on the odor intensity of exhaust from a 2-stroke cycle diesel engine.

reduced by minimizing the uncontrolled volume of fuel in the injector tip. Reduction in odor intensity for a type 6–71 engine operating on No. 2 diesel fuel at various engine conditions for three injectors with decreasing uncontrolled volumes of fuel in the injector tip as reported by Ford[64] are shown in Fig. 18. The odor measurements were carried out using a 20 to 1 dilution ratio and the odor intensity was determined based on the Turk numbers. These results indicate the diesel odor intensity can be significantly reduced by effective injector design for one type of diesel engine. It remains to be seen if this technique for odor reduction will be useful in other types of diesel engines.

Based on the discussion in the preceeding paragraphs it may be concluded that progress is beginning to be made on understanding the chemical nature of the odorous constitutents that comprise diesel exhaust. The results presented above also underscore the immense complexity of the diesel odor problem. Effective injector design has been shown to reduce diesel odor in one class of engines, but general control techniques to eliminate diesel odor emissions are yet to be developed.

D. Polynuclear Aromatic Hydrocarbon Emissions

1. Background Material

Polynuclear aromatic hydrocarbons (PNA) have been found in the exhaust gas of automotive engines. Haffman and Wynder[108] have identified over 20 different PNA's in the "tars" collected from the exhaust of spark-ignition engines, and a number of these compounds are known carcinogens. Unfortunately, the biological activity of such a complex mixture is very difficult to assess due to the interactions among the carcinogens and the carcinogenic promoters and inhibitors which were also found in the mixture.[108] Wynder and Hoffman[109] reported that the carcinogenic activity of these exhaust "tars" on mouse skin was in the range of 1/25th to 1/50th of that of the same concentration of PNA carcinogens in cigarette smoke. This result, taken along with the large dilution of automobile exhaust, has led some individuals to minimize the potential carcinogenic hazards associated with exhaust emissions of PNA. However, the potential health hazards associated with emissions of PNA from combustion devices must be considered in more detail in the future, and the purpose of the subsequent pargraphs is to briefly review the results of research dealing with PNA emissions from automotive engines.

2. Compression-Ignition Engines

Ray and Long[110] and Scott and co-workers[100,106] have analyzed the particulate phase associated with diesel exhaust. Ray and Long[110] collected their samples from a CFR diesel engine and Scott and co-workers[100,106] obtained their samples from multicylinder 2- and 4-cycle engines. These investigators separated the particulate samples into benzene-soluble and benzene-insoluble fractions and reported that the benzene-soluble fraction contained both unburned fuel and PNA, and that the benzene-insoluble fraction was composed primarily of carbonaceous materials.

Both the 2- and 4-stroke cycle engines were found to emit total particulate concentrations in the range 100 to 500 μg/liter of exhaust. In general, the 4-stroke cycle engine emitted larger percentages of benzene-soluble particulates than the 2-stroke engine. However, at full load, the 4-stroke cycle engine was found to emit very small quantities of benzene-soluble particulate material and large quantities of smoke.

A number of PNA were found in the benzene-soluble fractions of the particulates emitted by both engines. However, the total concentration of PNA was found to be less than 0.1% of the total particulate sample. The studies showed that the polycyclics emitted by the 4-stroke cycle engine were in the range of one-half to one-tenth of the polycyclics emitted by the 2-stroke cycle engine. Additional studies with various fuels indicated that the fuel boiling range is not the most important factor affecting particulate emissions and that the aromatic hydrocarbon content of the fuel could not be correlated with PNA content of the exhaust.

Begeman and Colucci[111] have recently reported the PNA emissions from a 2-stroke cycle diesel engine operated on a simulated driving cycle for city buses. The authors reported that approximately 85% of the benzene-soluble organic matter collected from diesel exhaust was unburned fuel, and that the volatility of the condensed fuel was lower than the volatility of the fuel supplied to the engine. The average emission rates of benzo(a)pyrene (BaP), a known carcinogen, was reported to be 62 μg/gal fuel, which, taking into account the engine speeds, exhaust volumes, and loads in the simulated bus driving cycle, was approximately one to ten times lower[111] than the results reported by Reckner and Squires.[103] Begeman and Colucci[111] also reported that the PNA emission rates for the diesel engine tested were lower than the average emission rates from a 25 spark-ignition car survey and one specific 364-in.^{3} displacement spark-ignition engine operating on premium fuel, but they were higher than the emission rates from one specific 283-in.^{3} displacement spark-ignition engine operating on regular fuel.[111,112]

Additional research designed to help clarify the potential health hazards associated with PNA emissions from diesel engines is required. Measurement of PNA emissions from a characteistic sample of both 2- and 4-stroke diesel engines that are commonly employed in buses and trucks would be very useful in evaluating the relative importance of PNA emissions from compression-ignition and spark-ignition engines. This suggested program would also define the relative PNA emissions from the various diesel engine designs. The effects of fuel composition including the PNA content in diesel fuels as related to PNA exhaust emissions should also be studied. Finally, significant reduction in both the odor and total hydrocarbon emission from 2-stroke cycle diesel engines have been accomplished by careful design of the fuel injector tips to minimize the uncontrolled volume of fuel in the injector tip, and the effects of the new injector design on PNA from this type of diesel engine should also be studied.

3. Spark-Ignition Engines

Much of the early research associated with the emission of PNA from spark-ignition engines was carried out as a collaborative effort between the General Motors Research Laboratories and the Sloan-Kettering Institute for Cancer Research.[108,109,113-115] McKee and McMahon reviewed the early work associated with PNA in vehicle emissions.[116]

Begeman and Colucci[117] measured the BaP in the exhaust of a typical spark-ignition engine operating on a driving cycle which simulated city driving and reported that under their test conditions that approximately 36% of the BaP in the engines exhaust could be directly attributed to the BaP initially in the fuel. Between 0.1 and 0.2% of the BaP in the fuel was reported to survive the combustion process and was recovered from the exhaust, 5% of the BaP was accumulated in the crankcase oil, and some of the original BaP was reported to be converted into other PNA and more polar compounds.

In a subsequent, more comprehensive article, Begeman and Colucci[111] studied the influence of A/F ratio, driving mode, and oil economy on the PNA emitted by spark-ignition engines. During the course of the studies, the contribution of PNA in gasoline to the PNA emissions, the effects of two emission control systems designed to reduce carbon monoxide and hydrocarbons in exhaust on PNA emissions, and the average emission of PNA from a twenty-five car sample of privately owned vehicles in simulated city driving were determined.

Benzo(a)pyrene emissions, a commonly measured PNA in experiments of this type, were determined in all of the experiments and benz(a)anthracene (BaA) emissions were measured in most of the experiments. Both BaP and BaA are known to be carcinogens. In general, the ratios of individual PNA from various sources or from a given source for all operating conditions would not be constant. However, previous evidence indicating that the emissions of PNA are directly interrrelated[118-120] allowed the authors to conclude that the measurements of one or two members of this class of compounds could be considered to be indicative of the total PNA in the exhaust.

In the first series of tests the effects of operating mode, A/F ratio, oil consumption, and fuel composition effects on PNA emissions from a 283-in.³ displacement engine operating with regular fuel were determined. The tests were conducted in a dynamometer test stand and a driving cycle designed to simulate city driving was employed. The reported values of BaA, BaP,

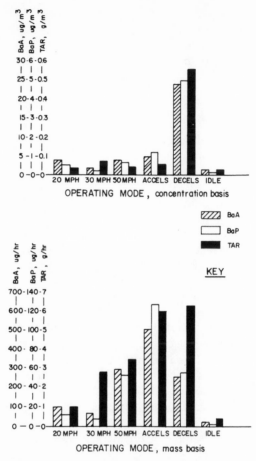

Fig. 19.   Effect of engine operating mode on PNA
hydrocarbon emissions from a spark-ignition engine.

and total tars, both on a concentration and mass basis as a function of engine
operating mode as reported in Ref. 111 are shown in Fig. 19. The results
presented in Fig. 19 clearly indicate that the concentrations of both BaA
and BaP are significantly greater in the deceleration mode than in any of the
other operating modes. However, on the basis of total mass emissions, the
PNA emissions were highest during accelerations when large total mass flow
rates were encountered. Based on these results, it can be concluded that the
operation of the engine in transient modes contributes significantly to total
PNA emissions during a typical city driving cycle. The effects of A/F ratio
on PNA emissions were also studied and it was found that both BaP and
BaA emissions were considerably higher for rich carburetion than for either

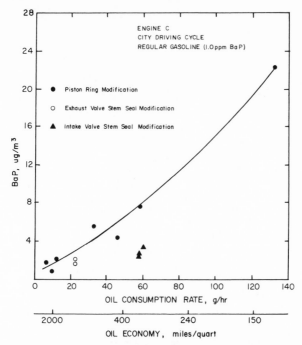

Fig. 20.   Effect of oil consumption on PNA hydrocarbon
emissions from a spark-ignition engine.

stoichiometric or lean carburetion. These results relating high PNA emissions
to transient engine operation and rich carburetion are in general agreement
with the work of previous investigators. [121,122]

High oil rate consumption in the test engines found to significantly
increase PNA emissions. In one series of tests with the 283-in.³ displacement engine operating on regular fuel in the simulated city driving cycle, the
effect of oil consumption on PNA was evaluated by systematically increasing
the oil consumption in a "normal" engine by removal of intake valve guide
seals, exhaust seals, or the expansion rings under one or more of the piston
oil control rings. The results of these experiments as reported in Ref. 111 are
shown in Fig. 20.

Based on these results it can be concluded that the BaP emissions are
approximately proportional to oil consumption when most of the oil is introduced into the chamber past the piston rings. Oil consumption due to leakage
into the exhaust system along the exhaust valve stems also follows this trend.
However, the BaP emissions were not significantly increased when the increased oil consumption was due to unsealed intake valve guides.

The correlation between PNA emissions and the oil consumption rate
in 4-stroke cycle spark-ignition engines may be of importance in determining

potential health hazards associated with 2-stroke spark-ignition engines in which fuel/oil mixtures are burned in the combustion chamber. The highest oil consumption rate reported in Ref. 111 of 146 miles per quart corresponds to approximately 29 parts gasoline per 1 part of oil, which is in the range of fuel/air mixtures (20/1 to 40/1) employed in 2-stroke cycle gasoline engines. These results may lead one to propose that PNA emissions from 2-stroke cycle engines would be high. Hunigen et al.[123] reported that BaP emissions from 2-stroke cycle engines were directly proportional to the concentration of the oil in the oil/fuel mixture, and Kuhn and Tomingas[124] reported BaP emissions in excess of 11,000 μg/gal of fuel, which is considerably higher than the results for 4-stroke cycle engines reported in Ref. 111, for 2-stroke cycle engines operating on a premixed fuel/oil ratio of 33 to 1.

Analysis of samples of crankcase oil obtained from the test engines indicated suprisingly high concentrations of PNA.[111] Based on these measurements it was reported that, when operating the engine under normal oil consumption conditions, the rate of entry of BaP into the crankcase was approximately 10 times greater than the rate of emission of BaP in the exhaust gas. Previous work by the authors[117] had shown that 5% of the BaP in the fuel burned found its way into the crankcase. However, in the present studies, it was reported that 10 to 40% of the BaP in the burned fuel was found in the crankcase. Based on these findings, the authors concluded that most of the BaP which enters the crankcase is formed in the combustion chamber by either pyrolysis of the fuel or oil, or is absorbed by the oil film left on the cylinder wall and is subsequently scraped into the crankcase.

Three different fuels with different initial concentrations of BaP were tested in order to ascertain if the correspondence between PNA in the exhaust could be directly related to the quality of BaP in the fuel as reported in Ref. 117. The PNA emissions were reported to be directly related to PNA in the gasoline, but other unidentified fuel composition factors also seemed to contribute to these emissions.[111] A series of tests also indicated that two emission control systems,[125,126] designed to reduce carbon monoxide and volatile hydrocarbons in the exhaust gases, were also effective in reducing the BaP emissions by factors ranging from 3 to 27 when compared to emissions from the same uncontrolled engines.

Finally, a fleet of 25 cars were tested by Begeman and Colucci[111] in order to determine the average rate of PNA emissions from the existing car population at the end of the 1963 model year. The cars were tested with the fuel left in the cars by the owners, and the test fleet was divided into two groups based on the model year, 1951–1959 and 1960–1963, in order to determine if on the average older cars emitted more PNA than newer cars. The average emission rates of the test fleet were 169 μg/gal, 660 μg/gal, and 3.1 g/gal of BaP, BaA, and tars, respectively, and in all cases the older group of cars was found to emit significantly more PNA than the newer group. The PNA emissions from the various cars differed widely due to the large variance of A/F ratios, oil economies, and possible fuel composition. However the authors[111] propose that the higher average emission recorded by the older group of cars is most probably due to the higher oil consumption rate of this group of cars.

Hangebrauch et al.[127] had previously reported PNA emissions from a group of 8 cars of two makes, including model years 1956–1964. They reported average BaP emissions of 169 $\mu g/gal$, which is in remarkably close agreement with the results of Begeman and Colucci,[111] and that the individual emission rates varied by a factor of ten with the older, high-mileage cars being the highest emitters. It should be noted that the results reported in Ref. 111 were for an auto population representative of approximately ten years ago. However, it is anticipated that the general trends obtained from this study may also be valid today.

Based on the discussions in the preceeding paragraphs, the potential health hazards associated with PNA emissions from combustion devices cannot be properly assessed until additional clinical data are assembled. However, based on our present limited knowledge of PNA formation mechanisms it can be concluded that these emissions from automotive sources can be minimized by utilizing fuels with minimum PNA impurities, operating spark-ignition engines at A/F ratios above stoichiometric, minimizing the consumption of lubricating oil, and employing carbon monoxide and volatile hydrocarbon control systems on automotive engines.

REFERENCES

1. Locklin, D.W., Weller, A.E., and Barrett, R.E., (edis). *The Federal R & D Plan for Air-Pollution Control by Combustion-Process Modifications,* Final Report Contract CPA 22–69–147, Battelle Memorial Institute, Columbus, Ohio, 1971, Ch. 2.
2. Vanpee, M., and Grard. F., The kinetics of the slow combustion of methane at high temperature, *Fifth Symposium (International) on Combustion,* Reinhold Publishing Corp., New York, 1955, pp. 484–490.
3. Brokaw, R.S., and Jackson, J.L., Effect of temperature and pressure on ignition delays for propane flames, *Fifth Symposium (International) on Combustion,* Reinhold Publishing Corp., New York, 1955, pp. 563–569.
4. Skinner, G.B., and Ruehrwein, R.A., Shock tube studies on the pyrolysis and oxidation of methane, *J. Phys. Chem.* **63** (1959) 1736–1742.
5. Kistiakowsky, G.B., and Richards, L.W., Emission of vacuum ultraviolet radiation from the acetylene–oxygen and the methane–oxygen reactions in shock waves, *J. Chem. Phys.* **36** (1962) 1707–1714.
6. Soloukhin, R.I., Quasi-stationary reaction zone in gaseous detonation, *Eleventh Symposium (International) on Combustion,* The Combustion Institute, Pittsburgh, Pennsylvania, 1967, pp. 671–676.
7. Higgin, R.M.R., and Williams, A., A shock-tube investigation of the ignition of lean methane and n-butane mixtures with oxygen, *Twelfth Symposium (International) on Combustion,* The Combustion Institute, Pittsburgh, Pennsylvania, 1969, pp. 579–590.
8. Asaba, T., Yoneda, K., Kakihara, N., and Hikita, T., A shock tube study of ignition

of methane–oxygen mixtures, *Ninth Symposium (International) on Combustion,* Academic Press, New York, 1963, pp. 193–200.

9. Miyama, H., and Takeyama, T., Kinetics of methane oxidation in shock waves, *Bull. Chem. Soc. of Japan* **38**, (1965) 37–43.

10. Glass, G.P., Kistiakowsky, G.B., Michael, J.V., and Niki, H., The oxidation reactions of acetylene and methane, *Tenth Symposium (International) on Combustion,* The Combustion Institute, Pittsburgh, Pennsylvania, 1965, pp. 513–522.

11. Seery, D.J., and Bowman, C.T., An experimental and analytical study of methane oxidation behind shock waves, *Comb, and Flame* **14** (1970) 37–48.

12. Snyder, A.D., Robertson, J., Zanders, D.L., and Skinner, G.B., Shock tube studies of fuel–air ignition characteristic, Monsanto Research Corp., *Tech. Rept. AFAPL-TR-65-93,* August 1965.

13. Cooke, D.F., and Williams, A., Shock tube studies of the ignition and combustion of ethane and slightly rich methane mixtures with oxygen, *Thirteenth Symposium (International) on Combustion,* The Combustion Institute, Pennsylvania, 1971, pp. 757–766.

14. Bowman, C.T., An experimental and analytical investigation of the high temperature oxidation mechanisms of hydrocarbon fuels, *Comb. Sci. Tech.* **2** (1970) 161–172.

15. Kozlov, G.I., On high temperature oxidation of methane, *Seventh Symposium (International) on Combustion,* Butterworth Scientific Publications, London, 1959, pp. 142–149.

16. Nemeth, A., and Sawyer, R.F., The overall kinetics of high temperature oxidation in a flow reactor, *J. Phys. Chem.* **73** (1969) 2421–2424.

17. Vandenabeele, H., Corbeels, R. and van Tiggelen, A., Activation energy and reaction order in methane oxygen flames, *Comb. and Flame* **4** (1960) 253–260.

18. Lifshitz, A., Scheller, K., Burcat, A., and Skinner, G., Shock tube investigation of ignition in methane–oxygen–argon mixtures, *Comb. and Flame* **16** (1971) 311–321.

19. Skinner, G.B., Lifshitz, A., Scheller, K., and Burcat, A., Kinetics of methane oxidation, private communication from K. Scheller, Sept., 1971.

20. Burcat, A., Lifshitz, A. and Scheller, K., Shock tube investigation of comparative ignition delay times for C_1–C_5 alkanes, *Comb. and Flame* **16** (1971) 29–33.

21. Burcat, A., Scheller, K., Crossley, R.W., and Skinner, G.B., Shock initiated ignition of ethane–oxygen–argon mixtures, *ARL Report No. 71-0032,* Aerospace Research Laboratories, Wright-Patterson AFS, Ohio, 1971.

22. Miller, R.E., Some factors governing the ignition delay of a gaseous fuel, *Seventh (International) Symposium on Combustion,* Butterworths Corporation, London, 1959, pp. 417–424.

23. Zimot, V.L., and Troshin, Yu. M., Ignition lag of hydrocarbon fuels at high temperature, *Combustion, Explosion and Shock Waves* **3** (1967)

24. Steinberg, M., and Kaskan, W.E., Ignition of combustible mixtures by shock waves, *Fifth Symposium (International) on Combustion,* Reinhold Publishing Company, New York, 1955, pp. 664–672.

25. Hawthorn, R.D., and Niton, A.C., Shock tube ignition delay studies of endothermic fuels, *AIAA J.* **4** (1966) 513–520.

26. Myers, B.F., and Bartle, E.R., Reaction and ignition delay times in the oxidation of propane, *AIAA J.* **7** (1969) 1862–1869.

27. Burcat, A., Lifshitz, A., Scheller, K., and Skinner, G.B., Shock tube investigation of ignition in propane–oxygen–argon mixtures, *Thirteenth (International) Symposium on Combustion,* The Combustion Institute, Pittsburgh, Pennsylvania, 1971, pp. 745–755.

28. Levinson, G.S., High temperature pre-flame reactions of *n*-heptane, *Comb. and Flame* **9** (1959) 63–72.

29. Mullaney, G.J., Peh, S.K., and Botch, W.D., Determination of induction times in one-dimensional detonations (H_2, C_2H_2, and C_2H_4), *AIAA J.* **3** (1965) 873–875.

30. White, D.R., Density induction times in very lean mixtures of D_2, H_2, C_2H_2, and

C_2H_4 with O_2, *Eleventh (International) Symposium on Combustion,* The Combustion Institute, Pittsburgh, Pennsylvania, 1967, pp. 147–154.
31. Gardiner, W.C., Observations of induction times in the acetylene–oxygen reaction, *J. Chem. Phys.* **35** (1961) 2252–2253.
32. Bradley, J.N., and Kistiakowsky, G.B., Shock wave studies by mass spectrometry. II. Polymerization and oxidation of acetylene, *J. Chem. Phys.* **35** (1961) 264–270.
33. Dove, J.E., and Moulton, D. McL., Shock wave studies by mass spectrometry. III. description of apparatus; data on the oxidation of acetylene and of methane, *Proc. Roy. Soc.* **A283** (1965) 216–228.
34. Hand, C.W. and Kistiakowsky, G.B., Ionization accompanying the acetylene–oxygen reactions in shock waves, *J. Chem. Phys.* **37** (1962) 1239–1245.
35. Stubbeman, R.F. and Gardiner, W.C., Shock tube study of the acetylene–oxygen reaction, *J. Phys. Chem.* **68** (1964) 3169–3176.
36. Stubbeman, R.F., and Gardiner, W.C., Delayed appearance of OH in the acetylene–oxygen reaction, *J. Phys. Chem.* **40** (1964) 1771–1772.
37. Miyama, H., and Takeyama, T., Delayed appearance of OH in acetylene–oxygen reaction, *J. Chem. Phys.* **42** (1965) 2636–2637.
38. Takeyama, T., and Miyama, H., A shock tube study of the acetylene–oxygen reaction, *Bull. Chem. Soc., Jap.* **38** (1965) 936–940.
39. Homer, J.B., and Kistiakowsky, G.B., Oxidation and pyrolysis of ethylene in shock waves, *J. Chem. Phys.* **47** (1967) 5290–5295.
40. Gay, I.D., Glass, G.P., Kern, R.D., and Kistiakowsky, G.B., Ethylene–oxygen reaction in shock waves, *J. Chem. Phys.* **47** (1967) 313–320.
41. Daniel, W.A., Flame quenching at the walls of an internal combustion engine, *Sixth Symposium (International) on Combustion,* Reinhold Publishing Company, New York, 1957, pp. 886–894.
42. Agnew, J.T., Unburned hydrocarbons in closed vessel explosions, theory versus experiment applications to spark ignition engine exhaust, *SAE Paper No. 670125,* 1968.
43. Daniel, W.A., Why engine variables affect exhaust hydrocarbon emissions, *SAE Paper No. 700108,* 1970.
44. Ellenberger, J.M., and Bowlus, D.A., Single wall quench distance measurements, Paper presented at the **1971** *Technical Session, Central States Section,* The Combustion Institute, Ann Arbor, Michigan, March 1971.
45. Friedman, R., and Johnston, W.C., The wall-quenching of laminar propane flames as a function of pressure, temperature and air–fuel ratio, *J. Appl. Phys.* **21** (1950) 791–795.
46. Green, K.A., and Agnew, J.T., Quenching distances of propane–air flames in a constant-volume bomb, *Comb. and Flame,* **15** (1970) 189–191.
47. Gottenberg, W.G., Olson, D.R., and Best, W.H., Flame quenching during high pressure, high turbulence combustion, *Comb. and Flame,* **7** (1963) 9–16.
48. Kurkov, A.P., and Mirsky, W., An analysis of the mechanism of flame extinction by a cold wall, *Twelfth Symposium (International) on Combustion,* The Combustion Institute, Pittsburgh, Pennsylvania, 1969. pp. 615–624.
49. Green, K.A., Quenching distances of propane–air flame at high pressures and several mixture ratios, *Ph. D. Thesis,* Drexel University, Philadelphia, Pennsylvania, June 1969.
50. Berlad A.L. and Yang, C.H., A theory of flame extinction limits, *Comb. and Flame* (1960) 325–333.
51. Shinn, J.N., and Olson., D.R., Some factors affecting unburned hydrocarbons in combustion products, *SAE Technical Progress Series, Vol. 6, Vehicle Emissions,* Society of Automotive Engineers, Inc., New York, 1964.
52. Daniel, W.A., and Wentworth, J.T., Exhaust gas hydrocarbons — Genesis and exodus *SAE Technical Progress Series, Vol. 6, Vehicle Emissions,* Society of Automotive Engineers, New York, 1964.

53. Daniel, W.A., Engine variables effects in exhaust hydrocarbon composition (a single cylinder engine study with propane as a fuel), *SAE Paper No. 670124,* 1967.

54. Wentworth, J.T., Piston and ring variables affect exhaust hydrocarbon emissions, *SAE Paper No. 680109,* 1968

55. Wentworth, J.T., and Daniel, W.A., Flame photographs of light load combustion point the way to reduction of hydrocarbons in exhaust gas, *SAE Technical Progress Series, Vol. 6, Vehicle Emissions,* Society of Automotive Engineers, Inc., New York, 1964.

56. Tabaczyski, R.J., Heywood, J.B., and Keck, J.C., Time-resolved measurements of hydrocarbon mass flow ratio in the exhaust of a spark ignition engine, *Fluid Mechanics Laboratory Publication No. 71-10,* Department of Mechanical Engineering, Massachusetts Institute of Technology, Cambridge, Massachusetts, April 1971.

57. Tabaczynski, R.J., Hoult, D.P. and Keck, J.C., High Reynolds Number Flow in a Moving Corner, *J. Fluid Mech.* 42 (1970) 249–255.

58. Yumlu, V.S., and Carey, Jr., A.W., Exhaust emission characteristics of four-stroke, direct injection, compression ignition engines, *SAE Paper No. 680420,* 1968.

59. Parez, J.M., and Landon, E.W., Exhaust emission characteristics of precombustion chamber engines, *SAE Paper No. 680421,* 1968.

60. Merrion, D.F., Effect of design revisions of two-stroke cycle diesel engine exhaust, *SAE Paper No. 680421,* 1968.

61. Marshall, W.F., and Hurn, R.W., Factors influencing diesel emissions, *SAE Paper No. 680528,* 1968.

62. Johnson, J.H., Sienicki, E.J., and Zeck, O.F., A flame ionization technique for measuring total hydrocarbons in diesel exhaust, *SAE Paper No. 680419,* 1968.

63. Hurn, R.W., Air pollution and the compression ignition engine, *Twelfth Symposium (International) on Combustion,* The Combustion Institute, Pittsburgh, Pennsylvania, 1969, pp. 677–687.

64. Ford, H.S., Merrion, D.F., and Hames, R.J., Reducing hydrocarbons and odor in diesel exhaust by fuel injector design, *SAE Paper No. 700734,* 1970.

65. Bascom, R.C., Broering, L.C. and Wulfhorst, D.E., Design factors that affect diesel emissions, *SAE Paper No. 710484,* 1971.

66. Bascom, R.C., and Hass, G.C., A Status report on the development of the 1973 california diesel emissions standards, *SAE Paper No. 700671,* 1971.

67. Barnes, G.J., Relation of Lean combustion limits in diesel engines to exhaust odor intensity, *SAE Paper 680445,* 1968.

68. Aaronson, A.E., and Matula, R.A., Diesel odor and the formation of aromatic hydrocarbons during the heterogeneous combustion of pure cetane, in a single-cylinder diesel engine, *Thirteenth Symposium (International) on Combustion,* The Combustion Institute, Pittsburgh, Pennsylvania, 1971, pp. 627–637.

69. Milks, D., Savery, C.W., Steinberg, J.L., and Matula, R.A., Studies and analysis of diesel engine odor production, *Paper No. EN-5E,* Presented at *Clean Air Congress of the International Union of Air Pollution,* Washington, D.C., December, 1970.

70. Bracco, F.V., A model for the diesel engine combustion and NO formation, Presented at the *Central States Section/Combustion Institute, Spring Meeting,* Ann Arbor, Michigan, March 23 and 24, 1971.

71. Fletcher, R.S, Chng, K.M., Heywood, J.B., and Bastress, E.K., Models of combustion and nitric oxide formation in direct and indirect injection compression-ignition engines, Presented at the *Central States Section/Combustion Institute,* Spring Meeting, Ann Arbor, Michigan, March 23 and 24, 1971.

72. Cornelius, W., Stivender, D.L., and Sullivan, R.E., A combustion system for a vehicular regenerative gas turbine featuring low air pollutant emissions, *SAE Paper No. 670936,* 1967.

73. Korth, M.W., and Rose, Jr., A.H., Emissions from a gas turbine automobile, *SAE Paper No. 680402,* 1968.

74. Lieberman, A., Compositions of exhaust from a regenerative turbine system, *J. Air Poll. Control Assoc.* **18** (1968) 149–153.
75. Cornelius, W. and Wade, W.R., The formation and control of nitric oxide in a regenerative gas turbine burner, *SAE Paper No. 700708*, 1970.
76. Lozano, E.R., Melvin, W.W., and Hochheiser, S., Air pollution emissions from jet engines, *J. Air Poll. Control Assoc.* **18** (1968) 392–394.
77. George, R.E., and Burlin, R.M., Air pollution from commercial jet aircraft in Los Angeles county, *Los Angeles Air Pollution Control District*, Los Angeles, Calif., April, 1960.
78. Smith, D., Sawyer, R.F., and Starkman, E.S., Oxides of nitrogen from gas turbines, *J. Air Poll. Control Assoc.* **18** (1968) 30–35.
79. Heywood, J.B., Fay, J.A., and Linden, L.H., Jet aircraft air pollutant production and dispersion, *AIAA Paper No. 70-115*, 1970.
80. Hammond, Jr., D.C. and Mellor, A.M., A preliminary investigation of gas turbine combustor modelling, *Comb. Sci. and Tech.* **2** (1970) 161–172.
81. Linden, L.H., and Heywood, J.B., Smoke emissions from jet engines, *Comb. Sci. and Tech.* **2** (1971) 401–411.
82. Hammond, Jr., D.C., and Mellor, A.M. Analytical calculations for the performance and pollutant emissions of gas turbine combustors, *AIAA Paper No. 71-711*, 1971.
83. Heywood, J.B., Gas turbine combustor modeling for calculating nitric oxide emissions, *AIAA Paper No. 71-712*, 1971.
84. Crowe, C.T., Pratt, D.T., Bowman, B.R., and Sonnichsen, T.W., Prediction of nitric oxide formation in turbojet engines by PSR analysis, *AIAA Paper No. 71-713*, 1971.
85. Edelman, R., and Economus, C., A mathematical model for jet engine combustor pollutant emissions, *AIAA Paper No. 71-714*, 1971.
86. Roberts, R., Aceto, L.D., and Kollrack, R.H., An analytical model for NO formation in a gas turbine combustion chamber, *AIAA Paper No. 71-715*, 1971.
87. Fletcher, R.S., and Heywood, J.B., A model for nitric oxide emissions from aircraft gas turbine engines, *AIAA Paper No. 71-123*, 1971.
88. Huls, T.A., Myers, P.S., and Uyehara, D.A., Spark ignition engine operation and design for minimum exhaust emission, *Progress in Technology, Vol. 12, Vehicle Emissions — Part II*, Society of Automotive Engineers, Inc., New York, 1967.
89. *Air Quality Criteria for Photochemical Oxidants*, U.S. Department of Health Education and Welfare, National Air Pollution Control Administration, Publication No. AP–63, Raleigh, North Carolina, 1970.
90. Leighton, P.A., *Photochemistry of Air Pollution*, Academic Press, New York, 1961.
91. Tuesday, C.S., *Chemical Reactions in the Lower and Upper Atmosphere*, Interscience, New York, 1961, p. 15.
92. Altshuller, A.P., Kopczynski, A.P., Lonneman, S.L., Becher, T.L., and Slater, R., Chemical aspects of the photooxidation of the propylene–nitrogen oxide system, *Environ. Sci. Tech.* **1** (1967) 899–914.
93. Jackson, M.W., Effect of some engine variables on composition and reactivity of exhaust hydrocarbons, *SAE Paper No. 660404*, 1966.
94. Altshuller, A.P., An evaluation of techniques for the determination of photochemical reactivity of organic emissions, *J. Air Poll. Control Assoc.* **16** (1966) 257–260.
95. McReynolds, L.A., Alquist, H.E., and Wimmer, D.B., Hydrocarbon emissions and reactivity as functions of fuel and engine variables, *SAE Paper No. 650525*, 1965.
96. Dimitriades, B., Eccleston, B.H., and Hurn, R.W., An evaluation of the fuel factor through direct measurements of photochemical reactivity of emissions, *J. Air Poll. Control Assoc.* **20** (1970) 150–160.
97. Jackson, M.W., and Everett, R.L., Effect of fuel composition on amount and reactivity of evaporative emissions, *SAE Paper No. 690088*, 1969.
98. Turk, A., *Selection and Training Judges for Sensory Evaluation of the Intensity and*

Character of Diesel Exhaust Odors, U.S. Public Health Service Publication No. 999-AP-32, 1967.

99. Rounds, F.G., and Pearsall, H.W., Diesel exhaust odor — Its evaluation and relation to exhaust gas composition, *SAE Trans.* **65** (1957) 608–627.

100. Linnell, R.H. and Scott, W.E., Diesel exhaust composition and odor studies, *J. APCA* **12** (1962) 510–515.

101. Colucci, J.M., and Barnes, G.J., Evaluation of vehicle exhaust gas odor intensity using natural dilution, *SAE Paper No. 700105,* 1970.

102. State of California Administrative Code, Title 17, Public Health Sanitation (Register 66, No. 39–30570, Exhaust Odor and Irritation), November 12, 1966.

103. Reckner, L.R. and Squires, R.E., Diesel exhaust odor measurements using human panels, *SAE Paper No. 680444,* 1968.

104. Springer, K.J., and Hare, C.T., Four years of diesel odor and smoke control technology evaluations — A summary, *ASME Paper No. 69-WA/APC-3,* 1969.

105. Vogh, J.W., Nature of odor components in diesel exhaust, *J. APCA,* **19** (1969) 773–777.

106. Reckner, L.R., Scott, W.E., and Biller, W.F., The combustion and odor of diesel exhaust, *Proc. Am. Petrol. Inst.* **45** (1965) 133–147.

107. Spindt, R.S., Barnes, G.J., and Somers, J.H., The characterization of odor components in diesel exhaust gas, *SAE Paper No. 710605,* 1971.

108. Hoffman, D., and Wynder, E.L., A study of air pollution carcinogenesis III. The isolation and identification of polynuclear aromatic hydrocarbons from gasoline engine exhaust condensate, *Cancer* **15** (1962) 93–102.

109. Wynder, E., and Hoffman, D., A study of air pollution carcinogenesis III. Carcinogenic activity of gasoline exhaust condensate, *Cancer* **15** (1962) 103–108.

110. Ray, S.K., and Long, R., Polycyclic aromatic hydrocarbons from diffusion flames and diesel engine combustion, *Comb. and Flame* **8** (1964) 139–151.

111. Begeman, C.R., and Colucci, J.M., Polynuclear aromatic hydrocarbon emissions from automotive engines, *SAE Paper No. 700469,* 1970.

112. Begemen, C.R., Carcinogenic aromatic hydrocarbons in automobile effluents, *SAE Technical Progress Series, Vol. 6, Vehicle Emissions,* Society of Automotive Engineers, Inc., New York, 1964.

113. Hoffman, D., and Wynder, E., Analytical and biological studies on gasoline engine exhaust, *National Cancer Institute Monograph, No. 9,* Washington, D.C., 1962, pp. 91–116.

114. Hoffman, D., and Wynder, E., Studies on gasoline engine exhaust, *J. Air Poll. Control Assn.* **13** (1963) 322–327.

115. Hoffman, D., Theisz, E., and Wynder, E., Studies on the carcinogenicity of gasoline exhaust, *J. Air Poll. Control Assn.* **15** (1965) 162–165.

116. McKee, H., and McMahon, W., Polynuclear aromatic content of vehicle emissions, *Am. Petrol. Instit.,* Technical Report No. 1, Project No. 21–2139, Aug. 28, 1967.

117. Begemen, C. R., and Colucci, J.M., Benzo (a) pyrene in gasoline partially persists in automobile exhausts, *Science* **161** (1968) 271.

118. Hangebrauck, R., vol Lehmden, D., and Meeker, J., *Sources of Polynuclear Hydrocarbons in the Atmosphere,* U.S. Dept. HEW, Public Health Service Publication No. 999-AP-33, 1967.

119. Sawicki, E., Stanley, T., Hauser, T., Johnson, H., and Elbert, W., Correlation of piperonal test values for aromatic compounds with the atmospheric concentration of benzo(a)pyrene, *Int. J. Air Water Poll.* **7** (1963) 57–70.

120. Sawicki, E., Hauser, T., Wobert, W., Fox, F., and Meeker, Jr., Polynuclear aromatic hydrocarbon composition of the atmosphere in some large american cities, *Am. Ind. Hdg. Assoc. J.* **23** (1962) 137–144.

121. Gofmekler, V.A., Maneta, M., Manusadshants, Z., and Stepanov, L., The correlation between the concentration of 3,4-benzpyrene and carbon monoxide in gas exhaust of autos, *Gigiena i Sanitariya* **28**, (1963) 3–8.

122. Griffing, M., Maler, A., and Cobb, D., *A New Tracer Technique for Sampling and Analysis of Exhaust Gas for Benzo(a)pyrene, Using Carbon-14,* Presented before the Petroleum Div., American Chemical Society, New York, Sept. 7–12, 1969.
123. Hunigen, E., Jaskulla, N., and Wettig, K., Die Herabsetzung krebsfördernder Schadstoffe in Automotoren-abgasen durch Kraftstoffzusätze und Schmierstoffauswahl, *Int. Clean Air Cong. Proc., Part I, London,* October 1966, pp. 4–7.
124. Kuhn, W., and Tomingas, R., Versuch zur Verhütung der Bildung von Verunreinigungen der Auspuffgase von Zweitaktmotoren und von Dieselmotoren durch innere Aktivierung im Motor, *Staub* **25,** (1965) 86–97.
125. Steinhagan, W.K., Niepoth, W.G., and Mick, S.H., Design and development of general motors air injection reactor system, *SAE Paper No. 660106,* 1966.
126. King, J.B., Schneider, H.R., and Tooker, R.S., The 1970 General Motors emission control systems, *SAE Paper No. 700149,* 1970.
127. Hangebrauck, R., Lauch, R., and Meeker, J., Emissions of polynuclear hydrocarbons from automobiles and trucks, *Am. Ind. Hyg. Assn. J.* **27** (1966) 47–56.

Chapter 4

The Kinetics of Pollutant Formation in Spark-Ignition Engines

H. K. Newhall

Mechanical Engineering Department *
University of Wisconsin
Madison, Wisconsin

I. INTRODUCTION

Consideration of chemical equilibrium thermodynamics within the context of the spark-ignition engine cycle leads to the expectation that fuel and air should be converted completely to carbon dioxide and water. Thus, the observed appearance in exhaust gases of significant quantities of such undesirable components as unburned hydrocarbons, nitric oxide (and carbon monoxide under chemically correct or fuel-lean conditions), points to a deviation from the equilibrium situation, and implies the significance of rate controlling chemical reactions during the course of engine cycle events. Such reactions are directly responsible for the emission of exhaust pollutants. For this reason the study of mechanisms and processes responsible for pollutant formation and/or destruction necessarily includes detailed consideration of the field of chemical reaction rate theory or chemical kinetics.

It has been common practice in the past to express combustion rates in terms of an overall or "global" mechanism consisting of an assumed single chemical reaction step in which fuel and oxidizer are converted directly to final products:

Fuel + Oxidizer→Products

Though neglecting completely the many simultaneous reaction steps and intermediate chemical species that participate in real combustion processes, such schemes in a limited number of cases have been useful in carrying out calculations where prediction of gross energy release rates rather than chemical details were the desired result. In contrast, the emissions problem

*Present Address: Chevron Research Company, Richmond, California

Table I. Elementary Chemical Reactions and Selected Rate Constants

Reaction	Rate constant[a]
1f. $H_2O + M \rightarrow OH + H + M$	$5.40 \times 10^{17} \exp(-123{,}600/RT)$
1b. $H + OH + M \rightarrow H_2O + M$	1.5×10^{16}
2f. $H_2 + M \rightarrow H + H + M$	$3.10 \times 10^{15} \exp(-110{,}000/RT)$
2b. $H + H + M \rightarrow H_2 + M$	$7.0 \times 10^{17}/T$
3f. $NO + M \rightarrow N + O + M$	$7 \times 10^{10} \sqrt{T}(150{,}000/RT)^2 \exp(-150{,}000/RT)$
3b. $N + O + M \rightarrow NO + M$	0.9×10^{15}
4f. $N_2 + M \rightarrow N + N + M$	$4.2 \times 10^{12} \sqrt{T}(224{,}900/RT) \exp(-224{,}900/RT)$
4b. $N + N + M \rightarrow N_2 + M$	6.1×10^{14}
5f. $N_2O + M \rightarrow N_2 + O + M$	$1 \times 10^{15} \exp(-61{,}000/RT)$
5b. $N_2 + O + M \rightarrow N_2O + M$	$1.82 \times 10^{13} \exp(-21{,}400/RT)$
6f. $NO_2 + M \rightarrow NO + O + M$	$5.4 \times 10^{21}(1/T) \exp(-74{,}000/RT)$
6b. $NO + O + M \rightarrow NO_2 + M$	2.0×10^{16}
7f. $O_2 + M \rightarrow O + O + M$	$6 \times 10^{13}(118{,}000/RT) \exp(-118{,}000/RT)$
7b. $O + O + M \rightarrow O + M$	1.0×10^{14}
8f. $OH + H \rightarrow H_2 + O$	$1.4 \times 10^{12} \exp(-6000/RT)$
8b. $H_2 + O \rightarrow OH + H$	$3.3 \times 10^{12} \exp(-8000/RT)$
9f. $OH + O \rightarrow O_2 + H$	$5.5 \times 10^{13} \exp(-1000/RT)$
9b. $O_2 + H \rightarrow OH + O$	$7.2 \times 10^{14} \exp(-16{,}900/RT)$
10f. $OH + H_2 \rightarrow H_2O + H$	$6.2 \times 10^{15} \exp(-6000/RT)$
10b. $H_2O + H \rightarrow OH + H_2$	$3.2 \times 10^{14} \exp(-21{,}100/RT)$
11f. $OH + OH \rightarrow H_2O + O$	$7.7 \times 10^{12} \exp(-1000/RT)$
11b. $H_2O + O \rightarrow OH + OH$	$8.3 \times 10^{13} \exp(-18{,}100/RT)$
12f. $CO + OH \rightarrow CO_2 + H$	$7.1 \times 10^{12} \exp(-7700/RT)$
12b. $CO_2 + H \rightarrow CO + OH$	$4.7 \times 10^{14} \exp(-27{,}250/RT)$
13f. $N_2 + O \rightarrow NO + N$	$7 \times 10^{13} \exp(-75{,}500/RT)$
13b. $NO + N \rightarrow N_2 + O$	1.55×10^{13}
14f. $N + O_2 \rightarrow NO + O$	$13.3 \times 10^{9} \times T \times \exp(-7080/RT)$
14b. $NO + O \rightarrow O_2 + N$	$3.2 \times 10^{9} \times T \times \exp(-39{,}100/RT)$
15f. $N + OH \rightarrow NO + H$	4.2×10^{13}
16f. $NO_2 + O \rightarrow NO + O_2$	0.58×10^{11}
16b. $NO + O_2 \rightarrow NO_2 + O$	$0.18 \times 10^{11} \sqrt{T} \exp(-47{,}000/RT)$
17f. $N_2O + O \rightarrow NO + NO$	$1.42 \times 10^{14} \exp(-28{,}000/RT)$
17b. $NO + NO \rightarrow N_2O + O$	$2.6 \times 10^{12} \exp(-63{,}800/RT)$
18f. $NO_2 + NO_2 \rightarrow NO + NO + O_2$	$4 \times 10^{21} \exp(-26{,}900/RT)$
18b. $NO + NO + O_2 \rightarrow NO_2 + NO_2$	$2.4 \times 10^{9} \exp(+1046/RT)$
19f. $NO_2 + N \rightarrow NO + NO$	3.6×10^{12}
19b. $NO + NO \rightarrow O_2 + N_2$	Rate-negligible

[a] Rate constant units: bimolecular reactions, cm^3/mole-sec; termolecular reactions, cm^6/mole^2-sec. Third body, M, assumed to be N_2. T, °K.

is concerned directly with intricate details of reaction processes related to chemical species appearing in relatively small concentrations. Therefore, the traditional concept of the global reaction must give way to consideration of accurate and detailed systems of simultaneously occurring "elementary chemical reactions." Such reactions represent real molecular processes, and as a result can be individually characterized by reaction rate constants whose values depend only on temperature, and are otherwise invariant regardless of the environment in which the reaction takes place. All reactions considered in the following discussion are of this type.

Table I is a listing of a large number of elementary reactions and rate constants pertinent to hydrocarbon–air combustion products. Sources of the reported rate constant values are disclosed in some detail in Ref. 1. Another excellent source of rate constant values is the series of volumes complied by Leeds University.[2]

In the spark-ignition engine a definite distinction can be made between the homogeneous chemical and physical processes occurring in bulk combustion gases and the heterogeneous processes occurring in the vicinity of combustion chamber surfaces. It has been demonstrated by previous investigators that the appearance of unburned hydrocarbons in engine exhaust gases is a result of these heterogeneous processes.[3] In contrast nitric oxide and carbon monoxide are pollutant species emanating from the more or less homogeneous processes associated with the bulk combustion gases. The present chapter will be devoted to the discussion of chemical reactions pertinent to the bulk combustion gases with particular emphasis on the formation and destruction of nitric oxide and carbon monoxide. The role of heterogeneous processes in unburned hydrocarbon emissions is treated in a subsequent chapter.

It should be noted that use of the term homogeneous as presently applied to bulk combustion gases does not necessarily imply the absence of gradients in temperature or in species concentrations. Rather it connotes the absence of marked discontinuities in such quantities as would be expected to appear at interfaces imposed by combustion chamber surfaces.

II. CHEMICAL THERMODYNAMICS OF THE ENGINE CYCLE

A. Chemical Equilibrium

The importance of chemical kinetics to the emissions problem arises through the departures from chemical equilibrium alluded to previously. It is therefore instructive to consider as an initial point of reference and also as a limiting case the behavior of a hypothetical engine cycle in which combustion gases proceed through a continuous sequence of shifting chemical equilibrium states. In this situation ignition and combustion of a parcel of the initial fuel–air mixture would lead instantly to formation of high-temperature combustion products containing significant quantities of such equilibrium dissociation products as atomic oxygen, atomic hydrogen, and hydroxyl radicals as well as the major species, H_2O, CO_2, and CO usually associated with lower temperature exhaust products. Additionally, the chemical equilibrium species distribution corresponding to peak engine cycle temperatures includes nitric oxide in concentrations of as much as one or two mole percent depending upon the fuel–air ratio employed.

Combustion is followed by the expansion process with attendant cooling of combustion gases. The consequent shift in chemical equilibria dictates recombination of atoms and radicals to form more stable low-temperature diatomic and polyatomic species. At the same time equilibrium considerations dictate the decomposition of nitric oxide in favor of N_2 and O_2 as temperatures fall. Carbon monoxide should, to a significant extent, be oxidized to CO_2. By the point of exhaust valve opening in the typical spark-ignition engine cycle, combustion gas temperatures are sufficiently low that the corresponding

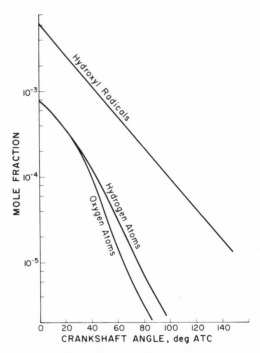

Fig. 1.   Theoretical equilibrium concentrations of atomic hydrogen, atomic oxygen, and hydroxyl radicals during the engine cycle.

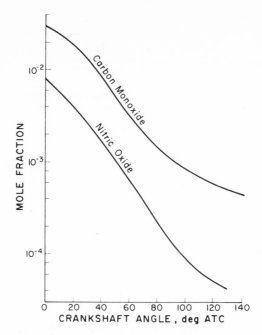

Fig. 2. Theoretical equilibrium concentrations of carbon monoxide and nitric oxide during the engine cycle.

equilibria would yield nearly complete decomposition of NO to N_2 and O_2 and for chemically correct and fuel-lean mixtures nearly complete oxidation of CO to CO_2. This behavior is illustrated by Figs. 1 and 2 which present calculated shifting chemical equilibrium species concentrations as would exist during the expansion process of a hypothetical chemical equilibrium engine cycle.

B. The Role of Chemical Kinetics

The preceding description of chemical equilibria clearly indicates that if equilibrium is to be maintained, pertinent chemical reactions must proceed rapidly relative to the time scale of engine cycle events. First, the combustion process must rapidly convert the initial fuel–air mixture to high-temperature, high-pressure equilibrium combustion products, Chemically this process is extremely complex involving perhaps as many as several hundred simultaneous chemical reactions and a large number of transient intermediate chemical species.

During the subsequent expansion and cooling process, equilibrium can be maintained only if recombination reactions are sufficiently rapid to promote recombination of oxygen atoms, hydrogen atoms, and hydroxyl radicals at the rate indicated by Fig. 1. Similarly Fig. 2 indicates that nitric oxide decomposition reactions and carbon monoxide oxidation reactions must proceed at high rates if the chemical equilibrium path is to be followed.

Thus emissions of the undesirable species NO and CO are directly related to the extent to which each of the above reaction processes is able to satisfy the dictates of rapidly shifting equilibria. In the following discussion each of these processes will be treated in some detail.

III. COMBUSTION OF HYDROCARBON FUEL–AIR MIXTURES AND THE APPROACH TO THE INITIAL HIGH-TEMPERATURE EQUILIBRIUM STATE

Formation of high-temperature combustion products from the initial hydrocarbon fuel–air mixture is extremely complex. It is estimated that as many as 200 simultaneous chemical reactions involving a large number of short lived intermediate species may be involved. For this reason, detailed treatment of the kinetics of this portion of the engine cycle process is probably beyond the scope of present analytical techniques.

A number of empirical correlations related to the overall rate of combustion may be used to establish the general rate of approach to the initial equilibrium state following ignition of the fuel–air charge. Probably the best known correlations of this type are those based on well-stirred reactor experiments.[4,5] Such experiments were originally designed for study of combustion processes under highly turbulent reaction rate limited conditions, and the results therefore provide an indication of the chemical rate of conversion of reactants to products as might occur locally in an engine combustion chamber.

Typical of results obtained in such experiments is the correlation for burning of octane with air under chemically correct conditions.[5]

$$\frac{N}{V p^{1.8}} = 48 \frac{\text{gram-moles/liter-second}}{\text{atm}^{1.8}} \tag{1}$$

where

N = moles of air in the fuel mixture consumed per second
V = reactor volume in liters
p = pressure in atmospheres

It can be shown that under conditions typical of those in an engine this reaction rate corresponds to a reaction time of approximately 0.1 msec, or, for an engine running at 2000 rpm, a crank shaft rotation of about one degree. Thus, the time lapse between the point of ignition and completion of major combustion reactions for a localized parcel of fuel–air mixture within the combustion chamber is apparently quite short relative to engine cycle events, and it seems probable that the principal products of combustion rapidly approach an initial high-temperature equilibrium state prior to the beginning of the expansion process.

IV. FORMATION OF NITRIC OXIDE

A. Fundamental Investigations

While chemical equilibrium considerations would predict the existence of nitric oxide in levels of as much as 1 or 2 mole % in peak temperature combustion gases, it is presently recognized that the formation of nitric oxide proceeds much less rapidly than do the initial combustion reactions, which convert fuel and air to combustion products. As a consequence there exists a significant time lapse between completion of major combustion reactions for a parcel of fuel–air mixture and the local formation of final equilibrium levels of nitric oxide. This behavior has been quantitatively observed in engine combustion chambers,[6] in high-pressure combustion vessels,[7,8] in shock-tube experiments,[9] and in steady-flow burner studies.[10]

The measurements described in Refs. 7 and 8 utilized the experimental apparatus shown schematically in Fig. 3. This system consists essentially of a cylindrical pressure vessel fitted with diametrically opposed quartz windows. By means of a discharge UV lamp and grating spectrometer set to the appropriate wavelength, spectral absorption due to formation of nitric oxide

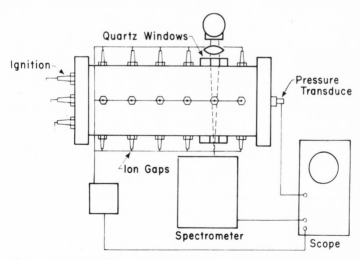

Fig. 3. Experimental system for ultraviolet absorption measurement of nitric oxide formation.[7]

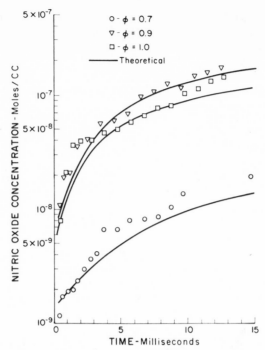

Fig. 4. Experimental and theoretical nitric oxide formation rates for combustion of H_2–air mixtures.[7]

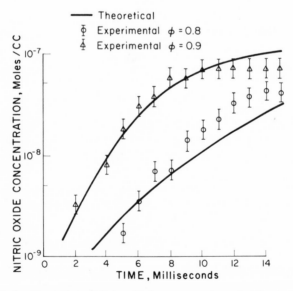

Fig. 5.   Formation of nitric oxide in propane–air combustion gases.[8]

as an axially propagating flame front enters the optical path can be measured and recorded. In this manner, the formation of nitric oxide in the immediate vicinity of the flame front is observed.

Resulting measurements of nitric oxide formation in combustion of hydrogen–air mixtures are plotted in Fig. 4.[7] Here the zero of time represents the instant of flame front arrival in the optical path of the spectroscopic system. Figure 5 presents similar but more recent results obtained for combustion of propane–air mixtures.[8] For both fuels, the nature of the nitric oxide formation process is similar. Nitric oxide is formed primarily in postflame combustion gases, and the approach to chemical equilibrium concentrations involves significant lapses of time, relative to the rate of events occurring in typical real engine cycles.

The solid lines of Figs. 4 and 5 represent independent theoretical calculation of the rate of nitric oxide formation for conditions corresponding to the experimental vessel conditions. These calculations were based on a simple mechanism employing the following elementary reactions.

$$N_2 + O \rightarrow NO + N \qquad (13f)^*$$
$$NO + N \rightarrow N_2 + O \qquad (13b)$$
$$O_2 + N \rightarrow NO + O \qquad (14f)$$
$$NO + O \rightarrow O_2 + N \qquad (14b)$$

leading to the rate equation

$$\frac{d[NO]}{dt} = k_{13f}[N_2][O] - k_{13b}[NO][N]$$
$$+ k_{14f}[N][O_2] - k_{14b}[NO][O] \qquad (2)$$

Numerical rate constants employed in the integration of Eq. (2) are those presented in Table I.

Since it was determined *a posteriori* from experimental results that the formation of nitric oxide is slow relative to the probable rates of relaxation of atoms and radicals to their post flame equilibria, post flame equilibrium concentration values for nitrogen and oxygen atoms were substituted into Eq. (2). Numerical evaluation of rate constants and atom concentrations was based on local time-varying post flame temperatures inferred from recorded pressure traces.

It is notable that excellent agreement exists between experimentally measured and theoretically predicted nitric oxide concentrations. This is particularly significant in view of the fact that the predictions involved solution of equations containing no adjustable parameters, the reaction rate constants being published values derived from independent investigations of elementary reactions. This result offers strong evidence that at pressures and temperatures pertinent to engine combustion, nitric oxide is formed primarily in equilibrated (with exception of NO), post flame combustion gases by a

*See Table I.

relatively simple mechanism involving the reactions of Table I. The calculation procedure involved in use of this mechanism is straightforward and within the means of most computer facilities.

Independent shock-tube studies of nitric oxide formation in shock ignited dilute H_2–O_2–Ar mixtures are in general agreement with the combustion vessel results presented in Figs. 4 and 5.[9] These shock-tube results are reproduced in Fig. 6 which shows nitric oxide formation versus time following ignition. Here nitric oxide concentrations in the vibrational ground state and first excited state are shown individually. The solid lines are based on theoretical rate calculations involving complete kinetics of H_2–O_2 oxidation as well as nitric oxide formation. Nitric oxide formation reactions included reactions (3), (15), and (17) of Table I as well as reactions (13) and (14) employed in obtaining the previous results of Refs. 7 and 8. On the basis of these calculations it was concluded that only reactions (13) and (14) contributed significantly to nitric oxide formation. Again the agreement between measured formation rates and those calculated on the basis of these reactions is excellent. The temperature sensitivity of these reactions is demonstrated by

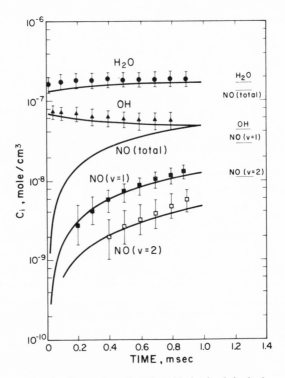

Fig. 6. Formation of nitric oxide in shock ignited H_2–O_2–Ar mixtures.[9]

comparison of the high rates of nitric oxide formation associated with the relatively high shock-tube experiment temperatures (2600–2800 °K), with the lower rates observed for the lower temperatures (2300–2500 °K), of the combustion vessel experiments.

Direct *in situ* measurements of nitric oxide formation within an engine combustion chamber are again broadly confirming of the foregoing observations.[6] Here recorded spectral intensities emitted by the reaction NO + O→NO$_2$ + h_ν were interpreted as an indication of nitric oxide concentration within the combustion chamber. Optically transmitting windows located at several points in the combustion chamber permitted evaluation and comparison of local nitric oxide formation rates at several locations. Typical results for an engine speed of 1200 rpm and fuel–air equivalence ratio of 0.9 are reproduced in Fig. 7. Here the measured rates of nitric oxide formation at two locations within the combustion chamber are presented. The estimated instant of flame front arrival in the optical path of the measuring system is marked by an asterik, and it is again apparent that nitric oxide formation occurs primarily in postflame combustion gases following flame front arrival.

The dashed lines in Fig. 7 are based on kinetic calculations similar in nature to those described with reference to the preceding combustion vessel experiments.[7,8] However, in the present case reaction (15) of Table I was included in the calculations as well as reactions (13) and (14). Agreement between experiment and theoretical calculations is not as favorable as was

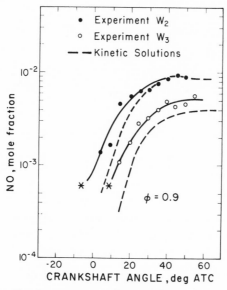

Fig. 7. Formation of nitric oxide in spark-ignition engine combustion chamber.[6]

the case for either the combustion-vessel or the shock-tube experiments. However, this can probably be ascribed to difficulties associated with control of experimental conditions within the engine combustion chamber, including the problem of cycle to cycle variation of the combustion process.

Measurements within a premixed flame stabilized on a flat flame burner indicate that nitric oxide formation in postflame combustion gases is again well described by reactions (13) and (14).[10] However, in direct contrast to the previously cited *in situ* measurements, a significant level of apparent "in flame" nitric oxide formation has been inferred from the burner experiments. It is not clear at present whether this finding can be ascribed to the difference in measuring techniques between the burner experiments and those previously described. The burner experiments, which were conducted at pressures ranging from 1–3 atm involved use of sampling probes. Axial sample probe position was adjustable and it was therefore possible to obtain a plot of nitric oxide concentration versus axial distance downstream of the stabilized flame front. Through knowledge of flow rates these data were converted to time rates of nitric oxide formation. The apparent quantity of "in flame" nitric oxide was inferred from extrapolation of axial nitric oxide profiles back to the flame front location.

In view of the previously cited *in situ* measurements, all of which indicate that nitric oxide formation occurs in significant quantities only in postflame combustion gases, the foregoing report of substantial in-flame nitric oxide should probably be considered suspect at the present time.

B. Elementary Reactions and Rate Constants for Nitric Oxide Formation

The preceding experimental findings provide a background for critical examination of the elementary chemical reactions related to nitric oxide formation. Of the reactions listed in Table I those participating directly in the formation of nitric oxide are the following:

$$N + O + M \rightarrow NO + M \tag{3b}$$
$$NO_2 + M \rightarrow NO + O + M \tag{6f}$$
$$O + N_2 \rightarrow NO + N \tag{13f}$$
$$N + O_2 \rightarrow NO + O \tag{14f}$$
$$N + OH \rightarrow NO + H \tag{15f}$$
$$NO_2 + O \rightarrow NO + O_2 \tag{16f}$$
$$N_2O + O \rightarrow NO + NO \tag{17f}$$
$$NO_2 + NO_2 \rightarrow NO + NO + O_2 \tag{18f}$$
$$NO_2 + N \rightarrow NO + NO \tag{19f}$$

As indicated by the preceding experimental findings only reactions (13f) and (14f) appear to contribute significantly to nitric oxide formation in hydrocarbon combustion products.

While several of the reactions (6f), (16f), (18f), and (19f) exhibit relatively large rate constants, all involve NO_2 as a reactant. Since NO_2 would be expected to occur only in trace quantities in high-temperature combustion products none of these reactions would be expected to contribute appreciably to formation of nitric oxide.

Reaction (17f) involving oxygen atom attack on N_2O has a relatively high rate constant. Again, however, due to the extremely low concentrations of N_2O that would be expected to exist in combustion gases, this reaction is probably not a significant source of NO.

Reaction (3b) forms nitric oxide by direct termolecular encounter of nitrogen and oxygen atoms and a third body. It can be shown that for levels of nitrogen atoms and oxygen atoms present in combustion gases reactions (13f) and (14f) are several orders of magnitude more rapid than (3b).

Reactions (13) and (14) have been studied independently by a number of investigators using diverse techniques ranging from shock tubes[11] to flow reactors.[12] Reasonable agreement among the observed rates has been attained, and the resulting numerical constants as appearing in Table I appear to be trustworthy.

In contrast the rate constant presented for reaction (15f) should be considered suspect at the present time, as its inordinately large value was derived from a single set of measurements based on a flow discharge technique.[13] Ordinarily atom–radical reactions of this nature would not be expected to be competitive with atom–molecule reactions such as (13f) and (14f), and the question of precision of the reported rate constant would be of little practical consequence. However as a result of the unusually large rate constant assignment there exist certain combustion regimes in which reaction (15f) might be dominant relative to (13f) and (14f). This is particularly true under rich mixture conditions with relatively low concentrations of O and O_2 for participation as reactants in (13f) and (14f).

In the theoretical calculations presented in Figs. 4 and 5 inclusion of reaction (15f) with (13f) and (14f) was found to have negligible effect for fuel-lean and chemically correct mixture ratios. However, for fuel-rich conditions inclusion of this reaction, with the rate constant value of Table I, resulted in the overestimate of the rate of nitric oxide formation. This finding is consistent with the preceding observation that the reported rate constant value appears to be inordinately large.

C. Application to the Engine Cycle

Results of the previously cited investigations have shown that reactions (13) and (14) are primarily responsible for nitric oxide formation in combustion gases under conditions pertinent to most practical combustion processes. The rates of these reactions, as expressed by Table I, are strongly temperature dependent, diminishing rapidly as temperatures fall below peak combustion levels. As a consequence in the engine combustion chamber nitric oxide formation is arrested at a point early in the expansion process, usually before reaching peak-temperature equilibrium concentrations. Engine exhaust emissions are thus dependent on the time available for nitric oxide formation in burned combustion gases prior to the beginning of expansion and also upon the rate of nitric oxide formation during this period of time.

Application of the preceding calculation techniques to mathematical simulation of the spark-ignition engine cycle has demonstrated that burned gases formed early in the combustion process (in the vicinity of the spark plug) ultimately attain relatively large local concentrations of nitric oxide, in some cases approaching local peak-temperature equilibrium levels.[14] In burned gases formed later in the combustion process (at points distant from the spark plug), nitric oxide formation reactions are more quickly arrested by the ensuing expansion, and as a result in these gases lower levels of nitric oxide develop.

This spatial concentration gradient in nitric oxide resulting from the variation of available reaction times across the combustion chamber is enhanced by the development of temperature gradients. It is well known that gases burned early in the combustion process achieve inordinately high temperatures as a result of compression by later burning portions of the charge. Thus, the early burned combustion gases not only have the longest time available for nitric oxide formation but also the highest temperatures and hence reaction rates.

The appearance of concentration gradients is confirmed experimentally by the previously cited data of Fig. 7. Here observations at two combustion chamber locations indicate formation of significantly larger quantities of nitric oxide at the location closer to the point of ignition.

Numerous accounts of the response of nitric oxide exhaust emissions to engine operating and design variables are contained in the literature. These are summarized in Chapter I and will not be reiterated here. General trends of this nature can be readily interpreted in terms of the foregoing kinetic considerations.

Engine ignition timing has a twofold effect. With advanced timing, increased combustion temperatures result and, hence, the rate of nitric oxide formation increases. Also, advanced ignition timing yields a longer period of time availble for nitric oxide formation prior to the beginning of expansion, so that for a given formation rate, higher nitric oxide concentrations will be achieved.

Observed effects of fuel–air ratio on nitric oxide emissions are also predicted kinetically. The rate of nitric oxide formation, as given by Eq. (2), depends both upon temperature and upon existing concentrations of atomic and diatomic oxygen. These two factors combine to yield a maximum rate

of nitric oxide formation for fuel–air mixtures somewhat lean of the stoichiometric value. For leaner mixtures the rate falls due to the attendant drop in
combustion temperature, while for richer mixtures the rate decreases due to
the accompanying deficiency in oxygen atoms and molecules. This behavior
is demonstrated by the experimental results plotted in Fig. 8. Here the spectroscopic measurements of nitric oxide formation[7] have been replotted to
show the instantaneous concentrations of nitric oxide versus fuel–air equivalence ratio for a number of elapsed times following flame front arrival. Also
plotted is the nitric oxide equilibrium profile. Each of the instantaneous
concentration curves appears similar to exhaust emissions data plotted against
equivalence ratio, exhibiting a peak at a fuel–air ratio slightly leaner than
chemically correct. It is notable that the chemical equilibrium curve does not
exhibit this behavior.

A method that has been proposed for control of nitric oxide emissions
from both mobile and stationary combustion sources involves addition of

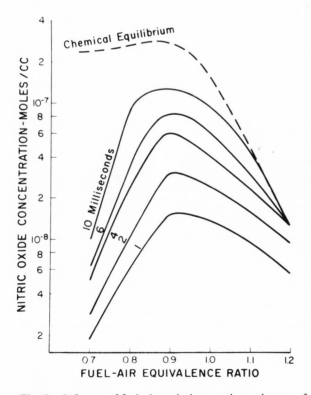

Fig. 8.   Influence of fuel–air equivalence ratio on the rate of
nitric oxide formation.

inert diluents to the initial fuel–air mixture. Methods proposed for practical application include the recirculation of burned exhaust gases into the fuel–air mixture[15,16] or the injection of water.[17]

In connection with such control techniques Fig. 9 presents a comparison between the measured rate of nitric oxide formation for a hydrogen–air–diluent mixture and that for a straight hydrogen–air mixture under otherwise similar conditions. These comparative results indicate that the diluent yields a significant reduction in the rate of nitric oxide formation. This effect appears to be pronounced during early parts of the formation process and tends to diminish as nitric oxide approaches chemical equilibrium levels late in the combustion process.

This finding would tend to indicate that the observed effectiveness of charge dilution in reducing emissions from practical combustion processes results from a reduced rate of nitric oxide formation rather than from a simple depression of the chemical equilibrium level.

The solid curves of Fig. 9 represent theoretical calculations based on reactions (13) and (14). The effect of the diluent on the rate of nitric oxide formation appears to be well represented by these calculations.

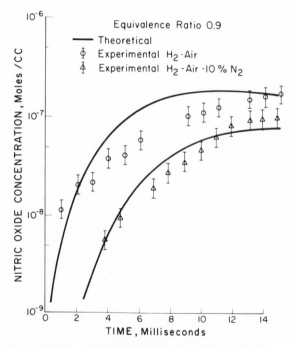

Fig. 9. Effect of reactant dilution on the rate of nitric oxide formation.[8]

V. EXPANSION AND EXHAUST PROCESSES

A. General

In the preceding discussion attention has been directed to the chemical rate processes of importance during the peak temperature portion of the engine cycle existing in the period between the time of ignition and the beginning of the expansion process. It has been pointed out that during this period carbon–hydrogen species rapidly approach peak-temperature equilibrium concentrations, while nitric oxide increases more slowly toward its equilibrium level. As the expansion process begins, the consequent cooling of combustion gases arrests the formation of nitric oxide, usually before peak-temperature equilibrium levels are achieved.

Previous discussion of equilibria indicated three important theoretical consequences of expansion and cooling:

1. Atoms and radicals formed at peak temperatures should recombine to form more stable low-temperature polyatomic species, for example H_2O and CO_2.
2. The ratio of carbon monoxide to carbon dioxide should decrease markedly.
3. Nitric oxide should decompose to N_2 and O_2.

Engine exhaust measurements of nitric oxide and carbon monoxide generally yield concentrations corresponding more closely to peak-temperature equilibria than to exhaust temperature values. It is therefore apparent that chemical reactions pertinent to the above processes are kinetically limited, and not sufficiently rapid to bring about the indicated equilibrium shift in the short time available during expansion.

Results of a detailed theoretical study of the chemical kinetics of the engine expansion process are broadly confirming of the above deduction.[1] Here rate expressions for 32 elementary chemical reactions were combined to yield 13 coupled nonlinear differential equations representing the time rates of change of each of 13 chemical species considered to be pertinent. These equations were integrated numerically, together with the coupled energy and cylinder volume-time equations, starting at the initial point of the expansion process. Results were obtained in the form of concentration-time histories for each of the 13 chemical species considered. These species encompass all of those appearing in significant quantities in peak-temperature equilibrium products of combustion and include atomic and diatomic nitrogen, oxygen, and hydrogen; the hydroxyl radical; carbon monoxide; carbon dioxide; nitrous oxide; nitrogen dioxide, and nitric oxide. Chemical reactions and corresponding rate constants included in the analysis are those listed in Table I.

B. Recombination of Atoms and Radicals

Computed concentration-time histories for atomic oxygen, atomic hydrogen, and hydroxyl radicals during expansion are presented in Figs. 10, 11, and 12. Also plotted in each figure is the concentration-time history that would prevail in a shifting, chemical equilibrium expansion. Engine operating conditions corresponding to these results are compression ratio 9 to 1 and speed 4000 rpm. Thus, the 5 msec point in these figures corresponds to a crankshaft angle of approximately 120° A.T.C. The fuel–oxidizer stoichiometry corresponds to a chemically correct mixture of octane with air.

According to the chemical equilibrium calculation each of the above species, O, H, and OH should recombine nearly completely during the course of expansion. The results of kinetic calculations reveal that during the initial stages of expansion the recombination reactions are sufficiently rapid to maintain equilibrium. However, as the process continues, the reduction in total pressure retards these reactions to the extent that hyperequilibrium concentrations of hydrogen atoms, oxygen atoms, and the hydroxyl radicals persist. This persistence extends to the end of the expansion process.

More fundamentally important than the behavior of individual species concentrations is the behavior of individual elementary reactions. A useful

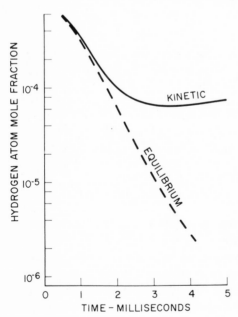

Fig. 10. Atomic hydrogen concentrations during expansion; fuel composition, C_8H_{18}; equivalence ratio, 1.0; engine speed, 4000 rpm; compression ratio, 9:1.[1]

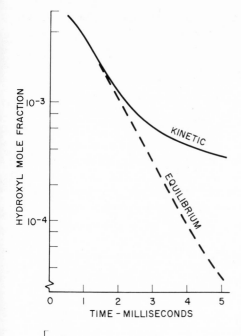

Fig. 11. Hydroxyl radical concentrations during expansion.[1]

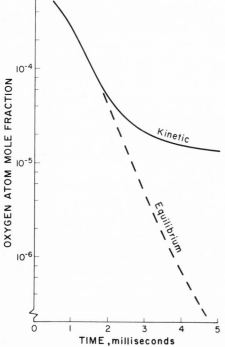

Fig. 12. Oxygen atom concentrations during expansion.[1]

indication of such behavior is the rate of an elementary reaction in both forward and reverse directions relative to the rates required for its continuous equilibration in the given system. This information can be obtained from the computed results by dividing product species concentrations by reactant concentrations, each raised to the appropriate stoichiometric exponent, and comparing the quotient so obtained with the equilibrium constant for the reaction. It is thus possible to formulate a quantitative measure of the degree of equilibration achieved by a particular elementary reaction. Such a formulation is given by Eq. (3). The numerator of the right-hand side of this equation represents the ratio of product species to reactant species each raised to its respective stoichiometric coefficient.

$$D_j = \frac{\prod_i [A_i]^{\nu_{ij}}}{Kc_j} \tag{3}$$

As an example for reaction (2) of Table I, Eq. (3) would take the form

$$D_2 = \frac{[H]^2/[H_2]}{Kc_2} \tag{4}$$

For a reaction, j, which is continuously equilibrated the "degree of equilibration," D_j, will equal unity. A deviation from unity would indicate that the reaction is not sufficiently rapid relative to the overall system process to remain equilibrated.

The degree of equilibration for each elementary reaction considered has been calculated for a number of points of a typical expansion, and the results are plotted in Fig. 13. Attention is restricted for the moment to those reactions

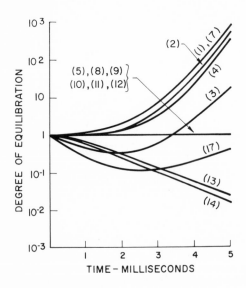

Fig. 13. Degree of equilibration for elementary reactions during expansion. Numbers in parentheses refer to reactions listed in Table I.

influencing recombination of atomic hydrogen, atomic oxygen, and the hydroxyl radical. These reactions include the three-body recombinations, (1b), (2b), and (7b)

$$H + OH + M \rightarrow H_2O + M \qquad (1b)$$
$$H + H + M \rightarrow H_2 + M \qquad (2b)$$
$$O + O + M \rightarrow O_2 + M \qquad (7b)$$

and the bimolecular atom-exchange reactions (8), (9), (10), and (11).

$$OH + H \leftrightharpoons H_2 + O \qquad (8)$$
$$OH + O \leftrightharpoons O_2 + H \qquad (9)$$
$$OH + H_2 \leftrightharpoons H_2O + H \qquad (10)$$
$$OH + OH \leftrightharpoons H_2O + O \qquad (11)$$

Inspection of the four atom-exchange reactions reveals that they are by themselves incapable of effecting a net recombination or reduction of the above atomic species. Therefore, recombination can occur only through the termolecular reactions. Figure 13 indicates that these reactions are initially equilibrated ($D_j = 1$); however, as expansion proceeds, a deviation from equilibration rapidly ensues. Clearly, the noted persistence of atomic species may be ascribed to failure of the termolecular reactions during expansion.

Figure 13 shows that in contrast to the termolecular reactions, the bimolecular atom-exchange reactions (8)–(11) are extremely rapid in both forward and reverse directions, and hence are continuously equilibrated. As a consequence the relative quantities of excess atomic species and hydroxyl radical are determined not by the relative rates of their respective recombination reactions but rather by the partial equilibrium established by the fast bimolecular reactions (8)–(11). The term "partial equilibrium" must be used with some care, for it can be demonstrated mathematically that the 4 bimolecular reactions (one of which is redundant under equilibrium conditions) are insufficient to describe a complete equilibrium. Rather the partial equilibrium should be interpreted as a transient distribution of atomic species dictated by the bimolecular reactions under constraints imposed by the lagging recombination reactions.

The partial equilibrium effect is clearly evidenced by Fig. 10 which shows that the concentration of atomic hydrogen reaches a minimum at an intermediate point in the expansion and then gradually begins to rise even though temperature decreases monotonically throughout expansion. This is a consequence of the shift of the partial equilibrium of reactions (8)–(11) as temperature decreases.

C. Oxidation of Carbon Monoxide

The computed concentration-time history of carbon monoxide is presented in Fig. 14. The behavior exhibited is similar to that of the atomic species. During the initial stages of expansion carbon monoxide is destroyed at a

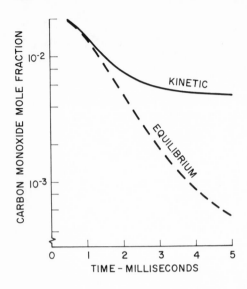

Fig. 14. Carbon monoxide concentrations during the expansion process.[1]

rate corresponding to a shifting chemical equilibrium expansion. However, as the process continues, an increasing deviation from equilibrium occurs; until at the end of expansion carbon monoxide concentration is as much as ten times the equilibrium value.

The elementary reaction primarily responsible for carbon monoxide oxidation is the bimolecular reaction involving the hydroxyl radical[12]

$$CO + OH \leftrightharpoons CO_2 + H \qquad (12)$$

This atom-exchange reaction is relatively fast in forward and reverse directions, and, as indicated by Fig. 13, it is continuously equilibrated throughout the expansion, a fact which is initially surprising in view of the noted nonequilibrium behavior of carbon monoxide.

If reaction (12) is at all times in equilibrium, then the relative levels of carbon monoxide and carbon dioxide will be controled solely by the existing levels of hydroxyl radicals and hydrogen atoms and in particular by the ratio of the two. This is expressed by Eq. (5).

$$\frac{[CO]}{[CO_2]} = \frac{1}{kc_{12}} \frac{[H]}{[OH]} \qquad (5)$$

It was previously noted that the distribution of excess atomic species and hydroxyl radicals is controled by the partial equilibrium established by the bimolecular reactions (8)–(11). From Figs. 10 and 11, it is apparent that as temperature falls this partial equilibrium favors a disproportionately large number of hydrogen atoms relative to hydroxyl radicals. For example, at the 4 msec points, hydroxyl concentration is approximately 4 times its equilibrium value, whereas hydrogen atom concentration is over 20 times its equilibrium value. Therefore during expansion the ratio of hydrogen atoms to hydroxyl radicals as dictated by the partial equilibrium is much greater than would be the case for complete equilibrium, and it is deduced from Eq. (5) that the ratio of carbon monoxide to carbon dioxide must be correspondingly greater than for complete equilibrium. As a consequence there results an excess of carbon monoxide.

Figure 15 presents a comparison of computed carbon monoxide concentrations for the end of expansion with measured engine exhaust concentrations.[18] Also included in this plot are chemical equilibrium concentrations of carbon monoxide corresponding to the initial point of the expansion and to the final point. It is apparent that over the range of fuel–air ratios considered, the kinetics calculations are in substantial agreement with reported exhaust measurements. One notable fact evident from the computed results and supported by the measurements is that the nonequilibrium behavior of carbon monoxide is most pronounced for fuel-rich mixtures.

D. Nitric Oxide Decomposition

According to chemical equilibrium calculation nitric oxide should decompose to diatomic nitrogen and oxygen during expansion. However, as shown in Fig. 16, the present analysis indicates that little if any decomposition actually occurs.

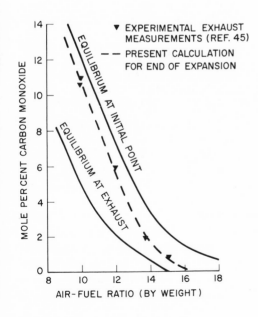

Fig. 15. Comparison of measured exhaust carbon monoxide concentrations with computed values.[1]

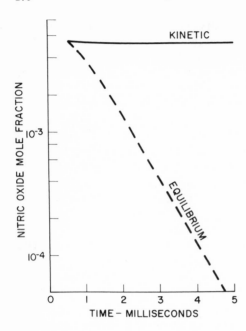

Fig. 16. Computed kinetic and equilibrium nitric oxide levels during expansion.[1]

The reactions influencing nitric oxide decomposition are the bimolecular atom-exchange reactions between nitric oxide and nitrogen atoms, (13b), between nitric oxide and oxygen atoms, (14b) (the reverse of the formation reactions), and between pairs of nitric oxide molecules, (17b).

$$NO + N \rightarrow O_2 + N \qquad (13b)$$
$$NO + O \rightarrow N_2 + O \qquad (14b)$$
$$NO + NO \rightarrow N_2O + O \qquad (17b)$$

The degree of equilibration for each of these reactions is plotted in Fig. 13, and it is evident that each deviates from equilibrium throughout most of the process. The rate of nitric oxide destruction attributable to each of the above elementary reactions was calculated for several points of the expansion and it was determined that none is sufficiently rapid to achieve appreciable decomposition. The most effective reactions are (13b) and (14b) both destroying nitric oxide at a rate 500 times that of reaction (17b). Reaction (14b) is endothermic and therefore much less favorable energetically than reaction (13b). However the large number of oxygen atoms relative to nitrogen atoms present in the system tends to compensate for this, and the two reactions proceed at about the same forward rate.

The computed behavior of nitric oxide as given by Fig. 16 agrees with previous experiments of the author.[19] These experiments involved spectroscopic instrumentation of an engine combustion chamber such that nitric oxide concentration within the chamber could be directly monitored during expansion. The results indicated that nitric oxide was essentially fixed in concentration throughout all observable portions of the expansion. Typical result of these experiment are displayed in Fig. 17.

VI. THE EFFECT OF PRESSURE OSCILLATIONS ON NITRIC OXIDE KINETICS

Unlike many combustion processes in present use, that occurring in the spark-ignition engine is not uncommonly accompanied by high amplitude pressure oscillations. These oscillations, usually referred to as "knock," are a result of sudden and often spontaneous combustion of portions of the fuel–air charge that normally would burn at a later time in the process. The resulting localized high rate of energy release induces high amplitude pressure waves, or shock waves, that subsequently undergo repeated reflections from combustion chamber surfaces. The result as viewed from any fixed point within the combustion chamber is a high amplitude pressure oscillation.

Pressure disturbances as occurring in shock waves are accompanied by related temperature disturbances. As a consequence, the pressure oscillations encountered under knocking conditions in the engine are accompanied by corresponding temperature oscillations. Since the temperature sensitivity of

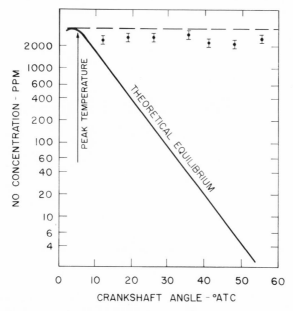

Fig. 17. Infrared measurements of nitric oxide during the engine cylinder expansion.[19]

the rate of nitric oxide formation is well documented it is to be expected that such temperature oscillations may be of significance from the standpoint of nitric oxide exhaust emissions.

From a theoretical point of view the effect of pressure (temperature) oscillations in nitric oxide formation can be qualitatively ascertained by considering a hypothetical square wave temperature oscillation symmetric about a mean combustion temperature, T_0. It can be shown very simply that if the variation of the rate of nitric oxide formation with temperature were linear, then the average rate of nitric oxide formation under oscillating conditions would be equal to the rate corresponding to the average temperature, and the oscillation would be of no consequence. However the rate of nitric oxide formation is generally nonlinear, varying exponentially with reciprocal temperature in accordance with an Arrhenius type of expression.

$$\text{Rate} = Ae^{-E/RT} \tag{6}$$

The general form of this expression is shown in Fig. 18. It is significant that an inflection occurs in this curve. For temperatures below the inflection the rate of reaction increases at an accelerating rate with increasing temperature while the opposite is true for temperatures above the inflection. As a consequence it can be shown that for temperatures below the inflection point the temperature oscillations will result in an effective nitric oxide formation rate in excess of that corresponding to the mean temperature T_0. On

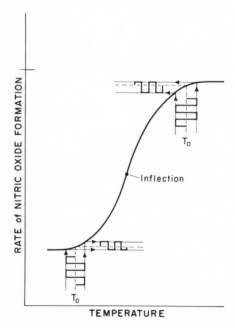

Fig. 18.   Qualitative representation of the theoretical influence of temperature oscillations on nitric oxide formation.

the other hand for mean temperatures above the inflection, oscillations will result in a reduction of the rate of nitric oxide formation relative to the mean temperature rate. This behavior is illustrated graphically in Fig. 18.

It can be shown that in air-breathing engines peak combustion temperatures are well below the inflection point of Fig. 18. Therefore it would be expected that in general temperature oscillations should directionally increase nitric oxide formation and emissions.

Very little basic experimental data related to the effect of engine knock on nitric oxide production has been published. Experimental nitric oxide levels occurring in oscillating combustion in a steady flow reactor have been published previously.[20,21] However the significance of these results is unclear due to the relatively low pressure and temperature levels and oscillation amplitudes employed.

Comparative measurements of emissions from automobile engines under knocking and nonknocking conditions have indicated increases of as much as 15% in nitric oxide as a direct result of knocking operation.[22,23] Typical of such results are those of Fig. 19 which is reproduced from Ref. (23). Here nitric oxide emissions from a vehicle operated on a chassis dynamometer with three different fuels are reported. Emissions obtained with a knocking fuel consisting of equal parts *iso*-octane and *n*-heptane are compared with those of two nonknocking fuels, one straight *iso*-octane and the other 50% *iso*-octane, 50% heptane with TEL added to suppress knock. It is apparent that both nonknocking fuels yielded similar emission levels while the knocking fuel gave a significant increase in nitric oxide. The similarity of results obtained with *iso*-octane and with the leaded reference blend are convincing evidence that the observed increase in nitric oxide under knocking conditions was the result of pressure oscillations rather than direct chemical effects due

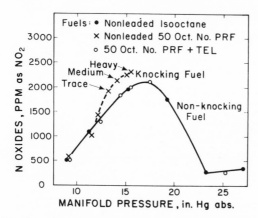

Fig. 19.   Experimental evidence for the effect of engine knock on nitric oxide emissions.[23]

to TEL or to fuel structure. Results of a more recent study[24] have been interpreted as indicating that nitric oxide emissions increase with increasing knock intensity. Unfortunately results of this study must be completely discounted since knock was induced by advancing the ignition timing of the engine. The very significant effect of ignition timing on NO emissions from engines under nonknocking conditions is well known and clearly would have masked variations in emission levels due to increased knock intensity.

REFERENCES

1. Newhall, H.K., Kinetics of engine-generated nitrogen oxides and carbon monoxide, *Twelfth Symposium (International) on Combustion,* Mono of Maryland, Baltimore, pp. 603–613.
2. Baulch, D.L., Drysdale, D.D., and Lloyd, A.C., *Critical Evaluation of Rate for Homogeneous, Gas-Phase Reactions of Interest in High Temperature Systems,* Vols. 1–5, May 1968–July 1970, Department of Physical Chemistry, The University, Leeds, England.
3. Daniel, W.A., and Wentworth, J.T., Exhaust gas hydrocarbons — Genesis and exodus, *Technical Progress Series, Vol. 6,* SAE, 1964, pp. 192–205.
4. Longwell, J.P., and Weiss, M.A., High temperature reaction rates in hydrocarbon combustion, *Ind. Eng. Chem.* **47** (1955) 1634–1643.
5. Clarke, A.E., Harrison, A.J., and Odgers, J., Combustion stability in a spherical combustor, *Seventh Symposium (International) on Combustion,* Butterworths, London, 1959, pp. 664–673.
6. Lavoie, G.A., Spectroscopic measurements of nitric oxide in spark ignition engines, *Combustion and Flame* **15** (1970) 97–108.
7. Newhall, H.K., and Shahed, S.M., Kinetics of nitric oxide formation in high pressure flames, *Thirteenth Symposium (International) on Combustion,* Mono of Maryland, Baltimore, 1971, pp. 365–373.
8. Shahed, S.M., and Newhall, H.K., Kinetics of nitric oxide formation in propane–air and hydrogen-air-diluent flames, *Combustion and Flame* **17** (1971) 2.
9. Bowman, T.C., Investigation of nitric oxide formation kinetics in combustion processes: The hydrogen–oxygen–nitrogen reaction, paper presented at the 1970 Fall Meeting, Eastern Section, Combustion Institute, Atlanta.
10. Fennimore, C.P., Formation of nitric oxide in premixed hydrocarbon flames, *Thirteenth Symposium (International) on Combustion,* Mono of Maryland, Baltimore, in press.
11. Wray, K.L., and Teare, J.D., Shock tube study of the kinetics of nitric oxide at high Temperatures, *J. Chem. Phys.* **36** (10) (1962) 2582–2596.
12. Kaufman, F., and Kelso, J.R., Thermal decomposition of nitric oxide, *Chem. Phys.* **23** (9) (1955) 1702–1707.
13. Campbell, I.N., and Thrush, B.A., Reactivity of hydrogen to atomic nitrogen and atomic oxygen, *Trans. Faraday Soc.* **64**, Part 5 (1968) 1265–1274.
14. Muzio, L.J., Starkman, E.S., and Caretto, L.S., The effect of temperature variations in the engine combustion chamber on formation and emission of nitrogen oxides, paper 710658, presented at the SAE Automotive Engineering Congress, Detroit, Jan. 1971.
15. Kopa, R.D., and Kimura H., Exhaust gas recirculation as a method of nitrogen oxides control in an internal combustion engine, Paper presented at 53rd Annual Meeting, Air Pollution Control Association, Cinnicinati May 1960.

16. Daigh, H.D., and Deeter, W.F., Control of nitrogen oxides in automotive exhaust," paper presented at the 27th Mid-year Meeting, API Division of Refining, San Francisco, May 1962.
17. Nicholls, J.A., El-Messiri, I.A., and Newhall, H.K., Inlet manifold water injection for control of nitrogen oxides-theory and experiment, *SAE Trans.* **78** (1970).
18. Huls, T.A., Myers, P.S., and Uyehara, O.A., Spark-ignition operation and design for minimum exhaust emissions, *SAE Trans.* **75** (1967).
19. Newhall, H.K., and Starkman, E.S., Direct spectroscopic determination of nitric oxide in reciprocating engine cylinders, *SAE Trans.* **76** (1968).
20. Seagrave, R.C., Reamer, H.H., and Sage, B.H., Oxides of nitrogen in combustion: Some microscopic measurements, *Combustion and Flame* **8** (March 1964) 11–19.
21. Seagrave, R.C., Reamer, H.H., and Sage, B.H., Oxides of nitrogen in combustion: Oscillatory combustion at elevated pressure, *Combustion and Flame* **9** (March 1965) 7–18.
22. Hanson, T.K., and Egerton, A.C., Nitrogen oxides in internal combustion engine gases, *Proc. Roy. Soc.* **A163** (1937) 90–100.
23. Gilbert, L.F., Hirschler, D.A., and Getoor, R.C., Oxides of nitrogen in engine exhaust gas, paper presented at the 132nd Meeting of the American Chemical Society, Sept. 1957.
24. Duke, L.C., Lestz, S.S., and Meyer, W.E. The relation between knock and exhaust emissions of a spark ignition engine, SAE Paper 700062 presented at the Automobile Engineering Congress, Detroit, Jan. 1970.

Chapter 5

Particulate Emission from Spark-Ignition Engines

G. S. Springer

Department of Mechanical Engineering
The University of Michigan
Ann Arbor, Michigan

I. INTRODUCTION

During the past fifteen years the gaseous components of automotive exhaust have been widely investigated to evaluate their role in air pollution. Relatively little effort has been devoted to the study of particulates emitted from automotive engines. The reason for the greater interest in gaseous emissions is understandable since gaseous components are emitted in considerably larger amounts than particulates, as illustrated in Fig. 1. Nevertheless, gasoline engines still introduce into the United States' atmosphere about one million tons of particulate matter annually,[1] contributing significantly to air pollution. Most notably, particulates affect (a) solar radiation and climate near the ground, (b) visibility, (c) material damage, (d) vegetation, and (e) health. In this paper, these effects are not examined in detail (interested readers are referred to the comprehensive summaries given in Ref. 2); instead, attention is focused on the emission of particulates from spark-ignition engines.

While there are precise definitions of gaseous emissions, there is no exact definition of what constitutes particulate matter in automotive exhaust. Here "particle" or "particulate" is defined as any dispersed matter, solid or liquid (except uncombined water), in which the individual aggregates are larger than single molecules ($\sim$0.0002 μ in diameter) but smaller than about 500 μ.[2,3] Spark-ignition engines exhaust many particles in this size range which are composed of various organic and inorganic substances.

In addition to the total amount of particulates emitted, the size and composition are also of significance in air pollution problems. For example, small particles may remain suspended in the atmosphere for several months while larger ones settle in seconds. The chemical composition may influence

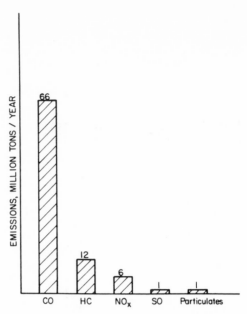

Fig. 1. Air pollution in the United States from automotive sources.[1]

photochemical reactions, the mineral content of vegetation, and human health. For these reasons, it is desirable to determine the physical and chemical properties of particulates, with special attention to

1. Physical characteristics
   a. Total amount (number and weight) emitted
   b. Size distribution
   c. Shape
2. Chemical composition
   a. Chemical composition of the total particulate matter emitted
   b. Chemical composition of different size particulates

Emission standards should be concerned with the foregoing properties. However, emission standards being considered by the U.S. Government and the State of California do not distinguish between particles of different sizes or composition but are concerned only with the total emission. For automobiles, the emission standard considered for 1975 is 0.1 gram per mile.

The properties listed above depend on several different parameters related to the fuel and the engine, including:

1. Fuel composition
2. Engine
   a. Make of engine

      b. Wear and maintenance
      c. Oil
3. Exhaust system
      a. Geometry and design
      b. Surface
      c. Pressure
      d. Temperature
4. Operating conditions
      a. Air–fuel ratio
      b. Speed (including accelerations and decelerations)
      c. Load
      d. Spark timing

The objective of this study is to survey and summarize the existing information on the amount and composition of particulates generated in spark-ignition engines and their dependence on the parameters listed above. It is noted that due to the large number of these parameters it is difficult to compare the results of various investigations obtained under widely different conditions. Certain general conclusions can be reached, however, by comparing existing data. For this reason, such comparisons are presented wherever expedient, the emphasis being on the trend rather than on the absolute values. When citing previous data no attempt was made to list all the conditions under which these were obtained. Only that information is given which is needed for the interpretation of the results. Readers interested in further details may consult the original references.

II. MEASUREMENT TECHNIQUES

Experimental systems and procedures have not yet been developed which are suitable for a complete determination of the physical and chemical properties of particulates, and for evaluating the effects of the fuel, the engine, the exhaust system, and the operating conditions on these properties. Instead, in the past different experimental techniques have been employed each providing information on certain aspects of the problem. In the following, the arrangements of the most commonly used experimental methods are briefly described, with emphasis on concepts rather than on experimental details.

In every experiment a sample of the gas (containing the particulates) is collected. The particulates from this sample are then separated and analyzed. For convenience, we divide the experimental procedures into two groups; (a) where the samples are taken from outside of the exhaust system (i.e., from the atmosphere, precipitation, vegetation, etc.), and (b) where the sample is extracted directly from the exhaust system.

A. Particulates Collected from Outside the Exhaust System

Particulate samples collected from outside the exhaust system do not provide direct information on the generation and emission of particulates from automobiles. From such samples one cannot separate, without undue difficulty, particulates generated by different sources (stationary or moving) or distinguish between particulates emitted by different types of automotive engines. From these samples none of the effects of the parameters listed previously (fuel, engine, and exhaust system parameters) can be assessed. Measurements performed outside the exhaust system do yield indirect evidence of the contribution of automotive sources to atmospheric particulates and, for this reason, several such measurements have been performed in the past. For example, it has been shown[4,5] that concentrations of lead, nickel, cadmium, and zinc in roadside soil and plants increase with traffic volume and decrease with distance from highways. Also, the lead content of precipitation can be correlated with the amount of gasoline consumed in a given geographical area (Fig. 2). A further and more direct evidence of the role of automotive emissions in air pollution was given by Daines et al.,[7] who demonstrated that near a highway the concentration of lead particles in the air depends upon both the traffic volume and the distance from the highway (Fig. 3).

The particulate content of the atmosphere was investigated by Robinson et al.,[8,9] Colucci and Begeman,[10] and Larsen and Konopinski.[11] Robinson

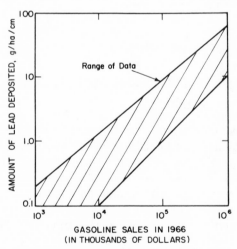

Fig. 2. The amount of lead deposited on one hectare by one centimeter of precipitation versus sales of gasoline in the locale of the collecting station (Lazrus et al.[6]).

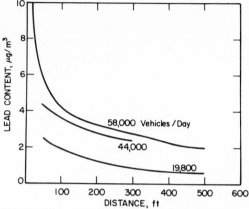

Fig. 3.   The concentration of lead in the atmosphere
as a function of traffic volume and distance from the
highway (Daines et al.[7]).

and his coworkers reported on lead particles in several locations across the
U.S. Colucci and Begeman sampled the atmosphere in Detroit and in Warren,
Michigan, Larsen and Konopinski sampled the atmosphere in the Sumner
Tunnel between Boston and East Boston. The former analyzed particulate
matter for lead, tar, vanadium, sulfates, and polynuclear aromatic hydrocarbons, the latter determined the lead and total particulate concentrations
at the inlet and outlet of the tunnel. As will be discussed in Section IV.A,
the results of these investigations agree, at least qualitatively, with those
obtained from sampling the exhaust gas directly.

Some of the shortcomings of the atmospheric sampling technique can
be overcome by collecting the exhaust in a large "bag" and by taking samples
from this bag.[12] For such systems approximately 8:1 air to exhaust gas
ratios are recommended to avoid condensation. In the past, 2000–5000 ft³
volume bags have been used.[12] With this method the exhaust from a given
engine can be isolated. Furthermore, some of the errors introduced by direct
sampling of the exhaust (most notably, errors due to anisokinetic sampling,
see Section II.B) can be avoided. The disadvantages of this method are that
some of the important parameters influencing the particle formation (e.g.,
temperature) cannot be varied, and it is difficult to eliminate from the bag
all background particles, and to reduce the humidity to acceptable levels
($\sim 10\%$).

B. Particulate Sampling from the Exhaust Stream

For a complete characterization of the particulates emitted from automotive engines a proportional sample of the exhaust is needed. Such a sample

can be analyzed for all the required physical and chemical properties described in the Introduction. Proportional samples are generally obtained in one of two ways: (a) The entire exhaust stream is thoroughly mixed with appropriately filtered ambient air. A small portion of this air–gas mixture is then withdrawn through a sampling probe into a particle size analyzer (Fig. 4). (b) The sample to be analyzed is taken directly from the exhaust gas stream (Figs. 5 and 6). In either of these methods extreme care must be exercised to ensure that the sample is representative of the conditions in the exhaust gas. To this end, attention must be paid to the following points:

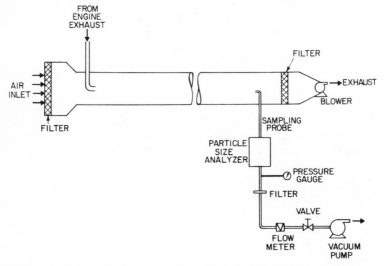

Fig. 4.   Schematic of apparatus for collecting particulates from exhaust diluted with air.

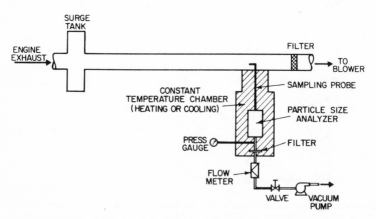

Fig. 5.   Schematic of apparatus for sampling the exhaust directly.

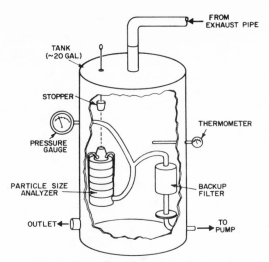

Fig. 6. Schematic of apparatus for sampling undiluted auto exhaust directly from the exhaust pipe (Lee et al.[15]).

1. The sampling must be done isokinetically, i.e., the gas velocity in the sampling probe must be equal to the velocity of the mainstream at the inlet of the probe. This will minimize bias in the sample in favor of either the smaller or larger particles. The flow rate required through the sampling probe for isokinetic sampling can be estimated from the known total flow rate and the velocity distribution in the sampled gas stream. The total particulate concentration ($\mu g/m^3$) may not be affected significantly if the flow rate through the probe differs only slightly from the value required for isokinetic sampling (Fig. 7). However, the size distribution might be more affected by changes in the sampling flow rate. Pulsations, such as occur in automotive exhaust, make isokinetic sampling difficult and should be minimized.

2. Condensation of the water vapor in the sampling line and in the measuring instruments must be avoided since the particulate matter cannot be separated readily from the condensate.

3. Losses of particulate matter in the sampling probe should be reduced as much as possible.

4. The sampling time must be sufficiently long to provide representative and reproducible results. This is particularly important when the testing is not done at a steady speed but under cyclic operation.

The above sampling requirements can be achieved reasonably well with the test method shown in Fig. 4. Dilution of the exhaust gas with air reduces

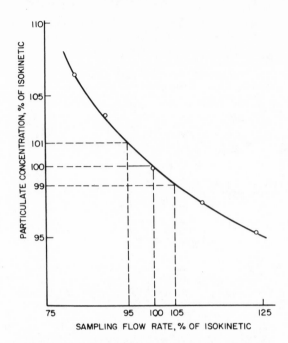

Fig. 7. Variation in particulate concentration as a function of variation in the flowrate through the sampling probe. 1970 Chevrolet 350 CID, V–8 engine, at 1800 rpm road load, leaded fuel. Exhaust pipe diam., 2 in.; circles, data, solid line, fit to data (Sampson and Springer[13]).

velocity fluctuations in the air–gas mixture, thereby facilitating isokinetic sampling. Dilution ratios of 8:1 to 10:1 are generally used. The entrained airstream also reduces the concentration of the water vapor, hence minimizing the danger of its condensation. This method introduces two difficulties: (a) the particles must be mixed uniformly with the air, and (b) the temperature of the exhaust gas cannot be varied over wide ranges.* The first of these may be overcome by proper design of the air duct. Variation in the particulate concentration ($\mu g/m^3$) across a duct can be reduced to about 10%.[14] The latter difficulty could only be eliminated by heating the entire mixing air stream.

The temperature of the sample can be varied within wide limits by the arrangement shown in Fig. 5, namely by taking the sample directly from the exhaust stream. Here the sampling probe and the instruments are enclosed in a constant-temperature chamber heated or cooled to the desired temperature. The highest attainable temperature is governed by the temperature

---

*Temperature control is desirable since the particulate concentration depends strongly upon temperature (see Section III.A).

limit of the particle analyzing instrument. The lowest temperature is that at which water vapor condensation becomes significant. In such a direct sampling method the sample may be significantly distorted (a) due to anisokinetic sampling rates resulting from the strong pulsations in the flow, and (b) due to variations of particulate concentration across the exhaust pipe. Pulsations can be minimized by placing a surge tank in the system (Fig. 5). For a 1970 Chevrolet 350 CID V–8 engine running at 1800 rpm and exhausting into a 2-in.-diam. pipe, a 2-ft³ surge tank reduces the magnitude of both the static and stagnation pressure fluctuations by a factor of ten.[13] Information regarding the variation in particulate concentration across an exhaust pipe was provided by Sampson and Springer,[13] who measured the concentration in the exhaust of the engine just described. The results of these measurements (Table I) indicate that the total particulate concentration ($\mu g/m^3$) varies only by about 4% across the 2-in.-diam. exhaust. The variation in size distribution may be larger, but data on this has not yet been reported.

The above shortcomings of the direct sampling method can be overcome by the system proposed by Lee et al.[15] and shown in Fig. 6. In this arrangement, the particle analyzer is placed into a container and thus pulsations in the flow and nonuniformities in the sample are reduced. The temperature of the entire system (container and instrumentation) may be regulated by appropriate heaters.

It is noted here that commercially available particle analyzers have been used in most experiments. Most analyzers can accommodate samples only up to $\sim$200–250°F and, therefore, require a cooling of the exhaust gas. Some newer units can withstand temperatures up to $\sim$1600°F.

In addition to the proportional sample, it is desirable to collect all the particulates, regardless of size, contained in the exhaust stream. Such measurements not only yield information on the total amount of particles emitted (including total weight and total chemical composition) but also provide a check on the accuracy of the proportional sample. The most convenient method of collecting all particulate matter is to introduce a suitable filter or filters directly into the exhaust stream. The ratio of the total amount of particulates collected to the amount in the gas stream depends upon the size and efficiency of the filter. Sufficient vacuum on the downstream side of the filter must be provided to reduce the back pressure on the engine created by

Table I. Particulate Concentration in Samples Taken at Various Locations in a 2-in.-Diam. Exhaust Pipe[a]

Probe location	(·)	(·)	(·)	(·)	(·)
Particulate concentration % of center	100	100.4	98.7	96.5	102.1

[a] From Sampson and Springer.[13]

Table II.   Comparison Between the Total Amounts of Particulates Collected from the Tailpipe and from a Bag into Which the Vehicle Exhausted[a]

Car[b]	Fuel	Total amount corrected (g/mile)		
		Tailpipe	Bag	Ratio
1	Leaded	0.025	0.149	5.9
2	Leaded	0.036	0.124	3.4
3	Leaded	0.054	0.189	3.5
4	Unleaded	0.0036	0.087	24.3
5	Unleaded	0.0029	0.069	24.0
6	Unleaded	0.0110	0.230	20.9
7	Unleaded	0.0024	0.120	50.0

[a] From Ter Hear et al.[12]
[b] Samples collected from 1966–1970 model cars during 7-mode Federal Cycles.

the filter. Details of one such recently proposed arrangement may be found in Refs. 14 and 16.

In principle, the experimental methods described above are suitable for studying in detail the formation and emission of particulates. In practice, the measurements are generally beset by difficulties limiting the range of data and introducing considerable experimental errors. The possible magnitude of errors is well illustrated by the interesting experiments of Ter Haar et al.[12] These investgators compared the total amounts of particulates collected from the tailpipe of an automobile and from a bag into which the exhaust discharged. Five to fifty times as many particles were collected from the bag than from the tailpipe (Table II). There are several possible explanations for this; the sample from the tailpipe was drawn at a constant rate and not isokinetically, the losses in the sampling probes were different, and last, but not least, the temperatures of the two samples were different. These and similar problems beset many of the experiments, thereby seriously affecting the accuracy of the results. This must be borne in mind when examining the data presented in Sections III and IV.

C. Test Procedures for Particulate Emissions

Although test procedures for gaseous exhaust emissions have been defined precisely,[17-19] similar procedures have not yet been established for particulate emissions. Consequently, the test conditions, the measuring techniques, and the instrumentation used vary widely. Experiments have been performed with engines mounted on dynamometers or with cars driven on chassis dynamometers. Measurements have been made with engines running at constant speed, accelerating, decelerating, or operating under predetermined operating cycles. The emission rates are different during cyclic operation than during steady speeds because large amounts of particulates, built up during cruising conditions, are exhausted during acceleration (see Sections III.A and IV.A). Therefore, the type of cycle used in a test has a significant bearing on the results. Prior to 1968 various cycles, simulating city-type driving, were used in particulate studies; more recently the 7-mode Federal Test procedure (often termed California Cycle), developed for gaseous emission tests for 1968–1971 model cars, was often employed. This cycle, described in detail in Refs. 17 and 18, is composed of seven different operating modes of idle, cruise, acceleration, and deceleration. A new cycle was recently instituted for 1972 model engines and vehicles,[19] and this could also be applied to particulates.

Before each test, the mileage could be built up under various operating conditions. For gaseous emission the mileage must be accumulated according to the schedule given in Ref. 18 (Federal Mileage Accumulation Schedule). For particulates this same schedule may be and has been used.

Standards for measuring techniques or instrumentation have not yet been specified.

III. PHYSICAL CHARACTERISTICS OF PARTICULATES

As stated in the Introduction, the physical characteristics of interest are the total weight, the size distribution, and the shape of the emitted particles. In recent years considerable effort has been made to evaluate these characteristics and to determine their dependence on various parameters such as fuel, engine, and operating conditions. The progress in these efforts is summarized below.

A. Total Particulate Matter Emitted

From the point of view of air pollution one of the most important parameters is the total amount of particulate matter emitted. This may be expressed in terms of the weight emitted per unit time of operation (g/min), per unit volume of exhaust gas (g/ft³), or per unit distance traveled (g/mile). Data have been reported in the literature using all three of these definitions. However, since the late 1960's the last definition (g/mile) has gained widest acceptance.

McKee and McMahon[20] were among the first to present total emission rates from automobiles. These investigators measured emission from various 1954–1957 model cars under steady operating conditions (Table III). After this study little work was reported concerning total emission until 1970 when Habibi et al.,[16] Moran and Manary,[21] Ninomiya et al.,[22] and Ter Haar and his coworkers[12] presented results on total particulate emissions from

Table III.   Total Particulate Emissions from Different Engines Operating on Leaded Fuel

Engine	Operating condition	Emission (g/mile)	Reference
1954 Chevrolet, 6 cylinder	30 mph cruise	0.15	McKee and McMahon[20]
1954 Oldsmobile, V–8	30 mph cruise	0.04	
1954 Plymouth, 6 cylinder	30 mph cruise	0.06	
1956 Oldsmobile, V–8	30 mph cruise	0.08	
1957 Chrysler, V–8	30 mph cruise	0.06	
1957 Ford, V–8	30 mph cruise	0.1	
Various 1966 models	7-mode Federal Cycle		Ter Haar et al.[12]
	4 cold cycles	0.51	
	4 hot cycles	0.24	
1966 "Popular Model," V–8	Simulated consumer test conditions	0.4	Habibi et al.[16]
1969 Ford, V–8	7-mode Federal Cycle		Ninomiya et al.[22]
	cold start	0.52	
	after 10 cycles	0.27	
1970 Chevrolet, V–8	Steady, 2250 ppm	0.15	Moran and Manary[21]

1960–1970 model cars and engines. By the time these measurements were performed it was well recognized that deposit accumulation and engine operating conditions play an important role in the emission rate. Therefore, many of these investigators performed experiments under various cyclic operating conditions. The results of these measurements are summarized in Table III. As can be seen, McKee and McMahon's emission rates are considerably lower than those obtained in later years. It is unlikely that earlier model engines emitted fewer particulates. Possibly, the low emission rates were due to the steady operation or to uncertainties in the measurements. The first of these possibilities is supported by the observation (see Table III) that under steady operation the emission rates are two to three times lower than under cyclic conditions.

The results in Table III also demonstrate that emission rates are considerably higher when the engine is started (cold start) then when it has been running continuously for some time. This is further borne out by the results of Ninomiya et al.,[22] who operated test vehicles in accordance with the 7-mode Federal Cycle. After start, the emission rate decreased until it stabilized after several cycles (Fig. 8).

The foregoing observations point to the significant role of temperature in the formation of particulates. However, little information is available on the effect of temperature, possibly because it is difficult to control the temperature of the exhaust gas (see Section II.B). One experimental result showing the effect of temperature on the total particulate emitted was reported by Sampson and Springer[13] and is reproduced here in Fig. 9. The total emission

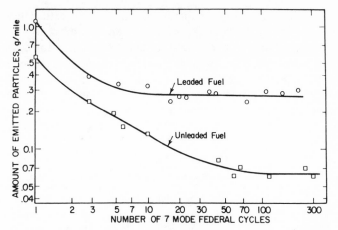

Fig. 8. The variation in particulate emition from 302 CID, V–8, 1969 Mustang with consecutive 7–mode Federal Cycles. (Ninomiya *et al.*[22])

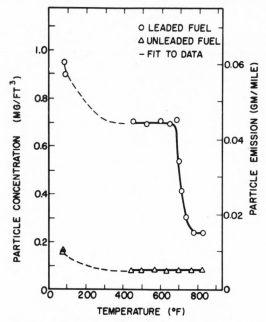

Fig. 9. The effect of temperature on the amount of particulates emitted. 1970 Chevrolet 350 CID, V–8 engine at 1800 rpm with road load and leaded fuel. (Sampson and Springer.[13])

rate appears to be extremely sensitive to the temperature of the exhaust gas, particularly above 600 °F and below 200 °F. Owing to its significance, the effects of temperature would merit further consideration.

The fact that operating conditions strongly affect particulate emission is true not only for cold start versus continuous operation but also for engines running at steady but different speeds, for acceleration and for deceleration. McKee and McMahon[20] showed that with increasing engine speed (from idle to 50 mph) the amount of particulate matter emitted (mg/min) increases considerably (Table IV). The emission rate further increases during acceleration. A similar effect was observed by Hirschler et al.,[23,24] who found that during acceleration the amount of lead emitted was several times higher

Table IV.   The Effect of Engine Speed on the Amount of Emitted Particulate. Leaded Fuel[a]

Engine	Particulate emission (mg/min)				
	Idle	30 mph cruise	50 mph cruise	50 mph accel.	30 mph decel.
1954 Chevrolet, 6 cylinder	6.4	78	77	71	68
1954 Plymouth, 6 cylinder	7.1	30	116	268	34
1954 Oldsmobile, V–8	18.0	20	50	293	77
1956 Oldsmobile, V–8	8.3	39	70	107	175
1957 Ford, V–8	8.6	52	245	206	572

[a] From McKee and McMahon.[20]

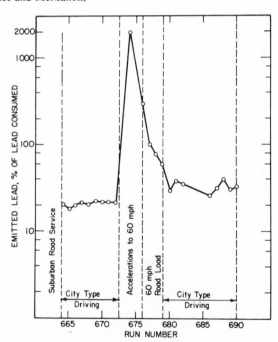

Fig. 10.   The effect of driving conditions on the amount of emitted lead (Hirschler and Gilbert[24]).

than the amount of lead consumed (Fig. 10). In contrast to this, at steady speeds only a fraction of the consumed lead was emitted. The increase in particle emission during acceleration is attributed to removal of particles from the exhaust system deposited during steady-state operation.

In addition to variations in operating conditions, the composition of the fuel used also influences particulate formation. In spite of the great importance of the fuel composition, relatively few investigators have explored this variable. The recent studies of Moran and Manary,[21] Ninomiya et al.,[22] Habibi et al.,[16] and Ter Haar et al.[12] illustrate the effects of fuel composition. Additives such as TEL, phosphorous, and commercial upper-cylinder lubricants, etc., cause substantial increase in the weight of particulates emitted (see Table V and Figs. 8, 11, 12). On the other hand, the total volume (rather than weight) of the emitted particulates appears to be considerably higher with unleaded than with leaded fuel (Fig. 13) and the mean particle diameter significantly smaller.[21] These results shed some light on the importance of fuel composition on particulate emission. Clearly, the weight

Table V.   The Effect of Fuel Composition on the Total Amount and on the Mass Median Equivalent Diameter of Emitted Particulates[a]

Fuel	Total amount emitted (g/mile)	MMED (μ)
Indolene HO 0 + 3 ml/gal TEL	0.16	$\ll 0.10$[b]
Indolene HO 0 + 3 ml/gal TEL + 1.0 theory EDB	0.15	0.65
Indolene HO 30 (3 ml/gal TEL, motor mix scavenger)	0.10	0.10
Indolene HO 15 (1.5 ml/gal TEL, motor mix scavenger)	0.03	1.5
Indolene HO 0 (Trace TEL)	0.003	—

[a] 1970 Chevrolet 350 CID, V–8 engine running at 2250 rpm. (From Moran and Manary.[21])
[b] Much less than 0.1 μ.

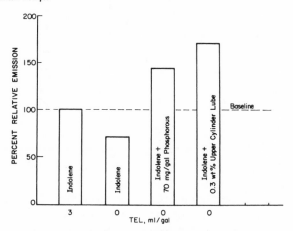

Fig. 11.   The effect of fuel on particulate emissions
(Ter Haar et al.[12]).

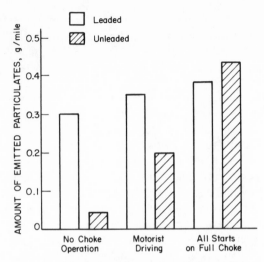

Fig. 12. Comparisons between particulate emissions with leaded and unleaded fuels (Habibi et al.[16]).

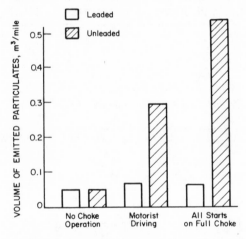

Fig. 13. Comparisons between the volume of particulates emitted with leaded and unleaded fuels (Habibi et al.[16]).

of emitted particulates is much lower with unleaded than with leaded fuel but the volume of very fine particles is larger. More research is needed to develop a complete understanding of the role of the fuel and fuel additives, as well as lubricants and lubricant additives, in the generation of particulates.

The physical characteristics of particulates depend also on the condition of the engine, as evidenced by the decrease in particulate emission rates from

Table VI.   The Effect of Engine Maintenance on Particulate Emission in a 1950 Ford 1/2–Ton Pickup Truck, 6–Cylinder Engine Using Leaded Fuel[a]

| Operating condition | Particulate emission (mg/min) and exhaust gas temp. (°F)[b] | |
	Before tune-up	After tune-up
Idle	14 (98 °F)	7 (126 °F)
30 mph cruising	36 (348 °F)	22 (410 °F)
Mild acceleration	226 (455 °F)	99 (478 °F)

[a] From McKee and McMahon.[20]
[b] Temperature measured at the end of the tailpipe.

a 1950 Ford 1/2-ton pickup truck after tune-up (Table VI). One must note, however, that the exhaust gas temperature increased after the tune-up; this increase may have accounted for some of the reduction in the emission.

The results discussed above demonstrate that the amount of particulate matter emitted depends strongly upon several factors and varies over a wide range. Therefore, from the information presently available, it is impossible to predict the emission rate from a vehicle. The generally recommended figure of 10–12 lb of particulates for each 1000 gal of fuel burnt[2] can be considered only a very rough estimate. It seems, however, that cars operating on leaded fuel do not satisfy the 0.1 g/mile emission, which is being considered as the standard for 1975.

B. Shape and Size Distribution of Particulates

Not all the emitted particulates are of equal importance to air pollution. Some particles settle quickly; hence their detrimental effects are less than those which remain suspended for longer periods. The settling time depends on the shape and size of the particle and, therefore, these characteristics must also be examined.

Particulate shape has not yet been investigated extensively. A number of micrographs have been presented in the literature[21,22] which show that particles may consist of single crystals and of aggregates, and may have a large variety of shapes, some being nearly spherical, some needlelike.

Owing to the large differences in shape, it is difficult to define simply and precisely the size of the particle. To overcome this difficulty the particle size is generally specified by a single dimension called "diameter." Unfortunately, some ambiguity exists regarding the definition of this term. The diameter is sometimes related to the settling velocity of the particle, in which case it is defined as the diameter of a sphere (of the same density as the particle) which would settle in still air with the same velocity as the particle.[3] Frequently, a "diameter" is deduced from observation of micrographs, from the intensity and direction of light scattered from the particle, or from particles

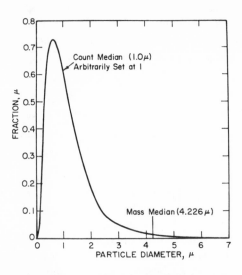

Fig. 14. Log-normal particle size distribution, assuming a count median diameter of 1 μ (from Ref. 3).

deposited on different stages of cascade impactors.[3] In addition to "diameter," the term "mass median equivalent diameter" (MMED) is also used. This is determined from the log-normal size distribution of particles (Fig. 14) such that half the mass lies on either side of the MMED.[3]

The most detailed investigations of particle size distributions were performed by Mueller et al.[25] and Moran and Manary.[21] The former tested 1961 and 1962 model cars operating at steady speeds (25, 45, 60 mph) and found 60–80% by weight of the particles to be less than 2 μ in diameter (Table VII). The latter performed measurements with a 1970 Chevrolet 350 CID V-8 engine running at 2250 rpm (Fig. 15). Moran and Manary's results confirm those of Mueller's, namely, that about 80% of the particles are less than 2 μ in diameter. These results, obtained at steady engine speeds, are also supported by the observations made by Ninomiya et al.[22] under the 7-mode Federal Cycle. From their data Ninomiya and his coworkers estimated that, with leaded fuel, about 40% of particulate mass emitted from a 1969 Ford 302 CID V-8 engine was associated with particles less than 0.1 μ in diameter. For unleaded fuel only about 10% was smaller than 0.1 μ. The results of

Table VII. Size Distribution of Particulates Emitted by One 1961 and Two 1962 Popular Low-Priced Model Cars, Operating at Steady Speeds (25, 45, 60 mph) Using Leaded Fuel[a]

Particle size (μ)[b]	>2	<2	<0.3
Size distribution by weight (%)	20–38	62–80	>43–55

[a] Mueller et al.[25]
[b] Diameter at unit density.

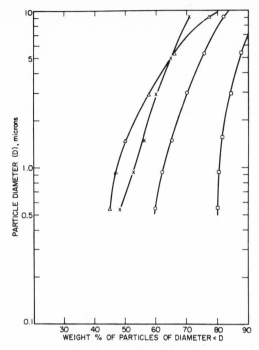

Fig. 15. The effect of fuel on the size distribution of emitted particles. Fuel used: △, Indolene HO 15 (1.5 ml/gal TEL, motor mix scavenger); ○, Indolene HO 30 (3 ml/gal TEL, motor mix scavenger); □, Indolene HO + 3 ml/gal TEL; ×, Indolene HO 0 + 1.0 theory EDB + 3 ml/gal TEL (Moran and Manary[21]).

Moran and Manary also show an increase in the weight of smaller particles with increase in the lead content of the fuel (Fig. 15).

The large number of small particles observed during steady operating conditions, and the increase in particle size and weight during acceleration suggest a possible mechanism for particle formation. Particles may originate in both the combustion chamber and the exhaust stream. However, owing to the prevailing high temperatures, only a small fraction of the emitted particles are likely to form in the combustion chamber. Most particles are expected to form and grow in the cooler exhaust system due to homogeneous and heterogeneous vapor phase condensation, enhanced by coagulation. Some of these particles are emitted directly, without settling, and these constitute most of the small particles in the exhaust. Some of the particles are deposited on the walls of the exhaust system, where they combine into larger particles. Many of these are removed when the flow rate is suddenly increased and these, together with rust and scale, account for the increase in the weight and size of particulates emitted during acceleration.

IV. CHEMICAL COMPOSITION OF PARTICULATES

Particulates emitted from automobiles are composed of various organic and inorganic substances including lead, carbon, alkaline earth compounds, iron oxides, tar, and oil mist. Of these, lead constitutes the largest weight fraction, at least when the fuel contains lead additives. Partly for this reason, and partly because of its toxicity, the lead in particulates has received the most attention, so much so that it has been investigated more thoroughly than any other aspect of particulate emission. It is expedient, therefore, to divide the discussion on the chemical composition into two parts, one dealing with lead and one with the remaining compounds.

A. Lead Content of Particulates

Owing to the high compression ratios generally used in American-built automobile engines, these engines require high (90–100) octane gasoline for satisfactory operation. Such high-octane ratings can be achieved either by adding tetraethyl lead or other organometallic compounds, or by increasing the aromatic content of the gasoline. Until recent years the first method was used almost exclusively, resulting in particulates of high lead content and contributing to the number of lead-containing particles in the atmosphere. This contribution is evident from the increase in the concentration of lead-containing particles in areas where traffic is heavy (Table VIII). As mentioned in Section II. A, this contribution is also manifested through the increased lead content of precipitation and of vegetation.

Table VIII. Concentration and Average Size of Lead-Containing Particles in Several Areas of the U.S.[a]

Area	Avg. daytime concentration ($\mu g/m^3$)	Avg. MMED (μ)
Los Angeles freeway	8.2–18.3	0.12
Downtown Los Angeles	2.9–9.0	0.26
Pasadena	0.9–4.3	0.24
San Francisco	2.8–6.1	0.25
Cincinnati	1.0–11.7	0.23
Chicago	0.5–7.8	0.30
Philadelphia	0.3–3.4	0.24
Cherokee	0.1–0.3	0.31

[a] Robinson et al.[8] and Robinson and Ludwig.[9]

Table IX. Average Composition of Emitted Particulate Lead Compounds[a]

	Weight % of compound[b]				
	PbCl·Br	α–NH$_4$Cl· 2PbCl·Br	β–NH$_4$Cl· 2PbCl·Br	2NH$_4$Cl· PbCl·Br	3Pb$_3$(PO$_4$)$_2$· PbCl·Br
City-type cycle, fuel + TEL only	68	24	6	2	—
City-type cycle, added sulfur[c]	70	30	—	—	—
City-type cycle, added phosphorus[d]	35	18	17	10	20
Constant speed, 60 mph, road load	60	20	20	—	—
Full-throttle accelerations	85	10	5	—	—

[a] Hirschler et al.[23]
[b] PbSO$_4$ and PbO·PbCl·Br·H$_2$O also occurred occasionally in concentrations of 5% or less.
[c] Sulfur content increased from 0.025 to 0.150 weight % by added disulfide oil.
[d] 0.4 theory phosphorous added as phosphorous additive A or B.

In discussing "lead" it must be recognized that the "lead" is in the form of different compounds, composed mostly of lead halides and complexes of ammonium halide and lead halide. A typical composition of the emitted lead compounds is given in Table IX. Unfortunately, such detailed chemical analysis is uncommon and the composition is generally unknown. Therefore, usually all lead compounds are simply referred to as "lead." This practice is also followed here.

The study of lead was pioneered by Hirschler and his coworkers,[23,24] who, in the late 1950's, investigated the lead content of particulates with the primary objective of evaluating the effect of fuel additives on the amount of emitted lead. Hirschler et al. estimated that lead compounds constituted 58 to 78% by weight of the fine (smaller than $\sim$5 μ) particles and 34 to 60% of the course (larger than $\sim$5 μ) ones. A word of caution is in order about the term "particle size" as used by these investigators. Hirschler et al. collected the particles from a precipitator and then washed them in a solution of trichlorethylene, which contained 5% of toluene and a dispersant. The lead compounds had negligible solubility in this solution, but some of the other compounds dissolved. Therefore, the "size" thus obtained is less than the actual size of the particle.

Following Hirschler, lead contents of particles were determined by Mueller et al.,[25] Ter Haar et al.,[12] Habibi et al.,[16] and Ninomiya et al.[22] The results of these investigations, summarized in Table X, show that in most cases the lead content of particles varies between 20–80% by weight. Habibi and his coworkers[16] reported somewhat higher values, Ter Haar et al.[12] reported somewhat lower ones. Although the reasons for the wide range of data are not readily evident, some plausible explanation for it can be offered. The fuel composition affects the amount of lead emitted, and differences in the fuel used may account for the spread in the results. Vari-

Table X. Lead Content of Emitted Particles (Leaded Fuel)

Engine	Operating condition	total particles	fine particles[b]	Reference
		Lead content (Wt %) of[a]		
1961–62 "popular model" cars	Steady speed (25, 45, 60 mph)		22–80	Mueller et al.[25]
1966 model cars	7–mode Federal Cycle			Ter Haar et al.[12]
	4 cold cycles	15–20		
	4 hot cycles	15–20		
1966 "popular model" cars	"No choke"	80–90		Habibi et al.[16]
	"Motorist driving"	80–90		
	"All starts on full choke"	80–90		
1968 Ford V–8	7–mode Federal Cycle	~50		Ninomiya et al.[22]

[a] Average values.
[b] Less than ~2 μ.

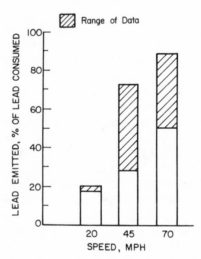

Fig. 16. The amount of lead emitted at various steady engine speeds, under road load conditions. Leaded fuel (Habibi et al.[16]).

ations in lead content may also be due to differences in the exhaust gas temperature (Fig. 9), accelerations (Fig. 10), and engine speed (Fig. 16). In connection with Fig. 16, it is worth noting the wide ranges of experimental uncertainties. Such large uncertainties are not uncommon in the measurements of particulates.

Since only a fraction (about 10 to 50 %) of the lead consumed is generally emitted, the question arises as to what happens to the remaining portion. According to Hirschler et al.[23,24] most of the lead is retained in the exhaust

Table XI.   Lead Retained by Automobile Exhaust Systems and Lubricating
Oil (Leaded Fuel)[a]

	1954 model		1953 model		1957 model	
Test mileage	26,996		19,345		9,800	
Total lead consumption (g)	5,300		4,154		1,950	
	Lead recovered					
	g	% of input	g	% of input	g	% of input
Ahead of muffler[b]	140	2.6	217	5.2	93	4.7
Mufflers	247	4.7	379	9.1	166	8.6
Tailpipes	5	0.1	11	0.3	2	0.1
Oil sludge	59	1.1	100	2.4	3	0.1
Oil filters	515	9.7	None used		None used	
Oil changes	160	3.0	436	10.5	187	9.6
Total	1126	21.2	1143	27.5	451	23.1

[a] Hirschler et al.[24]
[b] Includes conbustion chambers, exhaust ports, manifolds, and exhaust pipe.

system ahead of the tailpipe, in the oil, and in the oil filters (Table XI). Evidence supporting this finding was also presented by Moran and Manary.[21]

The ratio of lead emitted to lead consumed provides valuable insight into the lead content of particulates. However, the most important and most direct evidences of lead as a pollutant are the rate at which lead is emitted and the size distribution of the lead-containing particles.

Lead emission rates from 1966–1969 model cars vary from about 0.1 to 0.5 g/mile (Table XII), the emission rate being somewhat influenced by the operating conditions. More lead is emitted from cold than from hot engines, just as more total particulates are produced at cold starts than under continuous operation (Section III.A). The condition of the car, expressed in terms of the miles traveled at the time of the test, does not seem to affect the results appreciably, as long as sufficient miles are accumulated to stabilize the deposits. It is interesting to compare the data in Table XII with those given

Table XII.   Lead Emission Rates (Leaded Fuel)

Engine	Operating condition	Emission rate (g/mile)	Reference
1966 models[a]	7–mode Federal Cycle		Ter Haar et al.[12]
	cold cycles	0.512	
	hot cycles	0.240	
1966 popular model V–8[b]	Federal Mileage Accum. Schedule	0.12–0.20	Habibi et al.[16]
1969 Ford V–8[c]	7–mode Federal Cycle	0.1	Ninomiya et al.[22]

[a] Average of 16 cars tested; mileage from 30,000 to 100,000.
[b] Mileage from 5,000 to 28,000.
[c] Mileage 6,000.

by Larsen and Konopinski.[11] From sampling the air in the Summer Tunnel, these investigators estimated lead emission rates of about 0.031 g/mile. This is reasonably close to the values obtained directly from the exhaust (0.1–0.2 g/mile). The lower rates yielded by atmospheric samples are understandable, since the larger particles settle quicker, thereby resulting in lower particle masses in the sample.

Information regarding "lead particle" size was first provided by Hirschler et al.[23,24] but, as discussed earlier, the sizes reported by them referred to the size of the lead only and not to the total particle. Actual size distributions of lead containing particles were investigated by Mueller et al.,[25] Habibi et al.,[14,16] and Moran and Manary.[21] Mueller and his coworkers estimated that as much as 60–80% of the lead is contained in particles less than 2 μ in

Table XIII. Lead Particulate Emission as a Function of Size, Mileage, and Operating Condition[a]

Operating condition	Mileage	Percent of emitted lead		
		$>9\,\mu$	$<1.0\,\mu$	$<0.3\,\mu$
Federal Mileage Accum. Schedule	5,000	27	45	30
	16,000	39	36	25
	28,000	57	19	11
Simulated consumer test conditions	32,000	57	25	16

[a] 1966 Popular model cars with 327 CID, V–8 engine, 3 g/gal lead as motor mix (Habibi et al.[16]).

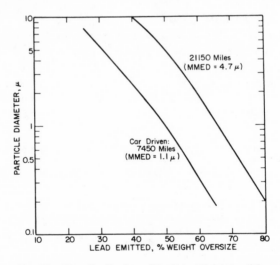

Fig. 17.   Size distribution of lead-containing particles.
Leaded fuel (Habibi[14]).

diameter. Habibi's measurements also show 50–60% of the lead residing in particles less than about 2 μ in size (Table XIII and Fig. 17). Moran and Manary's[21] results further support this conclusion. The fraction of lead in larger particles appears to increase as the mileage on the car increases (Table XIII). This phenomenon may be attributed to aggregates of particles deposited on the exhaust system during extended operation and exhausted subsequently.

Habibi's measurements indicate the mass median equivalent diameters of lead-containing particles to range from 1 to 4 μ. Samples taken from the atmosphere yield mass median equivalent diameters of 0.2–0.3 μ (Table VIII). As in the case of data taken from the Sumner Tunnel, the atmospheric sample again gives somewhat lower values than the sample taken directly from the exhaust. The faster settling of larger particles in the atmosphere biases the data in favor of smaller particles and this may account for the differences in the two measurements.

B. Chemical Composition of Particulates

A knowledge of the detailed chemical composition of particulates would provide a significant step toward the understanding of the particulate formation process. However, to date the general chemical composition has received little attention. The few results available show that besides lead, the major constituents are bromine, chlorine, and carbon (Table XIV and Fig. 18). In addition, particles contain metallic elements (largely iron and zinc) and various organic compounds. According to Moran and Manary[21] iron (rust) is mostly associated with larger particles, suggesting that some of these come from the surface of the exhaust system. Moran and Manary concluded that the fraction of organic materials is much higher in smaller particles than in larger ones. This is illustrated by one of the size distributions

Table XIV. Average Chemical Composition of Particulate Matter Emitted from a 1953, a 1954, and a 1957 Model Car (Leaded Fuel)[a]

Element	Range of composition (%)	
	Fine particles[b]	Coarse particles[c]
Lead	58–74	34–60
Chlorine	5.6–16.8	4.5–11.2
Bromine	8.2–21.3	9.2–23.6
Iron	0–1.3	1.2–11.2
Sulfur	0–1.9	0.2–2.3
Carbon	3.5–10.5	4.6–12.1

[a] Hirschler and Gilbert.[24]
[b] Less than ~5 μ.
[c] Greater than ~5 μ.

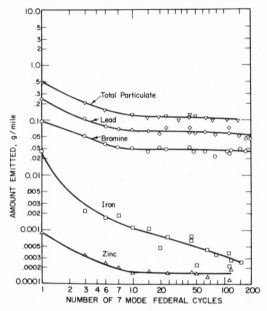

Fig. 18. The variation in the chemical content of
particulates with consecutive 7–mode Federal cycles.
302 CID, V–8, 1969 Torino. Leaded fuel (Ninomiya
et al.[22]).

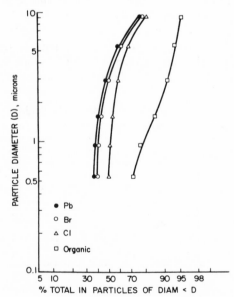

Fig. 19. Chemical composition of different size
particles. Fuel; indolene HO 15 (1.5 ml/gal TEL,
motor mix scavenger) (Moran and Manary[21]).

reproduced from Ref. 21 (Fig. 19). The presence of TEL and scavengers in the fuel changes the amount of lead, bromine, etc., in the particulate, but does not seem to affect the total amount of organic material present.

V. CONCLUDING REMARKS

The foregoing account indicates the strong activity in recent years in the field of particulate emission from spark-ignition engines, particularly within the last fifteen years. However, in spite of the many excellent studies, understanding of the phenomena of particulate formation and emission is extremely limited. The instrumentation and experimental methods must be perfected further to provide reliable and accurate data. Much work must still be done to understand the effects of the fuel, the engine, the operating conditions, and the exhaust system on the generation and emission of particulates. Such knowledge is needed in order to develop effective techniques for the reduction and control of emitted particulate matter.

ACKNOWLEDGMENTS

The author wishes to thank Messrs. J.T. Ganley and R.E. Sampson for their many helpful comments and suggestions on the manuscript. The support of this work by the National Institute of Health, under grant number AP A1012–01, is gratefully acknowledged.

REFERENCES

1. *The Sources of Air Pollution and Their Control,* PHS-Pub–1548, (revised) 1968, U.S. Department of Health, Education and Welfare, Division of Air Pollution, Washington, D.C.
2. *Air Quality Criteria for Particulate Matter,* Publication No. AP–49, Jan. 1969, U.S. Department of Health, Education and Welfare, Public Health Service, National Air Pollution Control Administration, Washington, D.C.
3. *Control Techniques for Particulate Air Pollutants,* Publication No. AP–51, U.S. Department of Health, Education and Welfare, Public Health Service, National Air Pollution Control Administration, Washington, D.C.
4. Motto, H.L., Daines, R.H., Chilko, D.M., and Motto, C.K., Lead in soils and plants: Its relationship to traffic volume and proximity to highways, *Environ. Sci. Tech.* 4 (March 1970) 231–238.
5. Lagerwerff, J.V., and Specht, A.W., Contamination of roadside soil and vegetation with cadmium, nickel, lead, and zinc, *Environ. Sci. Tech.* 4 (July 1970) 583–586.
6. Lazrus, A.L., Lorange, E., and Lodge, Jr., J.P., Lead and other metal ions in United States precipitation, *Environ. Sci. Tech.* 4 (Jan. 1970) 55–58.
7. Daines, R.H., Motto, H., and Chilko, D.M., Atmospheric lead: Its relationship to traffic volume and proximity to highways, *Environ. Sci. Tech.* 4 (April 1970) 318–323.
8. Robinson, E., Ludwig, F., DeVris, J.E., and Hopkins, T.E., *Variations of Atmospheric Lead Concentrations and Type with Particle Size,* Final Report, Project Number PA–4211, Nov. 1963, Stanford Research Institute, Stanford, California.

9. Robinson, E., and Ludwig, F., *Size Distributions of Atmospheric Lead Aerosols,* Final Report, Project Number PA–4788, April 1964, Stanford Research Institute, Stanford, California.

10. Colucci, J.M., and Begeman, C. R., The automotive contribution to air-borne polynuclear aromatic hydrocarbons in Detroit, *J. Air Poll. Control Assoc.* **15** (March 1965) 113–122.

11. Larsen, R.I., and Konopinski, V.J., Sumner Tunnel air quality, *Arch. Environ. Health* **5** (Dec. 1962) 597–608.

12. Ter Haar, G.L., Lenane, D.L., Hu, J.N., and Brandt, M., Composition, size and control of automotive exhaust particulates, *J. Air Poll. Control Assoc.* **22** (Jan. 1972), 39–46.

13. Sampson, R.E., and Springer, G.S., *Effects of Temperature and Fuel Lead Content on Particulate Formation in Spark Ignition Engine Exhaust,* Publication 72–1, April 1972, Fluid Dynamics Laboratory, Mechanical Engineering Department, The University of Michigan, Ann Arbor, Michigan.

14. Habibi, K., Characterization of particulate lead in vehicle exhaust-experimental techniques, *Environ. Sci. Tech.* **4** (March 1970) 239–253.

15. Lee Jr., R.E., Patterson, R.K., Crider, W.L., and Wagman, J., Concentration and Particle Size Distribution of Particulate Emissions in Automobile Exhaust, *Atmospheric Environment* **5** (April 1972), 225–237.

16. Habibi, K., Jacobs, E.S., Kunz, Jr., W.G., and Pastell, D.L., *Characterization and Control of Gaseous and Particulate Exhaust Emissions from Vehicles,* Presented at the Air Pollution Control Association-West Coast Section, Fifth Technical Meeting, San Francisco, October 1970. Report available from E.I. DuPont De Nemours & co. Wilmington, Delaware.

17. Control of air pollution from new motor vehicles and new motor vehicle engines, *Federal Register* **31**, Part II (March 1966) 5170–5178.

18. Control of air pollution from new motor vehicles and new motor vehicle engines, *Federal Register* **33**, Part II (June 1968) 8304–8324.

19. Control of air pollution from new motor vehicles and new motor vehicle engines, *Federal Register* **35**, Part II (Nov. 1970) 17288–17313.

20. McKee, H.C., and McMahon, Jr., W.A., Automobile exhaust particulates — source and variation, *J. Air Poll. Control Assoc.* **10** (Dec. 1960) 456–462.

21. Moran, J.B., and Manary, O.J., *Effect of Fuel Additives on the Chemical and Physical Characteristics of Particulate Emissions in Automotive Exhaust,* Interim Technical Report to the National Air Pollution Control Administration, submitted by the Dow Chemical Company, Midland, Michigan, July 1970.

22. Ninomiya, J.S., Bergman, W., and Simpson, B.H., Automotive particulate emissions, Presented at the *Second International Clear Air Congress,* International Union of Air Pollution Prevention Association, Washington, D.C., Dec. 1970. Reports available from Automotive Emissions Office, Ford Motor Company, Dearborn, Michigan.

23. Hirschler, D.A., Gilbert, L.F., Lamb, F.W., and Niebylski, L.M., Particulate lead compounds in automobile exhaust gas, *Ind. Eng. Chem.* **49** (July 1957) 1131–1142.

24. Hirschler, D.A., and Gilbert, L.F., Nature of lead in automobile exhaust gas, *Arch. Environ. Health* **8** (Feb. 1964) 297–313.

25. Mueller, P.K., Helwig, H.L., Alcocer, A.E., Gong, W.K., and Jones, E.E., Concentration of fine particles and lead in car exhaust, in *Symposium on Air Pollution Measurement Methods,* American Society for Testing and Materials, Special Technical Publication Number 352, 1962, pp. 60–77.

Chapter 6

Diesel Engines Combustion and Emissions

N. A. Henein

Mechanical Engineering Sciences Department
Wayne State University
Detroit, Michigan

I. INTRODUCTION

This chapter analyzes diesel combustion and emissions with emphasis on high-speed transportation engines. The contribution of these engines to air pollution in the USA is presented in Table I as the relative percentage of the emissions from other sources.[1] It may be seen that the two highest percentage emissions from the diesel engine are particulates and nitrogen oxides. Its other emissions that are not shown in Table I are the oxygenated hydrocarbons (including aldehydes), the odor constituents, and noise. The diesel emissions most offensive to the public are the exhaust odor and the visible smoke particulates.

The present Federal regulations call only for the control of exhaust smoke. With the growing public concern for cleaner air, more stringent regulations are

Table I. Contribution of the Transportation Diesel Engines to the Air Pollution Problem in the USA[a]

	$\dfrac{\text{Diesel}}{\text{Gasoline}} \times 100$ percentage	$\dfrac{\text{Diesel}}{\text{ECCS}^b} \times 100$ percentage	$\dfrac{\text{Diesel}}{\text{All sources}} \times 100$ percentage
Carbon monoxide	0.3	0.3	0.02
Gaseous hydrocarbons	3.7	3.4	1.9
Nitrogen oxides	14.3	5.5	4.8
Sulfur oxides	50.0	0.5	0.4
Combustion particulates	125.0	15.2	—
Total particulates	125.0	5.0	1.75

[a] Calculated from Ref. 1.
[b] ECCS, All energy conversion combustion systems.

expected to be made to control most of the undesirable emission species. The present chapter deals with all the combustion-generated emissions except noise.

A review of the literature on transportation engine emissions shows that most of the work has been related to the spark-ignition (SI) engine in which a flame propagates in a homogeneous fuel–air mixture. Few studies have been directed to the mechanisms of emission formation in other systems, such as diesel engines and gas turbines, in which the mixture is heterogeneous. For example, the unburned hydrocarbon emissions in the SI engine are mainly attributed to the wall quench phenomenon which has been studied both experimentally[2,3] and theoretically.[4] The corresponding phenomenon in the diesel engine or gas turbine has not yet been well defined. Some of the difficulties that arise in the theoretical analysis of the diesel engine emissions are related to the very complex turbulent gas flow patterns in the engine and to the multiplicity of the combustion mechanisms involved in the hetero-geneous fuel–air mixture formed in the different regions of the spray. Moreover, the cyclic nature of the engine operation and the instability of the charge constituents greatly hinders the experimental determination of the mechanisms of emission formation.

Previous diesel combustion studies have taken many approaches. Some of these were made on simplified systems or models in order to understand the fundamental, physical, and chemical phenomena which take place during the combustion process. These studies were concerned with droplet evapo-ration and combustion under steady- and transient-state conditions, droplet autoignition, spray evaporation, and combustion in constant volume bombs and engines. Many of the studies on engines dealt with the spray in a lumped form. The combined changes in the pressure and volume were used to calculate the rate of energy exchange between fuel and air. Engine photographic studies have proved to be very useful in analyzing some fluid dynamic aspects and the development of combustion. Detailed information on these studies may be found in the literature cited in Refs. 5–8. Other studies on spray pene-tration and evaporation and mixing were based on the continuum mechanics approach; these are discussed in Refs. 9–11. In the present chapter, the main findings of these studies will be used for the development of some combustion and emission models in the actual engine.

In high-speed transportation diesel engines, the time period available for the whole combustion process is very limited, and the engine should be designed to produce enough turbulence for the proper mixing of fuel, air, and other reacting gases. Present diesel engines can be divided into two major groups according to the method of producing turbulence in the cylinder:

1. *Direct-Injection or Open-Chamber Engines.* In this type of engine, turbulence is induced by the flow of air in a tangential direction to the cylinder during the intake stroke. The combustion chamber is open to all the fuel which is directly injected into the bulk of the air.

2. *Indirect-Injection or Divided-Chamber Engines.* In this type of engine, turbulence is produced by the flow of gases from one division in the chamber to the other, before and/or after combustion. Fuel is injected into the air in the prechamber. The products of combustion develop turbulence in the main chamber as they are discharged through a narrow passageway between the two chambers.

In the direct-injection engine, turbulence is generated prior to combustion, and is sometimes referred to as primary turbulence.[12] In the indirect-injection chamber, turbulence is combustion-generated, and is sometimes referred to as secondary turbulence.[12] Some combustion chambers have a combination of the two types. The mechanisms of combustion and emission formation are different in each type.

In the present chapter the combustion and emission characteristics of the direct- and indirect-injection engines will be studied. In order to help the reader analyze any particular combustion system, emphasis will be placed on the qualitative analytical approach rather than on detailed mathematical formulation. The mathematical formulations can be found in the references cited. Combustion and emission models will be developed for some of the systems and used to discuss the experimental results. A comparison between the emission characteristics of the different systems will be included.

Methods of emission control in diesel engines through basic design modifications or other devices will be studied. Areas for future research and development will be discussed and summarized in the concluding remarks.

Many of the combustion and emission phenomenon apply to both direct- and the indirect-injection engines. These will be discussed in Sections II and III, and will not be repeated in Section V, which deals with the indirect-injection engines.

II. DIESEL IGNITION AND COMBUSTION

A. Preignition Processes

Unlike the gasoline engine, in which combustion starts by an electric spark at one location, the combustion in the diesel engine starts by ignition nuclei at numerous locations in the combustion chamber. The time taken for the preignition process to produce these ignition nuclei, after the start of injection, is known as the ignition-delay period. The length of the ignition delay affects the rest of the combustion process, the mechanical and thermal stresses, noise, and exhaust emissions. Many formulas have been developed for the ignition delay and are summarized in Refs. 6–7.

The preignition processes in diesel engines may be divided into physical and chemical processes. The physical processes are:

1. Spray disintegration and droplet formation.
2. Heating of the liquid fuel and evaporation.
3. Diffusion of the vapor into the air to form a combustible mixture.

The chemical processes are:

1. The decomposition of the heavy hydrocarbons into lighter components.
2. The preignition chemical reactions between the decomposed components and oxygen.

It is difficult to draw a distinct line separating the physical and chemical processes since they overlap. The chemical processes start after the fuel vapor makes contact with the air. At the very early stages of combustion, however, the mass of fuel vapor which undergoes chemical reaction is too small to cause a detectable pressure rise. Therefore, the first part of the ignition delay can be considered to be dominated by the physical processes which result in the formation of a combustible mixture. The second part of the ignition delay can be considered to be dominated by the chemical changes which lead to autoignition. Autoignition is usually detected by the pressure rise due to combustion, or combustion luminescence. The pressure rise is due to the exothermic reactions. Luminescence may be due to the excitation of the formaldehyde molecules or to thermal radiation from the high-temperature carbon atoms in the flame.[13, 14] In general, illumination occurs after the start of pressure rise in direct-injection (DI) engines.[7]

In a detailed study of the ignition delay in DI diesel engines, it has been found that the preignition chemical reactions, rather than the physical processes, are the rate-controlling processes.[15]

B. Preignition Chemical Reactions

Very little is known about the detailed mechanisms of the reactions leading to the autoignition of hydrocarbon fuels in air. Even the reactions leading to the autoignition of simple hydrocarbons such as methane are not known. However, a study of the published data on the autoignition of liquid fuels in compression-ignition engines indicates that two types of lumped reactions take place prior to ignition. Garner et al.[16] found that, for four different liquid hydrocarbons, peroxides and aldehydes were formed during the ignition delay and reached their peak concentration just before the start of pressure rise due to combustion, as shown in Fig. 1. Also, they found that for high cetane number fuels, the peak concentration of these intermediate compounds occurred earlier than with lower cetane number fuels. Therefore we may consider that the preignition reactions take place in two stages as

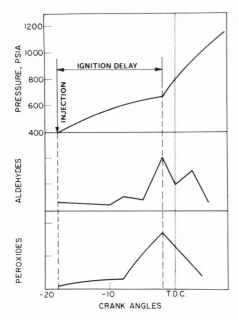

Fig. 1. Concentration of intermediate compounds during the combustion process (Garner[16]).

Table II. A Model for the Preignition Chemical Reactions

		Chain reactions	
	$C_nH_m + \left(n + \dfrac{m}{4}\right)O_2$	Initiating	....
Very slow reactions	....	Propagating	....
	....	Branching	....
			Intermediate compounds
	....	————	
Very fast reactions	Intermediate compounds	————	....
	....	Breaking	$n\,CO_2 + \dfrac{m}{2}H_2O$

illustrated in Table II. First, very slow reactions occur and form intermediate compounds such as peroxides and aldehydes. Then, once a critical concentration of these intermediate compounds has been reached, very fast chain reactions occur and lead to autoignition and the final products of combustion. If the critical concentration is not reached in any part of the spray, for example in the very lean areas of the spray or in the rich area in the core, partial oxidation products such as aldehydes and carbon monoxide will be left after the reaction.

A knowledge of an overall or global activation energy for the preignition reactions is believed to be necessary for developing mathematical models for emission formation in diesel engines. The overall or global activation energy for the preignition chemical reactions may be calculated from the ignition-delay data.

According to the above chemical reaction model, the ignition delay may be considered to end once the critical concentrations for the intermediate compounds are reached. If we deal with the combustion reactions as one lumped reaction between the fuel and oxygen, the following relation applies:

$$[F] + [O_2] \rightarrow [IC] \rightarrow \text{Final products} \tag{1}$$

Hence, the rate of formation of the intermediate compounds can be given by an Arrhenius-type relationship in terms of the fuel and oxygen concentration:

$$\frac{d[IC]}{dt} = k[F]^n[O_2]^m \tag{2}$$

where k = reaction velocity constant; $[F]$ = fuel vapor concentration; $[O_2]$ = oxygen concentration; $[IC]$ = intermediate compounds concentration; n and m = order of the reaction with respect to the fuel and oxygen, respectively; and t = time.

The reaction velocity constant may be given in terms of the global activation energy E_a as in Eq. (3). The activation energy is defined, according to the collision theory of reaction, as the minimum amount of translational energy a molecule must have in order to react upon impact. According to transient state theory of reaction, which is based upon interatomic attraction and repulsion forces, the activation energy is the minimum energy the reactants must acquire before the chemical change takes place.

$$k = c_1 e^{-E_a/R_0 T} \tag{3}$$

where c_1 = constant; E_a = apparent or global activation energy; R_0 = universal gas constant; and T = absolute temperature.

Since the ignition of the hydrocarbon–air mixtures occurs in a narrow limit of the fuel–air ratio, it can be assumed that the fuel and oxygen concentrations are constant and that Eq. (2) can be reduced to

$$\frac{d[IC]}{dt} = c_2 e^{-E_a/R_0 T} \tag{4}$$

where c_2 = constant.

The critical concentration of the intermediate compounds which is enough to start the reaction and cause a detectable pressure rise may be considered constant and to depend upon the combustion chamber. The ignition delay, ID, can, therefore, be expressed as follows:

$$ID = c_3 \, e^{E_a/R_0 T} \tag{5}$$

where c_3 is a constant which depends on the design of the combustion chamber. Equation (5) indicates that the activation energy, E_a, can be obtained by plotting log ID as a function of $1/T$.

The global activation energy for the preignition reactions of diesel No. 2, CITE (Compression-Ignition Turbine Engine), and gasoline fuels were found to be 5250 Btu/lb-mole, 10,430 Btu/lb-mole, and 14,780 Btu/lb-mole, respectively.[7] The cetane numbers of the above fuels are 57.5, 37.5, and 18, respectively. It is interesting to note that the global activation energy increases with the decrease in cetane number of the fuel. Low cetane number fuels are more stable and resist autoignition better than high cetane number fuels. In another study[15] it has been found that the global activation energy does not depend upon the physical properties of the fuel such as density and volatility, but is directly related to its ignition quality as measured by the cetane number.

III. DIRECT-INJECTION ENGINES (DI)

In DI diesel engines, particularly the transportation type, fuel is injected in swirling air. The optimum swirl may vary from one engine to another depending upon the design of the combustion chamber and the characteristics of the injection system.[17] In some of these engines, the ratio of the air swirl to engine speed was found to be between 7 and 13.[18]

Photographs by Watts and Scott[17] showed that the core of the spray is not significantly deflected by the air swirl as the spray top passes across the combustion chamber toward the walls. Since the spray consists of a very large number of different size droplets, the relatively large droplets may be assumed to be concentrated in the core. The smaller droplets and fuel vapor will be carried away from the core by the swirling air.

A. Fuel Evaporation

Fuel evaporation may be due to flash evaporation or due to heat transfer to the liquid fuel from the gases or wall surfaces. Flash evaporation is caused by the sudden drop in the fuel pressure as soon as it is injected under the effect of very high pressures, which range from about 3000 psi to 8500 psia.[19] These pressures usually exceed the critical pressures of the majority of the fuel constituents.[20] For example, the critical pressures of n-octane, n-decane, and n-dodecane are 361 psi, 311 psi, and 272 psi, respectively. Actual diesel fuels contain heavier hydrocarbons than those mentioned above. Their critical pressures decrease with the increase in the number of carbon atoms in the molecule. Upon injection, the fuel pressure drops to the cylinder air pressure which is in the range of 350 to 850 psia.[19] Higher pressures may be reached in supercharged engines.

The process of flash evaporation in the combustion chamber of diesel engines has not been the subject of any major study. This is probably due to the small amount of flashed vapor in comparison to the total amount of fuel injected. This small amount, however, may have a role in starting the preignition reactions, particularly in the light components of the actual fuels. Further studies in this area may prove useful for a better understanding of the preignition reactions in DI engines.

In many engines, particularly those with turbocharging, fuel evaporation in the cylinder may occur under supercritical conditions. Most of the previous studies on droplet evaporation have been concerned with subcritical evaporation which occurs in low compression-ratio naturally-aspirated diesel engines. Recent studies on supercritical evaporation have been limited to pure liquid compounds.[21] In diesel engines the droplets consist of a very large number of compounds with quite different thermodynamic properties. One major problem in the analysis of actual petroleum fuels is that of finding an average property for distillates. One of the methods proposed for such studies is that of the "characterization factor."[22] However, there is no way to check the validity of using this method for the study of supercritical evaporation of actual fuels in engines.

As the fuel droplets travel from the injector toward the walls, they gain heat by convection and decrease in size. The small size droplets which are carried away by the air are surrounded by relatively large air masses and are quickly heated. The high evaporation rate of the smaller droplets is due to two factors. The first is the high coefficient of heat transfer h, as given by Ref. 23, for droplets with no relative velocity with respect to the air:

$$hd/k = 2 \tag{6}$$

The second factor is the higher temperature potential between the surrounding air and the liquid. In the areas where the small droplets are concentrated, the mass of the fuel per unit air mass is lower than that in the core, where the droplets are larger and more crowded. The drop in the local air temperature in the region of the smaller droplets is therefore less than in the region of the larger droplets.

At the spray core, the mass-average temperature is lower because of the heat lost from the air to the concentrated fuel. El Wakil[24] suggested that adiabatic saturation conditions may prevail in the core and that the fuel may exist in the core as a liquid. Recent experimental results of Burt and Troth[25] showed that the rate of heat transfer to the fuel spray increases linearly with the air temperature and with the decrease in fuel volatility. These experimental observations are not consistent with the concept of adiabatic saturation in the spray core.[25] Further studies are needed in this area.

Watts and Scott[17] estimated that the fuel sprays take about 4 crankshaft degrees to reach the wall. This is equivalent to about 0.4 msec at a speed of 1800 rpm. During this time, some of the small droplets will be completely evaporated and the large droplets will be partially evaporated. The total amount of vapor that will be present in the combustion chamber before the jet hits the walls depends mainly upon the droplet size distribution in the spray.

A limited number of studies have been made on the droplet size distribution in fuel sprays in diesel engines. Most of these studies were performed more than thirty years ago. Future basic studies are needed in this area to utilize the modern electronic and holographic techniques.

The droplet size distribution in the spray has been found to be a function, among other factors, of the size of the nozzle hole, the ratio of its length to its diameter, the injection pressure, and the air pressure in the cylinder. The results of Sass[26] and Mehlig[27] for the droplet size distribution in fuel sprays, obtained under test conditions near those in naturally-aspirated diesel engines, showed that the droplet sizes vary from about 1μ to 50μ.

The time required for the complete evaporation of such droplets has been calculated for single compound droplets by using a mathematical model.[8] Under typical air conditions for naturally-aspirated diesel engines, it has been found that after 0.4 msec the fuel droplets of diameters less than 11μ for n-hexadecane, and 13μ for n-heptane will be completely evaporated. Most droplets of this size will be carried away from the core of the spray by the swirling air.

The concentration of the fuel components may be different in the various parts of the spray. Lamb[28] found that for multicomponent sprays the concentration of the higher-boiling component in the remaining drops increases as the spray evaporates. He also found that at higher air temperatures this trend toward increased concentration of the higher-boiling component was more pronounced. The fuel vapor which will be carried away by the swirling air will therefore consist of a large percentage of lighter compounds. The fuel droplets in the core will contain a larger percentage of the higher-boiling compounds.

B. Fuel–Air Distribution in the Spray

A study of the photographic films of Scott[29] showed that at the beginning of combustion the spray has a shape similar to that schematically drawn in Fig. 2. The average distance between the droplets is expected to change with the location in the spray, and is greatest near the leading edge (downwind), where the smaller droplets are concentrated. The average local fuel–air ratio, and consequently the combustion mechanisms, are therefore expected to vary from one location to the other.

If we plot the local equivalence ratio, and take into consideration all the fuel present — whether it is in the liquid or vapor phase — we get a distri-

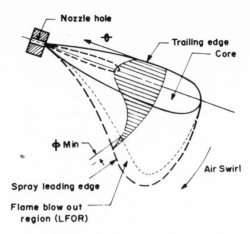

Fig. 2.   Schematic diagram for a fuel spray
injected in swirling air.

bution similar to that shown in Fig. 2. This distribution varies with the radial distance from the nozzle hole. However, at the leading edge (downwind) of the spray and for all radii, the fuel–air ratio always approaches zero and increases as we move toward the core of the spray.

It can also be assumed that most of the droplets which are carried away from the core will be completely evaporated before the start of ignition,[8] or that the mixture near the leading edge of the spray consists of premixed fuel–vapor and air.

C. A Model for Combustion and Emission Formation in DI Engines[30]

The present model will deal at first with a spray which does not impinge on the walls. Later, the behavior of the fuel deposited on the walls will be studied. The spray may be divided into several regions.

1. Lean Flame Region (LFR)

The concentration of the vapor between the core and the leading edge of the spray is not homogeneous, and the local fuel–air ratio may vary from zero to infinity. Ignition nuclei will be formed at several locations where the mixture is most suitable for autoignition. Photographic studies on spray combustion in diesel engines[17, 29] show that ignition starts near the leading edge (downwind) of the spray. This is illustrated in Fig. 3. Once ignition starts, small independent nonluminous flame fronts propagate from the ignition nuclei and ignite the combustible mixture around them; as an alternative, this area may undergo microvolume two-stage combustion.[15]

This mixture, on a mass average basis, is lean. The region in which these independent flames propagate or microvolume combustion occurs, will be

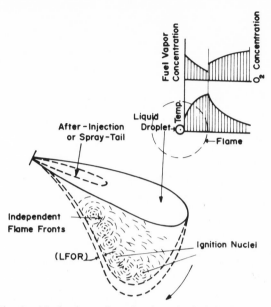

Fig. 3.   Mechanisms of combustion of a fuel spray
injected in swirling air.

referred to as the "lean flame region" (LFR). In this region, combustion is
complete and nitric oxide is formed at high local concentrations. Under very
light loads, the temperatures may not be high enough to produce high nitric
oxide concentrations at this early stage of the combustion process.

2. Lean Flame-Out Region (LFOR)

Near the far leading edge of the spray (downwind), the mixture is too
lean to ignite or support combustion. This region will be referred to as the
"lean flame-out region" (LFOR). Within this region, one would expect some
fuel decomposition and partial oxidation products. The decomposition
products consist of lighter hydrocarbon molecules. The partial oxidation
products contain aldehydes and other oxygenates. It is believed that this
region is one of the main contributors to the unburned hydrocarbons concentration in the exhaust. It corresponds to the quench zone in SI engines
as far as unburned hydrocarbon formation is concerned. The width of the
LFOR depends upon many factors including the temperature and pressure
in the chamber during the course of combustion, the air swirl, and the
type of fuel. In general, higher temperatures and pressures extend the
flames to leaner mixtures, [2, 19] and thus reduce the LFOR width. Increases
in temperature and pressure occur during the combustion of the rest
of the spray. Thus, the width of the LFOR region varies and depends
upon all factors which affect the cylinder gas pressure and temperature. Such

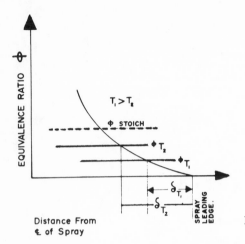

Fig. 4.   Effect of gas temperature on the
width of the LFOR.

factors include the overall fuel–air ratio, turbocharging, and coolant temperature. The effect of increasing the gas temperature on reducing the width of the LFOR is illustrated in Fig. 4. This would result in a reduction in the unburned hydrocarbons and aldehydes emission formation.

The anticipated effect of the increase in air swirl, binary diffusivity, and ignition delay (with all other parameters kept constant) is suggested in Fig. 5. Before ignition starts, the small droplets and vapor will be carried farther away from the centerline of the spray in the downwind direction, resulting in a wider LFOR.

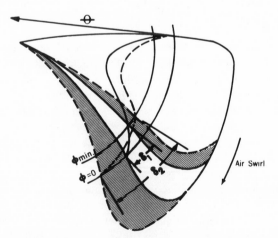

Fig. 5.   Effect of air swirl, binary diffusivity, and
ignition delay on the width of the LFOR.

At the borders between the LFOR and LFR, primary reactions take place and the initial hydrocarbons are attacked and reduced to CO, H_2, H_2O, and the various radical species (H, O, OH). Other unburned hydrocarbons, with less carbon atoms than the initial fuel molecules, are formed. Fristrom and Westenberg[31] suggested that a paraffin molecule in a lean fuel flame undergoes a reaction according to the following scheme:

$$OH + C_nH_{2n+2} \rightarrow H_2O + C_nH_{2n+1} \rightarrow C_{n-1}H_{2n-2} + CH_3 \qquad (7)$$

Fristrom and Westenberg[31] indicated that since hydrocarbon radicals higher than ethyl are thermally unstable, the initial radical C_nH_{2n+1} usually splits off CH_3, forming the next lower ethylenic compound as in Eq. (7). The unsaturated hydrocarbons formed are attacked rapidly by oxygen atoms in this region, forming oxygenated hydrocarbons. If the reaction is completed, as in the LFR, recombination reactions occur and produce CO_2 and H_2O. However, if the flame is quenched as in the LFOR, then unburned hydrocarbons, oxygenated hydrocarbons, CO and other intermediate compounds are left without being completely burned.

3. Spray Core

Following the ignition and combustion in the LFR, the flame propagates toward the core of the spray. In the region which is between the LFR and the core of the spray, the fuel droplets are larger. They gain heat by radiation from the already established flames and evaporate at a higher rate. The increase in temperature will also increase the rate of vapor diffusion, due to the increase in molecular diffusivity. These droplets may be completely or partially evaporated. If they are completely evaporated, the flame will burn all the mixture within the rich ignition limit. The droplets which are not completely evaporated will be surrounded by a diffusion-type flame as illustrated in Fig. 3. The rate of combustion of these droplets depends upon many factors which govern the rate of evaporation and diffusion of the fuel vapor to the flame, and the rate of diffusion of oxygen to the flame. The combustion in the core of the jet mainly depends upon the local fuel–air ratio. Under part-load operation, this region contains adequate oxygen; combustion is expected to be complete and to result in high NO production. The temperature of the flame depends upon both the mixture temperature before the start of the droplet combustion and the heat of combustion which is a function of the heavy compounds of fuel used. The flame zone temperature is a major factor affecting the NO formation.

Near full-load conditions, incomplete combustion occurs in many locations in the fuel-rich core. Fristrom and Westenberg[31] suggested that in fuel-rich saturated-hydrocarbon flames the initial reaction is simply the H-atom stripping, i.e.,

$$H + C_nH_{2n+2} \rightarrow H_2 + C_nH_{2n+1} \tag{8}$$

Recombination between the hydrocarbon radicals may occur, and hydro-carbons heavier or lighter than the original fuel molecules may be formed. This may explain the observations made by Barnes[32] and Milks et al.[33] that in diesel engines with a single-compound hydrocarbon fuel, the exhaust emissions contained molecules with more and with less carbon atoms than the original fuel. In addition to unburned hydrocarbons, carbon monoxide, oxygenated compounds, and carbon may be formed near full-load. Nitrogen oxides are formed at low concentration under these conditions.

4. Spray Tail

The last part of the fuel to be injected usually forms large droplets due to the relatively small pressure differential acting on the fuel near the end of the injection process. This is caused by a combination of decreased fuel injection pressure and increased cylinder gas pressure. Also, the penetration of this part of the fuel is usually poor. This portion is referred to as the "spray tail." Under high-load conditions, the spray tail has little chance to get into regions with adequate oxygen concentration. However, the temperature of the surrounding gases is fairly high (near the maximum cycle temperature), and the rate of heat transfer to these droplets is very high. These droplets therefore tend to evaporate quickly and decompose. The decomposed products contain unburned hydrocarbons, and a high percentage of carbon molecules. Partial oxidation products include carbon monoxide and aldehydes.

5. After-Injection

Under medium and high loads, many injection systems produce "after-injection." When this occurs the injector valve opens for a short time after the end of the main injection. In general, the amount of fuel delivered during after-injection is very small; however, it is injected late in the expansion stroke, under a relatively small pressure differential, and with very little atomization and penetration. This fuel is quickly evaporated and decomposed, resulting in the formation of CO, carbon particles (smoke), and unburned hydro-carbons. An extensive study of the effect of after-injection on exhaust smoke is given in Ref. 34.

6. Fuel Deposited on the Walls

Some fuel sprays impinge on the walls. This is especially the case in small, high-speed, DI engines because of the shorter spray path and the limited number of sprays. The rate of evaporation of the liquid film depends on many factors including gas and wall temperatures, gas velocity, gas pressure, and properties of the fuel. Previous studies by El Wakil et al.[24] showed that the liquid droplets reach an equilibrium temperature during evaporation and remain at this temperature until they are completely evaporated. For the liquid film, the local wall temperature may be lower than the equilibrium temperature corresponding to the gas pressure. Therefore, the rate of evaporation of the liquid film is expected to be less than the corresponding rate for the droplets, and it is expected that this liquid film will be the last to be evaporated. The vapor concentration is maximum on the liquid surface and decreases with increased distance from the surface. It we assume that the surrounding gas has a high relative velocity and contains enough oxygen, the flame will propagate to within a small distance from the wall. Combustion of the rest of the fuel on the walls will depend upon the rate of evaporation and mixing of fuel and oxygen. If the surrounding gas has a low oxygen concentration or the mixing is not appropriate, evaporation will occur without complete combustion. Under this condition, the fuel vapor will decompose and form unburned hydrocarbons, partial oxidation products, and carbon particles.

As the piston moves on the expansion stroke, the gases flow outward radially in an inversed squish motion to fill the expanding space between the piston top and the cylinder head. In shallow-bowl combustion chambers, most of the combustion process takes place in the bowl. In deep-bowl types, the reversed squish flow is significant and will help the mixing of incomplete combustion products and air during the expansion stroke. In both cases, the swirl motion will continue but, due to frictional losses, at a lower rotational velocity than that on the compression stroke. The combination of the inversed squish and swirl produce heterogeneous eddies and tends to draw the fuel vapor and the partial oxidation products, including carbon which has been formed from fuel deposited on the bowl walls. Photographic films by Scott[29] show eddies and heavy smoke clouds outside the bowl from the areas where the sprays impinge the wall. These eddies are heterogeneous, and upon combustion they burn with a luminous flame due to the presence of carbon particles.[35]

Based upon this model, the progression of combustion in the spray and the pressure and temperature development in the cylinder may be represented as shown in Fig. 6. The emission formation expected in the different regions of the spray without fuel impingement on the walls may be summarized as shown schematically in Fig. 7.

D. Heat Release Rates in DI Engines

Many studies have been made by Lyn,[35, 36] Austen,[37] and Grigg[38] to determine the heat release rate from pressure traces obtained in open-chamber diesel engines. Figure 8 shows sample traces of the gas pressure, rate of heat release, and cumulative heat release. The details of the work done to obtain these traces are given in Ref. 36. The negative heat release

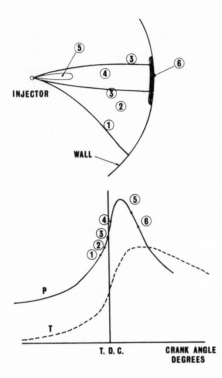

Fig. 6.   Progression of spray combustion, and cylinder pressure and temperature.

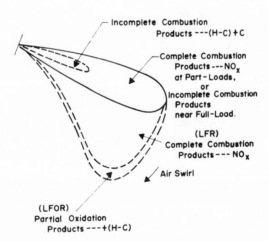

Fig. 7.   Emission formation in a fuel spray injected in swirling air.

after the start of injection, at 22° BTDC, is due mainly to the heat transfer from the hot air to the liquid fuel. During the ID the spray is formed and the droplets in the LFR and LFOR of the spray are the first to evaporate, forming a nonhomogeneous fuel–vapor–air mixture. As mentioned in Section II. A, the complete evaporation of these droplets takes place in much less time than the ID period.

Lyn observed that ignition occurred at 13° BTDC and the rate of burning reached its peak at about 10° BTDC; that is, it took three crank angle degrees to reach the peak. The end of the ID, 13° BTDC, can be considered to coincide with the start of combustion in the LFR. It is interesting to note that the flame observed, up to 10° BTDC, was of very low luminosity; this indicated that the burning was essentially confined to the premixed part of the jet.

Figure 8 shows that at the point of maximum heat release rate, point *B*, the cumulative heat release is about 5% of the total computed heat release. This reflects the approximate amount of fuel burned soon after ignition as a percentage of the total amount of fuel in this particular case.

The first appearance of the orange-colored luminous flame was observed at 10° BTDC, but did not spread to surround the tip of the jet until 7° BTDC. During the rest of the combustion process (near the core of the jet) the flame

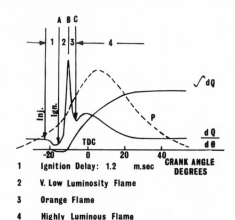

1	Ignition Delay: 1.2 m.sec
2	V. Low Luminosity Flame
3	Orange Flame
4	Highly Luminous Flame

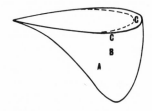

Fig. 8.   Progression of spray combustion and heat release in a DI engine.

was observed to have high luminosity due to the presence of carbon particles. High luminosity is a characteristic of diffusion-type flames.

E. Effect of Some Design and Operating Variables on DI Engine Emissions

The concentration of the different emission species in the exhaust is the result of their formation, as discussed in the above model, and their elimination or further formation in the cylinder and exhaust system.

$$
\begin{array}{c}
\text{Exhaust} \\
\text{emissions}
\end{array}
=
\begin{Bmatrix} \text{Emission} \\ \text{Formation} \end{Bmatrix}
+
\begin{Bmatrix} \text{Further} \\ \text{formation} \end{Bmatrix}
-
\begin{Bmatrix} \text{Elimination} \end{Bmatrix}
+
\begin{Bmatrix} \text{Further} \\ \text{formation} \end{Bmatrix}
-
\begin{Bmatrix} \text{Elimination} \end{Bmatrix}
$$

$$
\begin{array}{ccc}
\text{in} & \text{in} & \text{in} \\
\text{the spray} & \text{the cylinder} & \text{the exhaust system}
\end{array}
$$

In general, "further formation" applies to the nitric oxide and "further elimination" applies to the incomplete oxidation products. The further formation and elimination reaction rates are functions of the oxygen (or oxydants) concentration, the local mixture temperature, mixing, and residence time.

1. Unburned Hydrocarbons

The unburned hydrocarbons in the diesel exhaust consist of either original or decomposed fuel molecules, or recombined intermediate compounds. A small portion of these hydrocarbons originate from the lubricating oil.

In the spray, the unburned hydrocarbons are related to the LFOR, spray core, fuel on the walls, spray tail, and after-injection. The mechanisms of formation and oxidation of the hydrocarbon molecules depend upon most of the engine operating variables. The effects of some of these variables are studied in the following sections.

a. Fuel–Air Ratio. In diesel engines, if the changes in volumetric efficiency are neglected, the mass of air per cycle is almost constant. The change in load is accomplished by controlling the amount of fuel injected. This produces variations in fuel distribution in the spray, amount of fuel deposited on the walls, cylinder gas pressure and temperature, and injection duration. In general, an increase in fuel–air ratio results in a decrease in the portion of the fuel in the LFOR as a ratio of the total fuel injected, and an increase in that portion in the core and on the walls.

An increase in fuel–air ratio affects the oxidation reactions in many ways. It results in longer periods of injection; and if injection timing and rate are kept constant, more fuel is injected later in the cycle. In general, this results in a shorter reaction time for the last part of the injected fuel. An increase in fuel–air ratio also results in lower oxygen concentration. These two factors tend to decrease the rate of the elimination reactions. However, higher gas temperatures are reached because more fuel is burned and because there is a drop in the percentage heat losses to the coolant.[39] This tends to increase the elimination reactions rate.

At very light loads and idling conditions, it can be assumed that the fuel does not reach the walls and that its concentration in the core is small. Under these conditions the unburned hydrocarbon emissions originate mainly from the LFOR. The increase in the local temperature of this region, due to subsequent combustion of the rest of the spray, is very small, and the elimination reaction rates are very slow. These reactions are further reduced due to the very low concentration of the fuel molecules as they diffuse in the air around this region. The ratio of the unburned hydrocarbon formed in this region to total fuel injection is the highest at idling. This ratio decreases with the increase in fuel–air ratio due to the factors discussed above.

At part loads, the increase in fuel–air ratio causes more fuel to be deposited on the walls, and produces higher concentrations in the core. The unburned hydrocarbons formed in these regions increase. However, there is sufficient oxygen in the mixture so that, with increased temperature, the oxidation reactions are promoted and the hydrocarbon emissions are reduced. The reduction in the unburned hydrocarbon emissions (corrected to the stoichiometric ratio) with the increase in fuel–air ratio or BMEP has been observed by Hurn,[40] Marshall and Hurn,[41] Hames et al.,[102] and Perez and Landen.[42] These are shown in Figs. 9, 10, and 11, respectively.

It should be noted that in the study of emissions in heterogeneous combustion systems, the observed concentration may be misleading, particularly at part loads. This is due to the effect of dilution by excess air. For example, some reported data show that the concentration of the incomplete combustion products increases with the increase in load from no load. After correcting these data for the effect of dilution, the opposite trend is observed, as discussed in the present model.

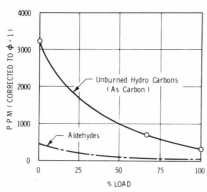

Fig. 9. Effect of load on unburned hyrocarbons and aldehyde emissions in a DI engine.

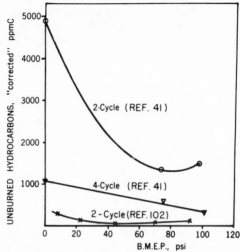

Fig. 10. Effect of load on unburned hydrocarbons in 2-cycle and 4-cycle DI engines, speed 1800 rpm.

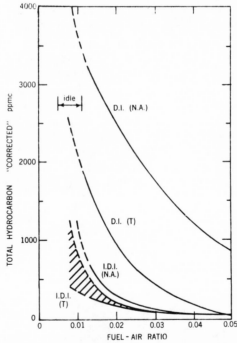

Fig. 11. Effect of fuel–air ratio on the unburned hydrocarbon emissions in DI and IDI engines. (N.A.): naturally aspirated; (T): turbocharged (Perez and Landen[42]).

The emissions data may be judged as a mass ratio of the fuel, as corrected to $\phi = 1$, or as specific emissions in mass per horsepower hour.[43] The specific emissions take into account the mechanical efficiency of the engine.

At full-load and overload conditions, an increase in fuel–air ratio results in the formation of more unburned hydrocarbon molecules in the core and near the walls. Under these conditions, the contribution of the LFOR to the total emission is very small. The oxidation reactions are also limited due to the lack of oxygen, in spite of the very high temperatures reached. This may explain the increase in the unburned fraction of the fuel at full load observed by Marshall and Hurn[41] and Hames et al.,[102] shown in Fig. 10.

The molecular structure of the unburned hydrocarbon emissions is expected to vary with fuel–air ratio. At idling and light loads the hydrocarbon emissions are related to the LFOR and are mainly composed of the original fuel molecules. This is because the molecules have little chance to decompose later in the cycle due to the relatively low gas temperatures reached under light loads.

At high loads the hydrocarbon emissions originate from the fuel molecules in the core and on the walls. Under these conditions, the temperatures reached in the cycle are fairly high and cause decomposition of some of the original fuel molecules. Since the fuel–air ratio in the core and near the walls is generally rich, there is a great possibility that some recombination reactions may occur between the hydrocarbon radicals and the intermediate com-

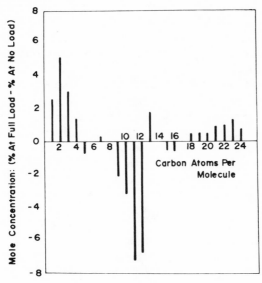

Fig. 12.   Effect of load on distribution of unburned hydrocarbon emissions in a DI engine.

pounds.[31] This results in higher concentrations of the heavier hydrocarbons. Figure 12 shows the variation in the mole concentration of the different molecules at full load and no load. The data for this figure were obtained from Ref. 40. It should be noted that at full load the molecules with 6 to 12 carbon atoms were reduced, but the molecules with 1 to 4, and 18 to 24 carbon atoms were increased. This implies that the medium-size molecules were decomposed to lighter molecules, and that the intermediate compounds were recombined to form heavier molecules.

The process of recombination of the hydrocarbon compounds and radicals may also result in compounds having a different structure than the original fuel. This has been reported by Bascom et al.[44] and Aaronson and Matula.[45] Aaronson and Matula found that a large number of aromatic hydrocarbons were synthesized during the heterogeneous combustion of pure cetane–air mixtures in an air-aspirating spray burner and in a Cooperative Fuel Research (CFR) diesel engine. In the CFR engine the aromatic hydrocarbons were formed from the straight chain fuel used because the lubricating oil did not contain any cyclic hydrocarbon compounds.

b. Turbocharging. At any fuel–air ratio, turbocharging, apart from its effect on spray formation, increases the average gas temperature over the whole cycle. With the same oxygen concentration, the increase in the average temperature causes the rate of oxidation reactions to increase and the concentration of the unburned hydrocarbon emissions to be reduced. This is enhanced by further oxidation reactions in the exhaust manifold and the turbocharger. With turbocharging, higher exhaust temperatures are also reached, the reaction time is increased, and mixing is improved. The reduction of the unburned hydrocarbon emissions with turbocharging has been observed by Perez and Landen[42] as shown in Fig. 11 for both the DI and indirect-injection IDI engines.

c. Injection Timing. An advance in injection timing was found by Khan and Grigg to increase the unburned hydrocarbon emissions.[46] Apparently, the longer ignition delay allows more fuel vapor and small droplets to be carried away with the swirling air, producing a wider LFOR. Another factor suggested by Khan and Grigg[46] is the increase in fuel impingement on the walls.

d. Swirl. An increase in swirl in the DI engine improves the mixing process and the hydrocarbon oxidation processes. Excessive swirl may, however, produce a wider LFOR, as discussed above, or an overlap of the sprays and an increase in unburned hydrocarbon emissions.

The swirl in DI engines may be changed by varying the ratio of the bowl diameter, d, to its depth, l. This ratio, d/l, is known as the aspect ratio. Due to conservation of moment of momentum, the deep-bowl pistons tend to have a higher swirl than the shallow-bowl pistons. The variation in the aspect ratio has been found by Watts and Scott[17] to have little effect on the fuel economy; it may, however, affect the emission formation. The data published on this are insufficient; however, the comparisons made by Bascom *et al.*[44] show that the unburned hydrocarbon emissions are higher in the deep-bowl than in the shallow-bowl (Mexican Hat) pistons.

    e. Other Factors. Other factors which affect the unburned hydrocarbon emissions are related to the injection system design, timing, and rate of injection. Merrion[47] reported the results of tests on injectors with different (sac) volumes between the needle seat and the nozzle holes. He found that the reduction in the sac volume greatly reduces the unburned hydrocarbon emissions. The fuel remaining in the sac seeps out through the injector holes during the expansion stroke. This fuel has very little chance to mix with the oxidants before the exhaust valves open.

2. Carbon Monoxide

    Carbon monoxide is one of the compounds formed during the intermediate combustion stages of hydrocarbon fuels.[31] As combustion proceeds to completion, oxidation of CO to CO_2 occurs through recombination reactions between CO and the different oxidants. If these recombination reactions are incomplete due to lack of oxidants or due to low gas temperatures, CO will be left.

    During the early stages of spray combustion in DI diesel engines, CO is believed to be formed at the borders between the LFOR and LFR. But since the local temperature is not high enough, very little oxidation reactions take place. Later on, during the combustion process, the local temperature may increase and promote the oxidation reactions.

    In the LFR, the carbon monoxide formed as an intermediate compound is immediately oxidized because the oxygen concentration and the gas temperature are adequate. In the core of the spray and near the walls, CO is formed at a high rate. The rate of its elimination depends mainly upon the local oxygen concentration, mixing, local gas temperature, and the available time for oxidation.

    The CO formed near the borders of the LFOR, as a ratio of the total fuel in the spray, depends upon the fuel–air ratio. At light loads this ratio is high because the gas temperature is low and very little oxidation reactions take place. An increase in load, or fuel–air ratio, results in lower CO emissions because of the increased gas temperatures and the elimination reactions. This has been observed by Perez and Landen[42] and by Marshall and Hurn[41] for the corrected CO concentration. The increase in fuel–air ratio beyond a certain limit may reduce the elimination reactions, despite the increase in temperature, due to the low oxidants concentration and the short reaction time. This results in high CO emissions with increase in load, as shown in Fig. 13.

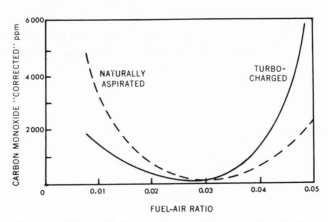

Fig. 13.   Effect of fuel–air ratio on carbon monoxide emission
in a DI engine (Perez and Landen[42]).

The effect of swirl on CO emissions was studied by Khan and Grigg[46] by using a shrouded inlet valve. They found that the optimum swirl for best economy is almost the same as that at which the CO concentration is minimum. This is believed to be due to the improved elimination reactions caused by the better mixing.

3. Aldehydes

Aldehydes are among the intermediate compounds formed during the autoignition process of hydrocarbon fuels in both diesel and gasoline engines. For diesel engines the work of Garner[16] was discussed in Section II. B. For gasoline engines under autoignition conditions (knocking), Withrow and Rassweiler[48] and Sturgis (49) noticed a high concentration of aldehydes ahead of the flame. This was not observed under no-knock conditions, indicating that high aldehyde concentrations are associated with the autoignition reactions.

In the fuel spray before ignition, the aldehydes concentration is expected to be high around the ignition nuclei in the LFR and low in the LFOR. After the small flame fronts are established they pass through the LFR and burn the aldehydes, leaving the aldehydes in the LFOR without oxidation. Like the unburned hydrocarbons, these aldehydes can be eliminated later in the cycle due to the temperature rise caused by the combustion of the rest of the spray. This, too, depends on the oxygen concentration. At light loads, the increase in fuel–air ratio enhances the oxidation reactions and results in a decrease in aldehydes emission concentration. This has been observed by Hurn,[40] Elliott,[50] Merrion,[47] and Marshall and Hurn.[41] The data of

Hurn is plotted in Fig. 9. At heavy loads the increase in fuel–air ratio, beyond a certain value, may result in higher temperatures but lower oxygen concentration. This may result in an increase in aldehydes emissions as observed by Merrion[47] and by Marshall and Hurn.[41]

4. Odor

Diesel odor has been the subject of many investigations during the last few years, particularly as is related to odor-producing species and instrumentation.[51-57]

In a recent study, Spindt et al.[55] identified many mono- and polyoxygenated partial oxidation products and certain fuel fractions as being the odor-producing compounds. These occur at very low concentrations in the exhaust.

According to Ref. 51 the principal components responsible for the characteristic oily-kerosene portion of diesel exhaust odor are alkyl benzenes, indanes, tetralins, and indenes. The contribution of the alkyl naphthalenes, which constitute a major portion of the mass of the oily-kerosene fraction, to the odor perception may be through a synergistic effect. The unburned fuel in the exhaust is very likely to contribute heavily to the oily-kerosene odor.

Somers and Kittredge[54] indicated that alkyl-substituted benzene and naphthalene type compounds have been related to the oily-kerosene odor quality and that oxygenated aromatic structures have been related to the smokey-burnt odor quality.

O'Donnell and Dravnieks[52] found from mass spectral data that in addition to the partial oxidation hydrocarbon compounds, sulphur species are among the more important odor contributors.

Many human panel techniques have been used to measure odor intensity. These techniques include a sniff box with different dilution ratios, natural dilution in a large building, or a Turk kit.[57, 58]

Data published by Merrion[47] showed that the odor intensity did not vary appreciably with engine speed, increased with load, and decreased with improving the design of the injection system.

Rounds and Pearsall[59] and Merrion[47] suggested that there may be some weak correlation between formaldehyde emission and odor intensity. Stahman et al.[56] found that with the use of some types of catalytic reactors the odor levels and the oxygenated compounds were reduced with no significant influence on the engine performance parameters.

Barnes[32] found that large differences in exhaust odor intensity were achieved by altering the intake atmosphere conditions to an engine. The engine used was a single cylinder, 4-stroke cycle diesel engine using n-heptane as a fuel. The artificial atmospheres supplied to the engine were comprised of oxygen plus an inert gas, or air. The inert gases used were argon, helium,

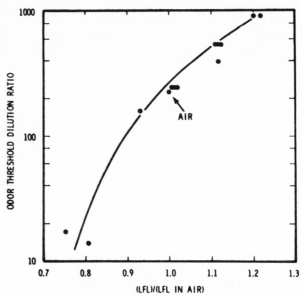

Fig. 14.   Relationship between lean flammability limit and
odor intensity (Barnes[32]).

nitrogen, and carbon dioxide. These mixtures of inert gases and oxygen have different lean ignitability limits. Barnes found that there is a correlation between the odor levels and the lean flammability limits (LFL) as shown in Figure 14.

The LFL ratio is defined as the ratio of percentage of fuel by volume at the LFL with the mixture to the percentage with air. A LFL ratio more than unity means that the fuel/oxidizer ratio is richer at the LFL than in air. Barnes[32] concluded that for such mixtures a larger portion of the injected fuel would be too lean to burn and would be partially oxidized. He observed that this was particularly true at idling and light loads.

In terms of the model developed in this chapter for the emissions in DI engines, the volume of the LFOR increases for the mixtures with high LFL ratio. This results in more fuel in that region and an increase in the concentration of the partially oxidized products.

Trumpy et al.[60] studied the effect of inlet temperature and equivalence ratio on odor production from premixed n-heptane–air mixtures in a motored spark-ignition CFR engine. The results are shown in Fig. 15a. The pressure traces showed that below the lower line no energy was released during the compression stroke, and the odor was that of n-heptane only. In the area between the two lines, smoke, odor, and eye irritation species were found in the exhaust. Under these conditions the pressure traces showed energy release

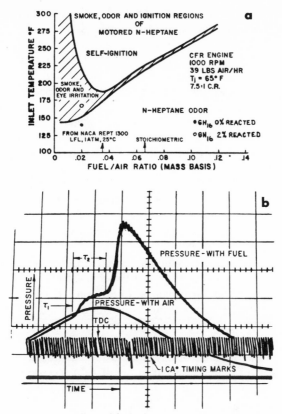

Fig. 15. (a) Effect of fuel–air ratio on odor in premixed
n-heptane–air mixture with and without self-ignition;
(b) pressure time trace showing self-ignition (Trumpy[60]).

indicating oxidation reactions. Above the upper line, autoignition occurred in two stages, as shown in Fig. 15b. This reduced the concentration of the partially oxidized products. Trumpy *et al.* concluded that the smoke, odor, and eye irritants are formed when the second-stage combustion is quenched.

5. Smoke Particulates

Different types of particulates are emitted from diesel engines under different modes and operating conditions. These particulates can be divided into the following:

1. Liquid particulates appearing as white clouds of vapor emitted under cold starting, idling, and low loads. These consist mainly of fuel and a small portion of lubricating oil, emitted without combustion;

they may be accompanied by partial oxidation products. These white clouds disappear as the load is increased.

2. Soot or black smoke is emitted as a product of the incomplete combustion process, particularly at maximum loads.

3. Other particulates include lubricating oil and fuel additives. The present study will be limited to the soot or black smoke.

The black smoke emission consists of irregularly shaped, agglomerated fine carbon particles.[61] According to current theories, these carbon particulates can be formed from the hydrocarbon fuels in the presence or absence of oxygen.

In the presence of oxygen, as in the LFR and in the core under light loads, pyrolysis of the fuel molecules may take place to form carbon. Carbon formation was observed during the early stages of combustion in the LFR by Alcock and Scott.[18] However, these carbon particles are completely burned due to the presence of enough oxygen.

In the spray core, especially under heavy loads, the oxygen concentration is low, the gas temperature is high, and the concentration of the high boiling point components is high, as discussed in Section III. A. These are easier to decompose than the lighter compounds. Pyrolysis of the molecules may take place and lead to the formation of acetylene and hydrogen.[62] The simultaneous condensation and dehydrogenation of acetylene results in solid carbon. Due to the presence of oxygen, partial oxidation reactions may take place and result in a high concentration of CO. According to Behrens[63] and Kassel[64] carbon may be formed if the ratio of carbon monoxide to carbon dioxide exceeds an equilibrium constant.

Smoke particles are also formed from the fuel deposited on the walls[29] by the same mechanism described for the spray core under heavy loads.

The carbon particles formed according to the previous mechanisms may be oxidized later in the turbulent flames formed during the combustion process. The extent of the oxidation reactions depends upon the oxygen concentration in the vicinity of the surface of the soot particles, temperature, and residence time.

The smoke intensity in the diesel exhaust is greatly affected by many parameters. By controlling them, smoke intensity may be reduced.

a. Effect of Fuel. The effect of a change in fuel properties on the injection and combustion process is not clearly understood. For example, the early work of Schweitzer[65] showed that increasing the fuel viscosity increases spray penetration; however, the recent work of Burt and Troth[25] showed no effect of the fuel properties on penetration and vaporization. This illustrates the need for further basic research on some of the most important physical phenomenon in diesel combustion.

Previous experimental results showed that higher cetane number fuels which are suitable for high-speed transportation engines have a tendency to

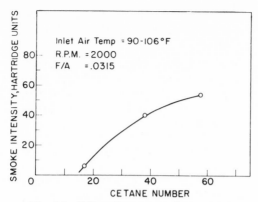

Fig. 16. Effect of cetane number on smoke
intensity in a DI engine.

produce more smoke.[34, 66, 67, 68] This is shown in Fig. 16 and is believed to be due to the lower stability of these fuels. This results in higher rates of carbon formation during the combustion process since more fuel is injected after the end of ID.[69] An interesting experiment was made by Rost[70] in which he measured the smoke intensity with regular gasoline and with additive treated gasoline of a cetane number equal to that of the regular diesel fuel. The results of his experiments showed that the smoke intensity increased as a result of the additive and was almost equal to that of the regular diesel fuel. Similar observations on the increase of smoke intensity with the increase in the cetane number are reported by Broeze and Stillebroer.[71]

Rost's experiment indicates that the fuel volatility has little effect on smoke reduction. The results of Golothan[66] showed that for a given cetane number less smoke is produced with more volatile fuels. Other results by Savage[72] showed an opposite trend. Thus, no definite conclusion can be drawn from the available experimental data on the effect of volatility on smoke. The change in the injection characteristics of each of the engines used with the more volatile fuel is believed to be one of the primary reasons for this disagreement.

   b. Injection Timing. Advancing the start of injection in DI engines with all other parameters kept constant results in longer delay periods, more fuel injected before ignition, higher temperatures in the cycle, and earlier ending of the combustion process. The residence time is therefore increased. All these factors have been found to reduce the smoke intensity in the exhust.[69] However, earlier injection results in more combustion noise, higher mechanical and thermal stresses, and higher NO concentrations.

   In a recent study, Khan[46] reported that very late injection reduces the smoke. The timing after which this reduction occurs is that at which the minimum ID occurs. He suggested that one of the factors that contributes to the reduction in smoke at the retarded timing is the reduced rate of formation due to the decrease in the temperature of the diffusion flames as most of these flames occur during the expansion stroke. Similar results are shown in Fig. 21.

   c. *Rate of Injection.*   Higher initial rates of injection have been found to be effective in reducing the exhaust smoke.[46] As discussed above, an early ending of the injection process will improve the elimination reactions.

   d. *Injection Nozzle.*   The size of the nozzle holes and the ratio of the hole length to its diameter have an effect on smoke concentration. A larger hole diameter results in less atomization and increased smoke.[46] An increase in the ratio l/d beyond a certain limit also results in increased smoke.[46]

   e. *Inlet Air Temperature.*   An increase in inlet air temperature results in higher gas temperature during the entire cycle. This affects the mixture formation as well as the chemical reactions which contribute to the carbon formation.

   The effect on the mixture formation is related to the spray characteristics — mainly penetration and atomization, evaporation, and diffusion. The increase in the gas temperature at a given pressure reduces the penetration.[73] This has been found to be more significant for more volatile fuels.[74] The reduction in gas density associated with the increase in temperature at constant pressure will result in smaller spray cone angles.[75] These factors result in high local concentrations of the fuel droplets near the nozzle. The increase in the rate of evaporation and diffusion with temperature will result in over-rich mixtures near the nozzle where mixing is not as effective as near the walls. Higher temperatures also increase the decomposition reactions. All these factors lead to increased smoke intensity. The results of Khan[69] showed an increase of smoke intensity with temperature at the different injection timings tested.

   In some less-volatile fuels the effect of increased intake air temperature may accelerate the oxidation reactions at a higher rate than that of the decomposition reactions. This produces a reduction in smoke intensity with the increase in temperature as explained in Ref. 34.

   f. *After-Injection.*   After-injection, sometimes called secondary injection, is different from dribbling. Dribbling is caused by fuel leakage past the valve seat when the valve is closed. It has a very bad effect on smoke intensity, unburned hydrocarbon emissions, carbon monoxide, and engine operation. During after-injection the needle valve is lifted off the seat as shown in Fig. 17.

   The start of after-injection, its duration, and the amount of fuel injected depend upon the load, speed, and coolant temperature. The effects of load and speed are related to the characteristics of the injection system, as discussed before. The effect of the coolant temperature has been examined in an experimental single-cylinder engine[39] in which ethylene glychol was the coolant.

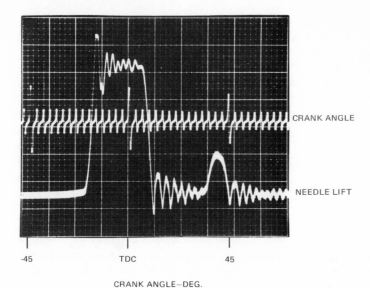

CRANK ANGLE

NEEDLE LIFT

-45   TDC   45

CRANK ANGLE–DEG.
Fig. 17.   Needle lift diagram showing after-injection.

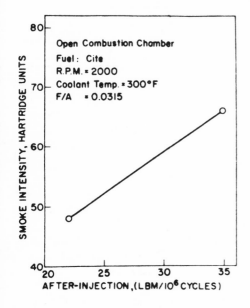

Fig. 18.   Effect of after-injection on smoke intensity.

The engine was under steady conditions for 1.5 hr during which after-injection was observed to increase with time. This is believed to be caused by the change in injection characteristics at the high coolant temperature. Figure 18 shows two data points for the after-injection and the corresponding smoke intensity at the beginning and end of this test. This figure clearly illustrates the marked effect of the increase in the amount of after-injection on the smoke intensity, as discussed earlier in the model.

The reduction of after-injection may be achieved by design modifications based upon theoretical studies and injection-system modeling. This has been the subject of some recent publications.[76-78]

g. *Other Approaches to Control Smoke Emissions.* Other approaches to reduce the diesel smoke include fumigation[56] and smoke-suppressant fuel additives. Barium-type additives were found to show little or no effect on power, odor, or gaseous emissions in a naturally-aspirated DI engine.[53] Catalytic reactors were found to have a little effect on smoke reduction.[56]

6. Nitric Oxide

Nitric oxide, NO, is formed during the combustion process at various concentrations in all the spray regions. Many mechanisms for NO formation in combustion systems have been proposed.[79-82] The widely accepted mechanism is that of Zeldovich[79]:

$$O_2 \rightleftarrows 2O \tag{10}$$

$$O + N_2 \rightleftarrows NO + N \tag{11}$$

$$N + O_2 \rightleftarrows NO + O \tag{12}$$

The chain reactions are initiated in Eq. (10) by the atomic oxygen which is formed from the dissociation of oxygen molecules at the high temperatures reached in the combustion process. According to this mechanism, the nitrogen atoms do not start the chain reaction because their equilibrium concentration during the combustion process is relatively low compared to the equilibrium concentration of atomic oxygen. Therefore, in diesel combustion, the local NO formation in the spray is related to the local oxygen atom concentration. This is a function of the local concentration of the oxygen molecules and the local temperature.

In the diesel engine, NO is not formed during the compression stroke even in very highly supercharged engines because of the relatively low temperatures reached. Although it is not formed in the LFOR during the early stages of combustion, raising the air temperature in this region may cause some NO to be formed here later in the cycle after the combustion of the rest of the spray.

Newhall and Shahed[83] found that in flames, NO is formed at higher rates with rich mixtures than with stoichiometric or lean mixtures. The final concentration, however, is maximum with a slightly lean mixture. The NO formation in the LFR may start at a lower rate than in the richer regions of the spray, but will reach much higher concentrations.

The results of Newhall and Shahed[83] also showed that most of the NO formation occurs during the post-flame reactions. It is expected that the LFR is one of the major contributing regions to NO formation since it is the first part of the spray to burn and has the longest post-flame residence time.

The temperature rise due to the combustion of the fuel in the core and on the walls may contribute to the NO formation in two ways. First, it increases the average temperature in the cylinder and results in higher concentrations of NO in the LFOR and LFR. Second, it may result in very high flame temperatures in the core. The NO formed in the core will also be influenced by the local oxygen concentration. For example, if the oxygen concentration in the core increases, such as when more holes are used to deliver the same amount of fuel, without spray interference, the NO formed in the core is expected to increase.

As the temperature decreases during the expansion stroke, the NO concentration does not decrease to the equilibrium concentration. Starkman and Newhall[84] and Lavoie et al.[82] found that in reciprocating engines the NO removal processes during the expansion stroke are very slow, and thus NO concentration remains nearly constant during expansion.

a. Effect of Fuel–Air Ratio. The effect of the overall fuel–air ratio on the NO emission and the related engine parameters is shown in Fig. 19 for a single-cylinder DI research engine. A detailed description of this engine is given in Ref. 39. The observed NO concentration in the exhaust increased with the increase in fuel–air ratio and reached a maximum at a ratio of about 0.048. It should be noted that the observed NO concentration does not represent the extent of NO formation by the combustion process because of the dilution with excess air. The NO concentration, corrected to the stoichiometric ratio, is plotted in Fig. 19 and shows that within the range of the experiments it decreased with the increase in fuel–air ratio.

The increase in fuel–air ratio resulted in an increase in the maximum mass-average gas temperature as indicated in Fig. 19 by the increase in the maximum pressure. It seems that this increase in the maximum mass-average gas temperature did not cause a corresponding increase in the NO formation. This suggests that the NO emissions in the DI engines are primarily related to the local combustion of the LFR. Also, the increase in NO concentration due to the subsequent combustion of the fuel in the core and on the walls may not be the main contributing factor in NO formation.

b. Effect of Type of Fuel. Figure 20 is a plot of the NO_x concentration for two fuels having cetane numbers of 35 and 59, as obtained by McConnell.[85] The injection timing was advanced for the low cetane number fuel to account for its longer delay period. This kept the start of combustion at the same point in the cycle for each fuel. With low cetane fuel, the ID is longer than with the other fuel and more fuel is present in the LFR when combustion starts. This

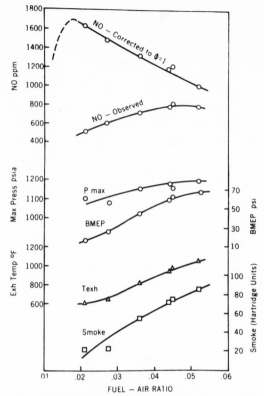

Fig. 19.   Effect of fuel–air ratio on nitric oxide emission
and other performance parameters.

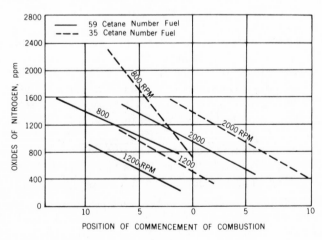

Fig. 20.

larger quantity of fuel produces a higher gas temperature upon combustion early in the cycle, and more NO is formed in the LFR.

The increase in NO formation with the decrease in cetane number of the fuel may be attributed to the effect of the preignition radicals on the mechanism of NO formation. This has been discussed by Fenimore.[86] Similar effects have been observed by Meguerian[87] with fuels of higher octane number during the combustion of homogeneous mixtures.

c. Effect of Injection Timing. The effect of injection timing on the NO emissions and other related engine parameters is shown in Fig. 21 for the single-cylinder DI research engine of Fig. 19. This figure shows that injection advance results in longer ID because the fuel is injected in air at lower temperatures and pressures. However, since the increase in ID in crank degrees is less than the injection advance, autoignition occurs earlier in the cycle. Longer ID causes more fuel evaporation and mixing in the LFR and high NO concentration. The NO formation in the other regions of the spray increases with injection advance due to the high temperatures reached.

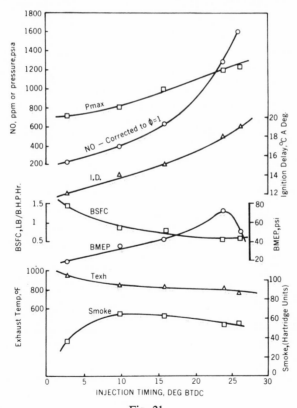

Fig. 21.

Late injection is an effective way to reduce NO emissions; however, this results in a loss in BMEP and fuel economy as shown in Fig. 21.

d. Effect of Swirl. The effect of swirl on the emissions in a DI engine was studied by Khan and Grigg[46] and discussed in Section III.E.2. They found that the shroud position which produces minimum CO also results in the highest NO concentration in the exhaust.

e. Effect of Turbocharging. Figure 22 shows the NO_x concentration with and without turbocharging at various fuel–air ratios, as obtained by McConnell.[85] Near the higher fuel–air ratios (60% air usage), turbocharging resulted in higher NO_x concentration. Figure 23 shows the same data as Fig. 22 but the NO_x concentrations are plotted versus the fuel delivery in $mm^3/cycle$.

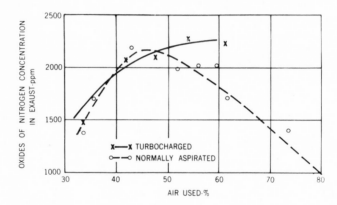

Fig. 22.

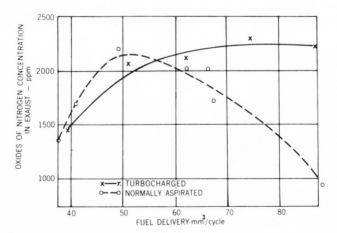

Fig. 23.

f. Effect of Intake Air Charge Dilution. One of the methods used to reduce the NO emissions in the diesel exhaust is to dilute the intake air charge with exhaust gases or water injection. This will be discussed in Section VII.

IV. THE M-SYSTEM

Meurer[88] developed the M-system with the idea of reducing both the preignition reactions of the autoignition process, and the decomposition of the fuel molecules associated with the heating of heterogeneous mixtures. In the M-system, the fuel is injected on a temperature-controlled surface of a spherical chamber in the piston. The rate of evaporation is controlled by the wall temperature and the air swirl. After the fuel vapor mixes with the air, it is ignited by several ignition sources formed by injecting a small percentage of the fuel into the chamber, away from the walls. More detailed information on this system may be found in Refs. 12 and 88.

The swirl and radial velocity components in the M-system are higher than those in the other types of open-chamber engines. The NO-emission characteristics of an engine of the M-system are given in Fig. 29 and will be compared with the other engines in Section VI.A.1.

V. INDIRECT-INJECTION ENGINES (IDI)

Indirect-injection engines, known as prechamber or divided chamber engines, have many different designs as detailed in Ref. 12. The performance of the IDI engine depends upon many factors including the ratio of the prechamber volume to the total clearance volume, the eddies and gas flow patterns in the prechamber and main chamber, size and direction of the throat, piston recesses, and the surface temperatures.

In the "swirl" type, the volume of the prechamber is about 50% or more of the total clearance volume. Swirl motion is produced in the prechamber during the compression stroke by the flow of air through a tangential throat. The swirl increases during the compression stroke and reaches its maximum a few degrees before top dead center (TDC). The swirl measured by Lyn[89] in a Comet Mark V, by using the Schlieren techniques, reached its maximum at seven degrees before TDC and was about 21 times the engine speed. The swirl helps to mix the fuel and air before the start of combustion in the prechamber. After ignition the flame moves toward the center of the prechamber due to the increase in the buoyant force acting on the hot products of combustion in the centrifugal field created by the swirl.[90] Nagao[91] found that the direction of injection with respect to the throat has a great effect on engine performance and smoke emissions. Other performance characteristics of the swirl chamber may be found in Ref. 12.

In other types of IDI engines, the volume of the prechamber and the area of the throat are smaller than in the swirl type. In these cases the turbulence produced in the prechamber during the compression stroke is milder than that in the swirl type. The main turbulence is produced after the start of combustion by the flow of the products from the prechamber into the main chamber where the combustion is completed.

The extent of combustion in the prechamber of any type of IDI engine depends mainly upon prechamber volume compared to the total clearance volume. Bowdon[92] found that in swirl chambers the ratio of the heat released in the prechamber to the total heat released was the same as the ratio of the prechamber volume to the total clearance volume.

The exhaust emissions in the IDI engines may be considered to take place in two stages: first, upon combustion in the prechamber; second, after the gases from the prechamber mix with the air in the main chamber. The extent of the reaction in each stage depends upon the concentration of the fuel and oxidants, temperature, mixing, and residence time. These factors change with the engine design and operating variables.

A. Effect of Design and Operating Variables

1. Effect of Load

Figure 24 shows the effect of load on the rates of heat input, heat release, and cylinder pressure. This figure is compiled from figures published by

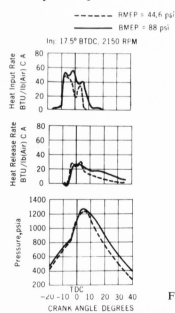

Fig. 24.   Effect of load on the rates of heat input, heat release, and pressure in an IDI engine.

Bowdon *et al.*[92] for a swirl-type IDI engine at constant speed and injection timing. During part-load operation, as the load is increased, more fuel is injected later in the cycle. It can be assumed that the oxygen concentration in the prechamber decreases as more fuel is injected. Therefore, the increase in fuel injection with load may reduce the extent of burning of the last portions of the fuel to be injected into the prechamber. The unburned hydrocarbons and CO formation in the prechamber may therefore increase with load, but their rate of oxidation in the main chamber likewise increases. This is due to the higher temperature reached in the main cylinder, during the expansion stroke, as can be seen from the pressure traces of Fig. 24. This results in lower H–C and CO emissions in the exhaust.

At fairly high loads, near the smoke limit, the rates of the oxidation reactions in the main chamber might not be high enough to eliminate the increased amounts of H–C and CO discharged from the prechamber. This is due to the small oxygen concentration and residence time, in spite of the very high temperatures reached. This may result in increased H–C and CO emissions near the smoke limit. Moreover, the high temperatures promote the decomposition reactions which result in increased carbon emissions.

Figures 25 and 26 show maps of the hydrocarbon and CO emissions, respectively, at various loads and speeds in a Comet Mark V IDI engine. These figures were obtained from Ref. 93. At any constant speed it is noticed that the concentrations of these emissions decrease with the increase in load from no-load. Their concentrations then increase as further increases in load bring it near the smoke limit. It should be noted here that the concentrations given in Figs. 25 and 26 are not corrected to the stoichiometric ratio and that those near no-load should be multiplied by a factor up to six in order to compare them with those at full load. The decrease in the unburned hydrocarbons and CO emissions with the increase in load from no-load has also

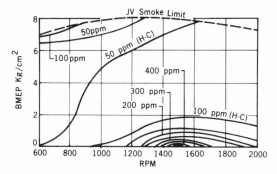

Fig. 25.   Unburned hydrocarbon emissions in an IDI
Comet V engine (Downs[93]).

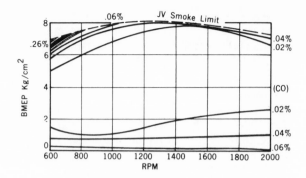

Fig. 26. Carbon monoxide emissions in an IDI Comet
V engine (Downs[93]).

been observed by Perez and Landen[42] for other types of IDI engines. This
has been shown in Fig. 11.

Both the prechamber and main chamber contribute to NO formation.
At light loads the NO emissions are mainly related to the combustion in the
prechamber, where a greater fraction of fuel is burned. This fraction decreases
with the increase in load.

At higher loads the mixture formed in the prechamber is rich, combustion
is incomplete, and the NO formation in the prechamber is not the primary
contributor to the NO concentration in the exhaust.

As the incomplete combustion products expand and pass through the
throat, they are mixed with each other. Because of the lack of oxygen, however, chemical reactions are not likely to occur at an appreciable rate in the
throat and the temperature may not increase. At very rich mixtures, the
flow through the throat may even cause cooling of the mixture as the pressure
decreases.

The increase in the temperature and NO formation in the main chamber
depends upon the amount of incompletely burned compounds discharged
from the prechamber, the oxygen concentration, and the timing of the discharge. The increase in load results in more unburned compounds to be
discharged in the main chamber, while producing a decrease in oxygen concentration. Higher loads (with constant injection timing) also result in a
later discharge in the main chamber and a relatively smaller temperature rise
after combustion due to the expansion work. The rate of expansion (piston
velocity) increases as the piston moves away from TDC in the first part of
the expansion stroke. Accordingly, it is expected that the increase in load
should result in lower NO concentrations. This is shown in Fig. 31 for the
NO mass emissions. An opposite trend is seen in Fig. 27 for the observed
NO emissions in the ppm for the engine of Fig. 25 and 26.

At very light loads and idling the NO mass emissions may decrease due to
the very low temperatures reached in the prechamber under these conditions.

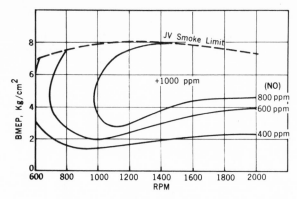

Fig. 27.   Nitric oxide emission in an IDI Comet V engine
(Downs[93]).

2. Effect of Speed

Figure 28 shows the effect of speed, at constant injection timing, on the
rates of heat input, heat release, and cylinder pressure for the swirl-chamber
engine of Fig. 24. The BMEP was constant within about $\pm 10\%$.

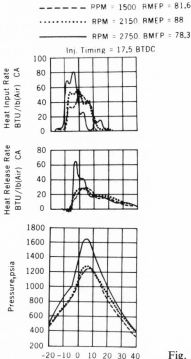

Fig. 28.   Effect of speed on the rates of heat input, heat
release, and pressure in an IDI engine.

For the same BMEP, more fuel is injected per cycle to account for the decrease in mechanical efficiency at higher speeds. At such speeds the rates of injection are higher, and in spite of the shorter delay period, the amount of fuel injected before TDC is larger. All these factors result in higher temperatures in the main chamber at higher speeds. Under the conditions of Fig. 28, therefore, the increase in speed is expected to improve the elimination reactions in the main chamber, to reduce the hydrocarbons and CO and to increase the NO concentration.

In IDI engines under actual running conditions, the change in speed may affect many of the other engine parameters, particularly the injection timing. The emission formation will therefore be a result of the variation in speed as well as of the other parameters. Figures 25, 26, and 27 show the variation in the H–C, CO, and NO emissions, respectively, at various speeds and at optimum injection timing for a swirl-chamber engine. Figure 31 shows the effect of speed on NO mass emissions in another IDI engine.

VI. COMPARISON BETWEEN THE EMISSION CHARACTERISTICS OF SOME TRANSPORTATION ENGINES

A. Diesel Engines

1. NO Emission

The NO emissions, in grams per horsepower-hour, are given for the three main types of combustion chambers used in transportation diesel engines in Figs. 29, 30, and 31. These figures are obtained from Ref. 94. Figure 29 is for an engine using the M-system; Fig. 30 is for a DI engine; and Fig. 31 is for an IDI engine. All three are 4-cylinder and water-cooled engines. The dotted lines in the figures indicate the smoke limit in each engine.

It is noted that the level of the NO emissions near the smoke limit in the IDI engine is lower than that in the DI or M-system engines. This may be caused by the relatively fuel-rich mixture during combustion in the prechamber, and the relatively lower maximum temperatures reached during combustion in the main chamber. Torpey[103] indicated that the major advantage of the IDI engine is its ability to run with retarded timings relative to the DI engine and yet have sufficiently high turbulence in the main chamber to give rapid combustion with low smoke level.

The NO emissions at a given BMEP in the IDI engine are not as sensitive to engine speed as in the other types. The M-system engine results in Fig. 29 show the sensitivity of the NO emissions to engine speed at any BMEP.

2. Emissions Based on a 13-Mode Cycle

A comparison between the brake specific emissions, on a 13-mode cycle, of different types of diesel engines was made by Marshall and Fleming[95];

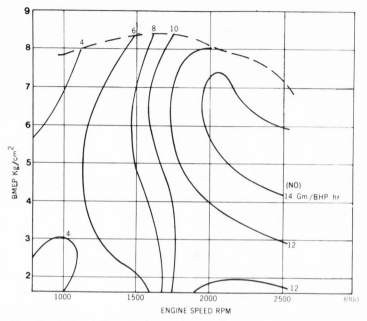

Fig. 29. NO emission characteristics of an M-system engine (Abthoff[94]).

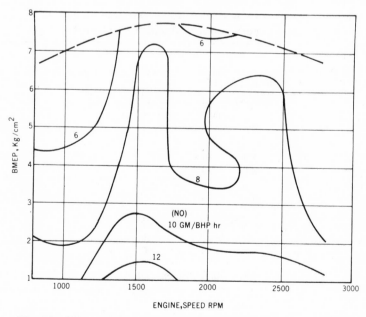

Fig. 30. NO emission characteristics of a DI diesel engine (Abthoff[94]).

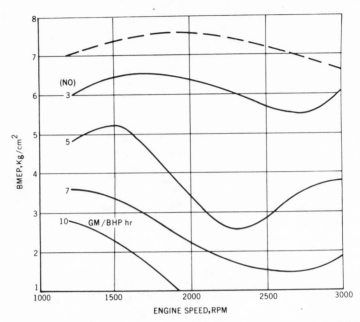

Fig. 31. NO emission characteristics of an IDI diesel engine (Abthoff[94]).

Table III. Comparison Between the Brake Specific Emissions in Different Types of Diesel Engines, Over a Modified 13-Mode Test Cycle

Engine and code identification	Power, bhp	Emissions, g/bhp-hr					
		Carbon monoxide	Nitric oxide (as NO_2)	Hydro-carbons (as CH_2)	Aldehydes (as HCHO)	Particu-lates	Odor
4-cycle, naturally-aspirated, DI							
A	34.5	10.0	8.7	1.9	0.5	0.6	4.5
B	53.6	4.2	6.7	3.1	0.4	0.6	3.6
C	57.1	5.8	19.8	3.0	0.4	0.1	4.0
D	58.1	4.9	16.7	3.1	—	—	4.4
E	68.1	5.6	9.5	2.3	0.3	0.2	4.5
F	73.1	5.7	6.9	.6	—	—	3.0
4-cycle, turbocharged, DI							
G	77.1	3.9	17.8	3.1	0.2	0.2	3.8
H	85.7	4.9	11.7	2.6	0.2	0.3	3.6
4-cycle, turbocharged, IDI							
I	88.6	2.3	6.1	0.3	0.1	0.4	3.3
2-cycle, DI							
J	46.7	6.1	14.7	0.7	—	—	3.3

the results are given in Table III. These results show that:

1. The IDI engine produced the lowest CO, NO, hydrocarbons, and aldehydes.
2. The odor levels were not greatly different for any of the engines tested.

B. Diesel and Gasoline Engines

Springer[96] made a comparison between the emissions from a 4-cylinder gasoline and a diesel-powered Mercedes 220 passenger car. The diesel engine had an IDI combustion chamber. The odor, smoke, and gaseous emissions from the diesel-powered car were measured by using methods developed for heavy-duty diesel trucks and buses. Comparative tests of the gasoline and diesel vehicles were also made by using the 1972 Federal Emissions Test Procedure for CO, hydrocarbons, and NO.

The results showed that the hydrocarbons, acrolein, and aldehydes seemed to correlate well with odor ratings. The 1972 procedure tests revealed that the diesel produced about 30% as much H–C, 5% as much CO, and about 50% as much NO_x as the gasoline.

Springer[96] reported that for these two engines, partial oxygenates such as acrolein, aliphatic aldehydes, and formaldehyde were substantially lower from the diesel than from the gasoline engine. The results of these tests should not be used to draw general conclusions in comparing the emissions in diesel and gasoline engines. More tests are needed before such conclusions are made.

VII. NO EMISSION CONTROL

Many controls may be used to reduce the NO emissions. These may, however, affect the combustion process, increase the concentration of other emission species, and affect engine performance.

The control of NO emissions in diesel engines should be made during the combustion process. The exhaust treatment reducing techniques used in gasoline engines employ a catalyst in the presence of CO.[97, 98] These techniques are not effective in diesel engines because of the presence of excess oxygen in the exhaust, even under full-load conditions.

As discussed earlier, the reduction in NO formation during the actual combustion process is mainly due to the drop in the maximum temperatures reached or to the reduced oxygen concentration. This may be achieved by any of the following methods or a combination of them.

A. Injection Timing

The retard of injection has been found to be very effective in reducing the NO emissions in both the DI and IDI diesel engines.[85, 93, 94, 99] This is

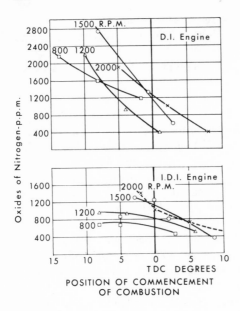

Fig. 32. Effect of injection timing on nitrogen oxides emission in DI and IDI engines (McConnell[85]).

mainly due to the reduction in the residence time and in the maximum temperatures reached in the different parts of the combustion chamber. McConnell[85] and others found that retarding the injection timing is more effective for the direct-chamber engines than for the indirect-chamber engines. This is shown in Fig. 32.

Springer and Dietzmann[53] found that retarding the injection timing in a DI engine had no apparent effect on odor or CO and increased the unburned hydrocarbons.

B. Water Addition

Diluting the charge by adding water affects both the emission formation and elimination in DI and IDI engines.

1. DI Engines

Adding water is expected to change the lean ignition limit and to increase the LFOR width and the H–C and CO formation. It may also reduce the elimination reactions due to the drop in the maximum temperatures. This would result in an increase in H–C and CO emissions.

Increasing inlet air humidity in a DI engine was found, by Springer and Dietzmann,[53] to significantly reduce NO and to increase smoke. Its effect on hydrocarbons and CO was minor except at full load.

Valdamanis and Wulfhorst[99] compared the effect of introducing water in the fuel as an emulsion to water induction with the intake air in a DI engine.

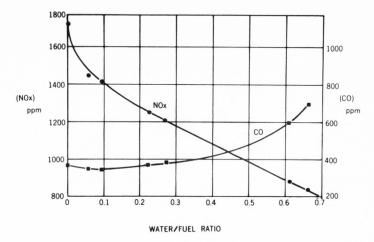

Fig. 33. Effect of water addition on nitrogen oxides and carbon monoxide in an M-system engine (Abthoff[94]).

As the ratio of water to fuel increased, the ID increased, and injection had to be advanced to obtain peak power. The increase in ID and injection advance for optimum power were greater with the emulsified fuel than with induced water. It is believed that the NO emission in these tests was affected by two factors: the effect of water on reducing the maximum temperatures and oxygen concentration, and the effect of injection advance on the maximum temperatures reached. These two factors resulted in an increase in NO emissions with the emulsified fuel and a decrease in NO emissions with the induced water.

The combined effects of water addition and injection advance (for optimum power) in Valdamanis tests affected the other emissions in different ways. It resulted in a reduction in smoke and an increase in unburned hydrocarbon emissions.

The effect of water injection on the emissions in the M-system was observed by Abthoff and Luther.[94] They found that water injection reduced the NO emissions and increased the CO emissions, as shown in Fig. 33.

2. IDI Engines

In IDI engines, water injection may result in more H–C and CO formation in the prechamber and less elimination in the main chamber.

The effect of water injection on a swirl-type IDI engine, Comet V, was studied by Torpey et al.[103]; the results are given in Fig. 34. In general, water injection decreased the NO, increased CO, H–C, and BSFC, particularly near fuel load.

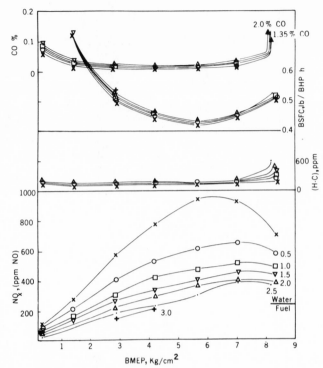

Fig. 34. Effect of water addition on the emissions of an IDI Comet
V engine (Torpey[103]).

C. Exhaust-Gas Recirculation

The reduction in NO emission in diesel engines by exhaust-gas recirculation is believed to be due to the increase in the heat capacity of the charge, as has been found in gasoline engines.[100, 101] The exhaust gas recirculation affects the combustion and emissions in the DI and IDI engines in a manner similar to water injection. The effect of percentage exhaust-gas recirculation on the NO and unburned hydrocarbon emissions and fuel consumption for a swirl-type IDI engine has been studied by Torpey,[103] and is shown in Fig. 35. The increase in percentage exhaust-gas recirculation reduced the NO emissions for all the BMEP range. Its effect on increasing H–C and BSFC was significant near full load.

Similar effects of the exhaust-gas recirculation on reducing the NO emissions have been observed by Abthoff and Luther[88] in another type of indirect-injection engine.

McConnell[85] studied the effect of reducing the oxygen concentration on the percent reduction in oxides of nitrogen by using different diluents. Figure 36 shows the effect of charge dilution by increasing the concentration of carbon dioxide, nitrogen, and exhaust gases on reducing NO emissions.

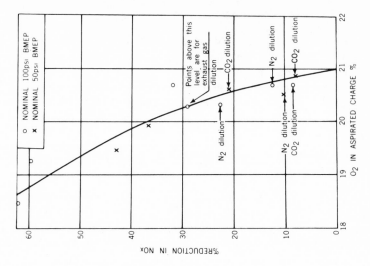

Fig. 36.   Effect of charge dilution on reducing NO emissions (McConnell[85]).

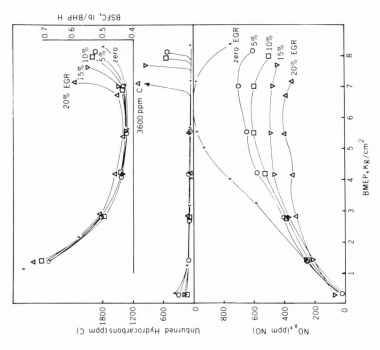

Fig. 35.   Effect of exhaust gas recirculation on the emissions of an IDI Comet V engine (Torpey[103]).

VIII. CONCLUDING REMARKS

The control of emissions in the diesel engine is expected to draw more attention in the future since this engine is very economical for heavy-duty transportation. It also has many applications in underground and enclosed spaces, where any reduction in the emissions is very desirable.

As discussed in the present chapter, the NO and other emission controls in diesel engines deal mainly with the combustion process itself rather than the postcombustion-stage techniques used in the exhaust system of gasoline engines. Therefore, it is believed that a better understanding of the combustion process itself may result in better control of the different emission species.

The combustion and emission formation models presented in this chapter are meant to help designers and research workers in assessing the potentialities of some chamber designs under the different operating conditions. The emphasis has been placed on the widely used types of combustion chambers. Due to the limited space, the discussions of some uncommon systems have been reduced. The author recognizes the complexity of the processes and the numerous design and operating parameters which affect the emissions in the diesel engine. However, an attempt has been made to emphasize the primary parameters which affect the combustion-generated emissions, and to explain general trends. Parameters related to a particular design may alter the trends discussed here. As more data become available, the models presented in this paper may be developed and improved.

The following areas are believed to be in need of future research for a better understanding of the mechanisms of combustion and emission formation in diesel engines:

(1) *Swirl in the combustion chamber.* In many cases the swirl has been measured by using vane-type rotors in a steady-state test rig. There are some doubts concerning the validity of this method for actual engine studies. Further studies using Schlieren photography and anemometer techniques may prove very useful.

(2) *Spray formation and droplet size distribution.* Most of the data related to diesel engines was published about 35 years ago. The use of the latest holographic techniques may lead to a better understanding of the behavior of the droplets and sprays.

(3) *Injection system improvement.* One of the areas which is in need of research is the elimination or reduction of afterinjection for the reduction of the smoke and incompletely burned emissions. Studies should be made with the help of injection-system modeling.

(4) *Multicomponent fuels.* It is noted that most of the previous fundamental work has been concerned with pure fuels. Actual fuels consist of a

very wide variety of components. More work is needed to study their evaporation, autoignition, and combustion.

(5) *Supercritical evaporation.* With the use of high turbocharging in diesel engines, the problem of supercritical evaporation is becoming more important.

(6) *Mechanisms of hydrocarbon fuels, autoignition, and combustion.* It is believed that the formation of the various emissions (particularly the aldehydes, carbon, and unburned hydrocarbons) is controlled by the mechanisms of oxidation. A change in temperature, pressure, or local fuel–air ratio may completely change the mechanism of oxidation and the resulting emissions. One of the areas which deserves further study is the mechanism of formation of carcinogenic and odoriferous aromatic hydrocarbons from straight-chain hydrocarbon fuels.

(7) *Odor measurement.* Efforts presently in progress need to be continued to develop a technique to identify odors and measure their intensity.

The above points show that most of the needed research is related to very complicated phenomena. Basic studies in these areas would provide a better understanding of the problem and would furnish data for mathematical modeling which has proved to be a very effective tool in research and development. These basic studies, together with research and development work on the actual combustion systems, are needed to enable better control of the diesel combustion and exhaust emissions.

REFERENCES

1. *The Federal R and D Plan for Air-Pollution Control by Combustion Process Modification,* Final Report prepared under Contract CPA 22–69–147 to Air Pollution Control Office, Environmental Protection Agency, by Battelle Memorial Institute, Jan. 11, 1971.
2. Friedman, R., and Johnson, W., The wall quenching of laminar propane flames as a function of pressure, temperature, and air–fuel ratio, *J. Appl. Phys.* **21** (8) (Aug. 1950) 791–795.
3. Daniel, W.A., Flame quenching at the walls of an internal combustion engine, *Sixth Symposium (International) on Combustion,* Reinhold Publishing Co., New York, 1957, pp. 886–894.
4. Kurkov, A.P., and Mirsky, W., An analysis of the mechanism of flame extinction by a cold wall, *Twelfth Symposium (International) on Combustion,* The Combustion Institute, Pittsburgh, Pa., 1969, pp. 615–624.
5. *Burning a Wide Range of Fuels in Diesel Engines,* Society of Automotive Engineers special publication, New York, PT–11, 1967.
6. Henein, N.A., and Bolt, J., Ignition delay in diesel engines, *SAE Automotive Engineering Congress,* Detroit, Mich., Jan., 1967, paper No. 670007.
7. Henein, N.A. and Bolt, J., Correlation of air charge temperature and ignition delay for several fuels in a diesel engine, *International SAE Automotive Engineering Congress,* Detroit, Mich., Jan. 1969, paper No. 690252.

8. Henein, N.A., A mathematical model for the mass transfer and combustible mixture formation around fuel droplets, *SAE Automotive Engineering Congress,* Detroit, Mich., Jan. 1971, paper No. 710221.

9. Adler, D., and Lyn, W.T., The evaporation and mixing of a liquid fuel spray in a diesel engine, *Diesel Engine Combustion Symposium Proceedings,* Institution of Mechanical Engineers, London, Vol. 184, Pt 3J, 1969–70, pp. 171–180.

10. Cotton, I., Hill, D.E., and McRae, P.R., Study of Liquid Jet Penetration in a Hypersonic Stream, *A.I.A.A..* 6 (1968) 2084–2089.

11. Wakuri, Y., *et al.,* Study of penetration of fuel spray in diesel engines, *J. Japan Soc. Mech. Engrs.* 3 (1960)

12. Obert, E.F., *Internal Combustion Engines,* 3rd Edition, International Textbook Company, Scranton, Pa. 1968.

13. Lewis, B., and Von Elbe, G., *Combustion Flames and Explosions of Gases,* Academic Press, New York, 1961, p. 150.

14. Jost, W., *Explosion and Combustion Processes in Gases,* McGraw-Hill, New York, 1946, p. 239

15. Henein, N.A., and Bolt, J.A., Kinetic considerations in the autoignition and combustion of fuel sprays in swirling air, *CIMAC 9th International Congress on Combustion Engines,* Stockholm, Sweden, May 1971, Paper No. A–7.

16. Garner, F.H., Morton, F., and Saundy, J.B., Preflame reactions in diesel engines, Part V, *J. Inst. Petrol.* 47 (1961) 175–193.

17. Watts, R., and Scott, W.M., Air motion and fuel distribution requirements in high-speed direct-injection diesel engines, *Diesel Engine Combustion Symposium, Proceedings,* Institution of Mechanical Engineers, London, Vol. 184, Pt 3J, pp. 181–191.

18. Alcock, J.F., and Scott, W.M., *Some More Light on Diesel Combustion,* Proceedings, Institution of Mechanical Engineers, (A.D.), London, Vol. 5, 1963, pp. 179–200.

19. Schmidt, Fritz, A.F., *The Internal Combustion Engine,* translated by R.W. Stuart Mitchell and J. Horne; Chapman and Hall, London, 1965, p. 78.

20. Lichty, L.C., *Combustion Engine Processes,* McGraw-Hill Book Company, 1967, p. 607.

21. Savery, W.C., and Borman, G.L., Experiments on droplet vaporization at supercritical pressures, *Paper presented to AIAA 8th Aerospace Sciences Meeting,* New York, Jan. 1970.

22. Maxwell, J.B., *Data Book on Hydrocarbons,* Van Nostrand, New York, 1950, p. 16.

23. Ranz, W.E., and Marshall, W.R., Evaporation from drops, *Chem. Eng. Prog.* 48 (3) (1952) 141–148.

24. El Wakil, M.M., Myers, P.S., and Uyehara, O.A., Fuel vaporization and ignition lag in diesel combustion, *SAE Trans.* 64 (1956) 713–729.

25. Burt, R., and Troth, K.A., Penetration and vaporization of diesel fuel sprays, *Diesel Engine Combustion Symposium, Proceedings,* Institution of Mechanical Engineers, London, Vol. 184, Pt 3J, 1969–70, pp. 147–170.

26. Sass, F., *Compressorless Diesel Engines,* Julius Springer, Berlin, 1299.

27. Mehlig, H., On the physics of fuel jets in diesel engines (Zur Physik der Brennstoffstrahlen in Diesel Maschinen), *Automob. Tech. Zeitschrift* 37 (1934).

28. Lamb, G.G., *Vaporization and Combustion of Multi-Component Fuel Droplets,* Semi-Annual Progress Report, Project Squid, Apr. 1, 1953 (NACA Report 1300, 1959, p. 31).

29. Scott, W.M., *Looking in on Diesel Combustion,* Society of Automotive Engineers Inc., New York, SP–345, 1968.

30. Henein, N.A., Combustion and emission formation in fuel sprays injected in swirling air, *SAE Automotive Engineering Congress,* Detroit, Mich., Jan. 1971, paper No. 710220.

31. Fristrom, R.M., and Westenberg, A.A., *Flame Structure,* McGraw-Hill, N.Y., 1965, p. 350.

32. Barnes, G.J., Relation of lean combustion limits in diesel engines to exhaust odor intensity, *SAE paper No. 680445,* 1968.

33. Milks, D., Savery, C.W., Steinberg, J.L., and Matula, R.A., *Studies and Analysis of Diesel Engine Odor Production,* Clean Air Congress of the International Union of Air Pollution, Washington, D.C., Dec. 1970.
34. Henein, N.A., and Bolt, J.A., The effect of some fuel and engine factors on diesel smoke, *SAE paper No. 690557,* Aug. 1969.
35. Lyn, W.T., Study of burning rate and nature of combustion in diesel engines, *9th Symposium (International) on Combustion,* Academic Press, New York, 1963, pp. 1069–1082.
36. Lyn, W.T., Calculations of the effect of rate of heat release on the shape of cylinder-pressure diagram and cycle efficiency, *Proceedings Institution of Mechanical Engineers,* (A.D.), London, Vol. 1, 1960–61, pp. 34–46.
37. Austen, A.E.W., and Lyn, W.T., Relation between fuel injection and heat release in a direct-injection engine and the nature of the combustion processes, *Proceedings, Institution of Mechanical Engineers,* (A.D.), London, Vol. 1, 1960–61, pp. 47–62.
38. Grigg, H.C., and Syed, M.H., The problem of predicting rate of heat release in diesel Engines, *Symposium on Diesel Engine Combustion, Proceedings,* Institution of Mechanical Engineers, London, Vol. 184, Pt 3J, pp. 192–202.
39. Bolt, J.A., and Henein, N.A., *Diesel Engine Ignition and Combustion,* The University of Michigan, Contract No. DA–20–018–AMC–1669(T), Final Report 06720–11–F, Feb. 1969, p. 542.
40. Hurn, R.W., Air pollution and the compression-ignition engine, *Twelfth Symposium (international) on Combustion,* The Combustion Institute, Pittsburgh, Pa., 1969, pp. 677–687.
41. Marshall, W.F., and Hurn, R.W., Factors influencing diesel emissions, *SAE paper No. 680528,* 1968.
42. Perez, J.M., and Landen, E.W., Exhaust emission characteristics of precombustion chamber engine, *SAE paper No. 680421,* 1968.
43. Johnson, J.H., Sienicki, E.J., and Zeck, O.F., A flame ionization technique for measuring total hydrocarbon in diesel exhaust, *SAE paper No. 680419,* 1968.
44. Bascom, R.C., Broering, L.C., and Wulfhorst, D.E., Design factors that affect diesel emissions, **1971** *SAE Lecture Series, Engineering Know-How in Engine Design,* SP-365, March, 1971.
45. Aaronson, A.E., Matula, R.A., "Diesel odor and the formation of aromatic hydrocarbons during the heterogeneous combustion of pure cetane in a single-cylinder diesel engine, *Thirteenth Symposium (International) on Combustion,* The Combustion Institute, Pittsburgh, Pa., 1971, pp. 471–481.
46. Khan, I.M., and Grigg, H.C., Progress of diesel combustion research, *CIMAC, 9th International Congress on Combustion Engines,* Stockholm, Sweedn, May 1971, paper No. A–18.
47. Merrion, D.F., Effect of design revisions on two stroke cycle diesel engine exhaust, *SAE Trans.* **77** (1968) paper No. 680422.
48. Withrow, L., and Rassweiler, G.M., *Ind. Eng. Chem.* **24** (1932) 528.
49. Sturgis, M.M., Some concepts of knock and antiknock action, *SAE Trans.* **63** (1955) 253–264.
50. Elliott, M.A., *Diesel Fuel Oils-Production, Characteristics and Combustion,* ASME, New York, 1948, pp. 57–120.
51. *Chemical Identification of the Odor Components in Diesel Engine Exhaust,* Arthur D. Little Inc., Final report to CRC and NAPCA, C–71475, CRC Project: CAPE–7–68 (1–69), HEW Contract No. CPA–22–69–63, June, 1970.
52. O'Donnell, A., and Dravnieks, A., *Chemical Species in Engine Exhaust and their Contributions to Exhaust Odors,* IIT Research Institute, Chicago, Ill., report No. IITRI C6183–5, Nov. 1970.
53. Springer, K.J., and Dietzmann, H.E., *An Investigation of Diesel-Powered Vehicle Odor and Smoke, Part IV,* Southwest Research Institute, San Antonio, Texas, Final Report No. AP–802, April 1971.
54. Somers, J.H., and Kittredge, G.D., *Review of Federally Sponsored Research on*

Diesel Exhaust Odors, U.S.E.P.A., Ann Arbor, Mich., Report No. 71–75, for presentation at the 64th Annual Meeting of the Air Pollution Control Association, Atlantic City, N.J., June 27–July 2, 1971.

55. Spindt, R.S., Barnes, G.J., and Somers, J.H., The characterization of odor components in diesel exhaust gas, *SAE paper No. 710605,* June. 1971.

56. Stahman, R.C., Kittredge, G.D., and Springer, K.J., Smoke and odor control for diesel powered trucks and buses, *SAE paper No. 680443,* 1968.

57. Stahman, R.C., and Springer, K.J., *An Investigation of Diesel Powered Vehicle Odor and Smoke,* National Petroleum Refiners Association, FL–66–46, Fuels and Lubricants Meeting, Philadelphia, Pa., Sept. 1966.

58. Turk, A., *Selection and Training of Judges for Sensory Evaluation of the Intensity and Character of Deisel Exhaust Odors,* U.S. Dept. of Health, Education and Welfare, PHS, Publication 999–AP–32. U.S. Dept. of HEW, Bureau of Disease Prevention and Environmental Control, National Center for Air Pollution Control, Cincinnati, Ohio, 1967.

59. Rounds, F.G., and Pearsall, H.W., Diesel exhaust odor, *SAE National Diesel Engine Meeting,* Chicago, Ill., paper No. 863, Nov. 1956.

60. Trumpy, D.K., Sorenson, S.C., and Myers, P.S., Discussion of the paper by G.J. Barnes, *Relation of Lean Combustion Limits in Diesel Engines to Exhaust Odor Intensity,* SAE paper No. 680445, 1968.

61. DeCorson, S.M., Hussey, C.E., and Ambrose, M.J., Smokeless combustion in oil-burning gas turbine, *paper presented at Combustion Institute,* Central States Sect., March 26–27, 1968.

62. Porter, G., *The Mechanism of Carbon Formation,* Advisory Group for Aero. R–D Memo. Ag 13/M9, Scheveningen, Netherlands Conf., May 3–7, 1954.

63. Behrens, H., Flame instabilities and combustion mechanism, *Fourth Symposium (International) on Combustion,* Williams and Wilkins, Baltimore, 1953, pp. 538–545.

64. Kassel, L.S., *J. Amer. Chem. Soc.* **56** (1934) 1838.

65. Schweitzer, P.H., *Penetration of Oil Sprays,* Pennsylvania State College Engineering Experimental Station Bulletin, No. 46, July 1937.

66. Golothan, D.W., Diesel engine exhaust smoke: The influence of fuel properties and the effects of using barium-containing fuel additive, *SAE paper No. 670092,* Jan. 1967.

67. Troth, K.A., Relationship between specific gravity and other fuel properties and diesel engine performance, *ASTM Symposium on Diesel Fuel Oils,* 1966.

68. McConnell, G., and Howells, H.E., Diesel fuel properties and exhaust gas—distant relations? *SAE Trans.* **76** (1967) paper No. 670091.

69. Khan, I.M., Formation and combustion of carbon in a diesel engine, *Diesel Engine Combustion Symposium, Proceedings,* Institution of Mechanical Engineers, London, Vol. 184, Pt 3J, pp. 36–43.

70. Rost, H., *M.T.Z.* **22** (1961) 458.

71. Broeze, J.J., and Stillebroer, G., Smoke in high speed diesel engines, *SAE J.* (March 1949) p.64.

72. Savage, J.D., *The Diesel Engine Exhaust Problem with Road Vehicle,* Diesel Engineers and Users Association, paper No. S. 302, June 1965.

73. Parks, M.V., Polonski, C., and Coye, R., Penetration of diesel fuel sprays in gases, *SAE paper No. 660747,* Oct. 1966.

74. Burman, P.G. and DeLuca, F., *Fuel Injection and Controls for Internal Combustion Engines,* Simmons-Boardman Publishing Corp., N.Y., 1962, p. 135.

75. Sitkei, G., *Beitrag zur Theorie der Strahlzerstäubung,* Acta Tech., Vol., 25, No. 1–2, 1969, pp. 81–117. Technical Translation-F–129, NASA., "Contribution to the Theory of Jet Atomization."

76. Becchi, G.A., Analytical simulation of the fuel-injection in diesel engines, *SAE paper No. 710568,* June 1971.

77. Wylie, E.B., Bolt, J.A., and El-Erian, M.F., Diesel fuel-injection system simulation and experimental correlation, *SAE paper No. 710569,* June 1971.

78. Rosselli, A., and Badgley, P., Simulation of the Cummins diesel injection system, *SAE paper No. 710570*, June 1971.

79. Zeldovich, Ya. B., Sadovikov, P. Ya., and Frank-Kamenetskii, D.A., *Oxidation of Nitrogen in Combustion*, Academy of Sciences, USSR, Moscow-Leningrad, 1947.

80. Eyzat, P., and Guibet, J.C., A new look at nitrogen oxides formation in internal combustion engines, *SAE Automotive Engineering Congress*, Detroit, Mich., Jan. 1968, paper No. 680124.

81. Heywood, J.B., Fay, J.A., and Linden, L.H., Jet aircraft air pollutant production and dispersion, *AIAA paper 70-115*, New York, Jan. 19, 1970.

82. Lavoie, G.A., Heywood, J.B., and Keck, J.C., "Experimental and Theoretical Study of Nitric Oxide Formation in Internal Combustion Engines, in *Combustion Science and Technology*, 1970, pp. 313–326.

83. Newhall, H.K., and Shahed, S.M., Kinetics of nitric oxide formation in high pressure flames, *Thirteenth Symposium (International) on Combustion*, The Combustion Institute, Pittsburgh, Pa., 1971, pp. 381–389.

84. Starkman, E.S., and Newhall, H.K., Direct spectroscopic determination of nitric oxide in reciprocating engine cylinders, *SAE Automotive Engineering Congress*, Detroit, Mich., Jan. 1967, paper No. 670122.

85. McConnell, G., Oxides of nitrogen in diesel engine exhaust gas: Their formation and control, *Proceedings IME* **178** Pt. 1, No. 38 (1963–64) pp. 1001–1014.

86. Fenimore, C.P., Formation of nitric oxide in premixed hydrocarbon flames, *Thirteenth Symposium (International) on Combustion*, the Combustion Institute, Pittsburgh, Pa., 1971, pp. 373–380.

87. Meguerian, G.H., *Nitrogen Oxide Formation, Suppression, and Catalytic Reduction*, American Oil Company, Research and Development Dept., PD 23, July 1971.

88. Meurer, J., Evaluation of reaction kinetics eliminates diesel knock — the M combustion system of M.A.N., *SAE Trans.* **64** (1956) 250–272; **72** (1962) 712–748.

89. Lyn, W.T., and Valdmanis, E., The application of high-speed schlieren photography to diesel combustion research, *J. Photogr. Sci.* **10** (1962) 74–82.

90. Lewis, G.D., Combustion in a centrifugal-force field, *Thirteenth Symposium (International) on Combustion*, The Combustion Institute, Pittsburgh, Pa., 1971, pp. 625–629.

91. Nagao, F., and Kakimoto, H., Swirl and combustion in divided chamber diesel engines, *SAE Trans.* **70** (1962) 680–699.

92. Bowdon, C.M., Samage, B.S., and Lyn, W.T., Rate of heat release in high speed indirect-injection diesel engine," *Diesel Engine Combustion Symposium, Proceedings*, Institution of Mechanical Engineers, London, Vol. 184, Pt 3J, 1969–70, pp. 122–129.

93. Downs, D., A European contribution to lower vehicle exhaust emissions, presented at the *Conference on Low Pollution Power System Development*, Eindhoven, Holland, 23–25, Feb. 1971.

94. Abthoff, J.,, and Luther, H., Die Messungen der Stickoxid-Emission von Dieselmotoren und ihre Beeinflussung durch Massnahmen am Motor, *ATZ* **71** (1969) 4, 124–130.

95. Marshall, W.F., and Fleming, R.D., *Diesel Emissions Reinventoried*, (Washington) U.S. Dept. of the Interior, Bureau of Mines, RI 7530, July 1971.

96. Springer, K.J., *Emissions from a Gasoline and Diesel Powered Mercedes 220 Passenger Car*, Report AR–813, Southwest Research Institute, San Antonio, Texas, June 1971.

97. Campau, R.M., Low emission concept vehicles, *SAE Automotive Engineering Congress*, Jan, 1971. paper No. 710294.

98. Meguerian, G.H., and Lang, C.R., NO_x reduction catalysts for vehicle emission control, *SAE paper No. 710291*, Jan. 1971.

99. Valdamanis, E., and Wulfhorst, D.E., The effects of emulsified fuels and water induction on diesel combustion, *SAE paper No. 700736*, 1970.

100. Ohigashi, S., Kurodo, H., Nakajima, Y., Hayashi, T., and Sugihara, K., *A New Method of Predicting Nitrogen Oxides Reduction on Exhaust Gas Recirculation*, SAE paper No. 710010, Jan. 1971.

101. Quador, A.A., Why intake charge dilution decreases nitric oxide emission from spark ignition engines, *SAE paper No. 71009,* Jan. 1971.
102. R. J. Hames, D. F. Merrion, and H. S. Ford, Some effect of fuel injection system parameters on diesel exhaust emissions, *SAE paper No. 710671, 1971.*
103. Torpey, P. M., Whitehead, M. J., and Wright, M., Experiments in the control of diesel emissions, The Institution of Mechanical Engineers, Symposium of Air Pollution Control in Transport Engines, Nov. 1971, Paper C 124/71.

Chapter 7

Diffusion and Fallout of Pollutants Emitted by Aircraft Engines*

S. L. Soo

Department of Mechanical Engineering
University of Illinois
Urbana, Illinois

I. INTRODUCTION

Emission of pollutants by aircraft engines has become an important topic because of:

1. Complaints of smoke, odor, and soiling caused by jet emissions near an airport.
2. Reduced visibility caused by the aircraft that depends on good terminal visibility at the airport.
3. Unsettled argument as to the effect of high-altitude emission on the earth's environment.

The Air Quality Act of 1967 (Public Law 90–148, 90th Congress, S 780, November 21, 1967) directed the Secretary of Health, Education and Welfare to conduct "a full and complete investigation and study of feasibility and practicability of controlling emissions from jet and piston aircraft engines and of establishing national emission standards with respect thereto."

As of 1970 the concensus[1-6] appears to be that pollution by jet aircraft still amounts to a small portion of the total emission in a metropolitan area. Yet in the immediate vicinity of an airport, the emission per unit area per unit time is already comparable to that of an urban industrial environment. Moreover, increases in air traffic will make pollution by aircraft increasingly important.

---

*For definition of symbols used in this chapter see the Notation section directly preceding the references.

The usually identified pollutants are: particulates, nitric oxide, hydrocarbons, carbon monoxide, and sulfur dioxide. These components have
various origins within the combustion process, produce different effects on
the environment, and are subject to different degrees of correction.

Most pollution control authorities tend to treat particulates and their
distribution in a given environment apart from gaseous pollutants. Particulates
serve as condensation nuclei for smog and therefore possess a special unde-
sirability. Particulates are visible because of light absorption, reflection and
scattering, and are more often the cause of complaints than are gaseous
pollutants.

It must be recognized that identification of particulates as a single group
is an over simplification. This is because among materials constituting particulates, some are nontoxic (such as soot) and others are toxic (such as lead);
consequently, they should be treated separately when dealing with the general
case of dispersion of particulates. The situation around a jetport is relatively
simple. The smoke from jet engines consists mainly of soot and its dispersion
is simple to compute and monitor.

Soot, hydrocarbons, and carbon monoxide arise from incomplete combustion and may be reduced significantly by engine design improvements
including higher combustion temperatures. However, reduction of these
components through an increase in temperatures tends to increase the formation of nitric oxide as long as air is used as an oxidant. Thus, trade-offs in
design will occur. Sulfur dioxide originates from sulfur in the fuel and can be
reduced by reducing the sulfur content of the fuel. In treating diffusion and
fallout, it appears reasonable to focus on particulates (soot) and nitric oxide
as typical species since they are the most difficult emissions to control and
their dispersion characteristics (gas–solid) are quite different. Other species
are more susceptible to correction by engineering design and fuel technology.

Consideration of air pollution will increasingly influence the planned
expansion of major airports and introduction of new airports. Model studies
of anticipated flight operation and meteorology in a given location have to
be made so that pollution will be kept at an acceptable level. This survey
identifies sources of pollution and delineates basic relations and procedures
for analyzing data from measurements, and predicting the level of pollution
from a specified set of operating conditions.

II. IDENTIFICATION OF SOURCES

Excellent accounts of the emission sources of the jetports are available.
References 1–7 constitute a beginners list.

Extensive measurements have been conducted at the Los Angeles International Airport by George et al.[1] They have investigated this source of

Table I.   Average Daily Pollutant Emissions in tons/day from Various Sources in Los Angeles County, 1969[1]

| | Motor vehicles | Power plants | | Jet aircraft |
		April 15–Nov. 15	Rest of year	
Particulates	43	1	6	11
Carbon monoxide	9282	(negligible)		24
Nitrogen oxides	624	135	145	7
Hydrocarbon	1677	4	6	61
Sulfur dioxide	31	30	115	3
Totals	11,657	170	272	106

air pollution since 1959. They calculated that the average total daily emission of particulates in the Los Angeles county in 1969 was 11 tons from jet aircraft (7 tons at the Los Angeles International Airport), 43 tons from motor vehicles, and 1 ton (April 15 to November 15) to 6 tons (rest of the year) from stationary power plants. Other species are given in Table I. They also reported a figure called lb/average flight. This average was 3–5 lb per engine in the pre-jumbo jet period. Their survey included time, range, and fuel consumption. It is clear that aircraft are presently, and will be increasingly, a significant source of air pollution. The acuteness of the problem of pollution is seen by comparing the 1969 data to the 1971 emission data on particulates at the Los Angeles Airport as shown in Table II. By this time, piston-engine population had dwindled to become an insignificant particulate contributor. In 1971 the overall average was 11 tons per day as compared to 7 tons per day in 1969. Table III also shows averages for various operating modes.

Since engine ratings differ widely, it is often desirable to express emissions in terms of pounds of emission per thousand pounds of thrust. This is termed specific emission rate.

Table II.   Particulate Emission at the Los Angeles Airport (1971)[a]

Type	Emission of particulates
Turbojet	560 tons per year
Turbofan	3260 tons per year
Turboprop	40 tons per year
Piston engines	21 tons per year
Engine run-up	110 tons per year
Service vehicles	20 tons per year
Vehicles entering and leaving airport	40 tons per year
Miscellaneous	40 tons per year
Total	4091 tons per year

[a]Furnished by Professor J.W. Patterson, Department of Environmental Engineering, Illinois Institute of Technology, Chicago, Illinois.

Table III. Average Particulate Emission of Jet Engines at Various Modes of Operation (1971)[a] (lb/min/engine)

	Turbojet	Turbofan
Taxi	0.392	0.3
Approach	0.425	0.734
Climb-out	0.683	0.834
Take-off	1.108	1.525

[a] Furnished by Professor J. W. Patterson.

Specific emission rates were reported by Bastress and Fletcher in 1969.[2] They measured exhaust emissions during various operating modes (idle, taxi, approach and climb-out, landing, and take-off) of seven classes of aircraft. The results in Table IV show that in terms of lbs/1000-lb fuel (called emission index), long-range jet engine aircraft tended to emit the least (0.3) particulates during idle and taxi, the greatest (0.6) during landing, take-off, and climb-

Table IV. Emission Indices for Aircraft Engines[2,4]

Engine class	Operating mode	Air/fuel ratio	Emission index (lb/1000 lb fuel)				
			CO	Organics	NO_x	Particulates	SO_2
Long-range jet	Idle and taxi	133	174	75	2.0	0.3	1.0
	Approach	103	8.7	16	2.7	1.1	1.0
	Land take-off climb-out }	75	0.7	0.1	4.3	0.6	1.0
Medium-range jet	Idle and taxi	132	50	9.6	2.0	0.6	1.0
	Approach	108	6.6	1.4	2.7	2.7	1.0
	Land take-off climb-out }	69	1.2	0.6	4.3	2.5	1.0
Piston transport	Idle	10.0	600	160	0	2	0.2
	Taxi	11.5	900	90	3	2	0.2
	Approach	12.0	800	60	5	2	0.2
	Land take-off climb-out }	9.5	1250	190	0	2	0.2
Turbine helicopter	Idle and taxi	111	118	11.5	2.0	1.0	1.0
	Approach	75	11	0.6	2.7	1.5	1.0
	Climb-out	63	4	0.3	4.3	1.5	1.0
Automotive	Average		300	55	27	4.5	2.3
Piston [4]						(0.4 lead)	

out, and the highest (1.1) during approach. Medium-range jets gave the highest particulates (0.6, 2.5, and 2.7, for the respective modes) among all jets. Turbohelicopters produced 1.0, 1.5, and 1.5 for the respective modes, and piston transports produced 2, 2, and 2, respectively.

Various emission-reduction techniques have been tried. It may be noted that low-smoke combustors for turbine engines increased nitrogen oxides (toxic) by 7% while reducing smoke by 50%; hence, it gives a better looking but more harmful plume.[4] Fuel additives for reducing smoke are usually not effective. TEL is still needed in high-performance piston engines. Curtailing taxi operation is one way to reduce emission. This may be accomplished by transporting passengers to the aircraft on the runway (Dulles International Terminal). Near air terminals, the pollutant concentrations were comparable to emission densities in adjacent urban communities. This is expected to become more severe in future years as both number of aircraft and number of flights increase.[2] Table IV also includes automotive piston engine data[4] for comparison.

In a study of the generation of pollutants, Heywood et al.[3] gave special attention to nitric oxide and soot. They calculated the mass of particulates and nitric oxide as a percentage of the fuel. For a J–57 engine, these particulate results were 1.4, 0.7, and 1.0 for take-off, approach, and idle, respectively. The nitric oxide results were 0.1, 0.4, and 0.2, respectively. At high temperatures, particulates were negligible but NO went up to 0.53, 0.45, and 0.24, respectively, for the three modes of operations. They reviewed the kinetics for the formation of carbon in gas-turbine combustors burning liquid droplets. Such studies have shown that carbon formation was reduced by leaning out the primary zone of the combustor and injecting air into the fuel cone. The burning of soot once formed was shown to be a complex process. Soot particles are usually of 0.05 μm to 0.06 μm and up to 0.125 μm with irregularly shaped clusters up to 0.8 μm. Significant oxidation occurs mainly in the 2200 °K to 2500 °K range.

In a review, Sawyer[4] showed that when compared to motor vehicles, aircraft contributed only a small fraction of the pollutants in our atmosphere. However, around airports, jet aircraft were major contributors of pollution. He stated that smoke from jet aircraft looks bad rather than being harmful to health. He explained that the specific emission rate index was used instead of concentration of exhaust species because of the air–fuel ratio differences between piston engines (nearly stoichiometric) and jet engines (four times theoretical air at full load to ten times at idle). He cited specific emission indices (in lb/1000 lb fuel) of particulates of 0.3, 1.0, and 0.6 at idle–taxi, approach, and landing–take-off–climb-out, respectively, for turbo jets. These were compared to emission indices of 4.5 for particulates of soot and 0.4 for lead in automobiles. His survey gave, in a combined plot, emission index versus equivalence ratio (fuel–air ratio to that of stoichiometric fuel–air ratio) juxtaposed with operating conditions: idle, approach, cruise, and take-off (see Fig. 1). He also presented data on nitric oxide which had an emission index generally below 3, but no trend could be identified. Naturally, this conclusion cannot be universal; for example, larger engines would operate differently from small engines. Moreover, he showed that smoke reduction by fuel additives tended to adversely[4] affect engine operation. A leaner primary combustion zone mixture ratio increased nitric oxide. In another survey,[5] the relationship of aircraft emissions to California air-quality criteria was explored. The particulate concentration was limited to 100 $\mu g/m^3$ as a crude guide, and no conclusion was drawn. The mobility of such a source in comparison to various stationary sources in ground installation was also identified. Other items discussed were the use of metallic smoke suppressant, the use of boronhydride fuels, and the problem of high-altitude aircraft emission. The only conclusion available seems to be that the probable effects of these additives are not known. Although his concern was the application of aerospace technology to air pollution problems, both topics are significant in their own right; whether a solution should be identified with aerospace technology appears to be superfluous.

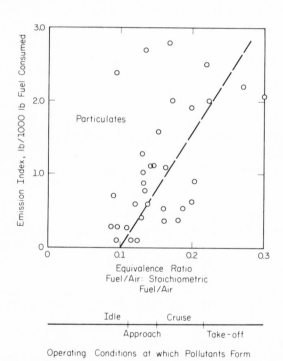

Fig. 1.   Emission index of particulates at various operating
conditions of aircraft from survey by Sawyer.[6]

It is seen that cumulative data such as in Tables I and II substantiate the need for pollution control but do not help future prediction and individual evaluation. Table III is an average indicator but does not show the merit of individual devices. This is especially true since turbofans are usually large engines. The emission index in Table IV must be viewed with some caution because it is the power output, or size of the jets rather than their range that makes possible a certain performance; higher efficiency usually comes with larger capacity. In the future, unless a better measure is devised, the specific emission rate should be included in all tabulations. This should help in the prediction of pollution level, especially if the emission indices are given for various engine output ranges such as cruising thrust.

III. DIFFUSION AND FALLOUT

The harmful effects of pollutants obviously depends on their spreading in a given environment.

In his survey, Fay[6] showed that the emission of air pollutants from jet aircraft constitutes only a small proportion of the total emission in a metropolitan area. In the immediate vicinity of an airport, however, the rate of emissions per unit area is comparable to that in an urban environment from sources other than aircraft. This is especially true for particulates. This might be changed greatly by the anticipated tripling of passenger movement over the next ten years. For instance, O'Hare Airport in Chicago is estimated to reach 900,000 landing–take-offs per year by 1980. This estimate is based on the current practice of 1 landing–take-off per 100 passengers. The John F. Kennedy Airport (10,000 landing–take-offs in 1965) already has a particulate concentration as high as those in the surrounding urban area.[7] This number is estimated to increase by a factor of five by 1980. Following the Los Angles study,[1] Fay took an arbitrary ceiling of 3000 feet in landing–take-offs as having a bearing on the ground level concentration and suggested that pollutants emitted above this level would not be mixed down to ground level due to extensive dilution. This appears to be an unnecessary prescription. Fay treated the problem as one of a moving source as did Sawyer.[4] He adopted the model of dilution by lateral mixing only for horizontal lateral width W_k. Fay gave the downwind concentration ρ_k of pollutant species k as

$$\rho_k = \frac{\dot{m}_k f}{u\,W_k} \tag{1}$$

where $\dot{m}_k$ is the mass rate of pollutant emitted per unit of vertical ascent and descent during a landing–take-off cycle at frequency f and wind speed u. Figure 2 is a diagrammatic representation of aircraft flight paths at one

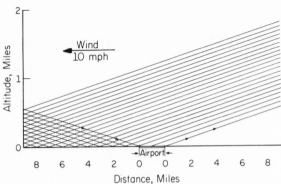

Fig. 2. Flight paths of deposition of pollutants for landing–take-off at 6-min intervals in 10-mph wind given by Fay.[4]

landing–take-off cycle every 6 min for a 2-hr period with the atmosphere moving downwind at 10 mph. For an estimated mass deposition rate of particulates of 16.6 g/m [based on (g/sec)/(m/sec descent)] from a four-engine (turbofan) aircraft, the mean annual concentration would be 60 μg/m^3 for $f = 10^6$ landing–take-offs per year, $u = 16$ km/hr, $W_k = 2$ km. The method here appears to be logical yet unnecessarily oversimplified. Like Sawyer, Fay indicated that aircraft pollutants originate within a vertical plane parallel to the wind direction in contrast to the case of ground sources in an urban area which originate in a horizontal or ground plane.

The nature of the moving source of an aircraft and its differences and similarities to a stack plume were treated rigorously by Soo[8] assuming Sutton diffusivities and simple geometry.[9] It was shown that proper identification of the coordinates of the moving source permits the use of the data and relations for ground source in predicting and correlating pollution by aircraft. Take the simple case of take-off in the absence of a head wind (see

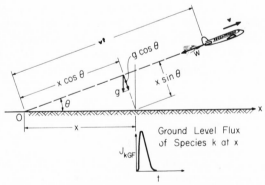

Fig. 3. Plume effect of an aircraft at take-off.[8]

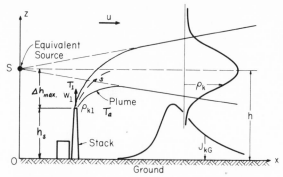

Fig. 4. Diffusion and transport of a stack plume.[8]

Fig. 3) at speed V, jet speed w_1, jet exit radius r_1, and density ρ_{k1} of species k, angle of ascent θ, diffusivity D'_{kz}, and gravitational acceleration g. Let us focus on the flux of pollutant species k at ground level at position x with respect to time t. The correspondence to the case of a stack plume (Fig. 4) is seen by the dimensionless correlations:

Aircraft	Stack
$x_k{}^* = \dfrac{D'_{kz}\,(Vt - x \cos \theta)}{V x^2 \sin^2 \theta}$	$x_k{}^* = \dfrac{D_{kz}\,x}{u h^2}$
$\tau_k{}^* = \dfrac{\tau_k\,g\,x \cos \theta \sin \theta}{D'_{kz}}$	$\tau_k{}^* = \dfrac{\tau_k\,gh}{D_{kz}}$

where D_k is the diffusivity of species k, τ_k is its relaxation time for momentum transfer and $\tau_k g$ is the terminal velocity. In Fig. 4 the stack has exit radius r_1, at exit velocity w_1, density ρ_{k1}. The plume discharges into a crosswind at speed u. The coordinate x is in the direction of u and coordinate z is opposite to the direction of gravity while y is normal to both. y is symmetric with respect to plane z–x containing the center plane of the plume with x measured from the equivalent source S, with h as the plume height. The diffusivities given in the Sutton diffusion equation are calculated from

$$
\left.
\begin{aligned}
D_{ky} &= \frac{2 - n}{4}\,C_{ky}{}^2\,u x^{(1-n)} \\[2mm]
D_{kz} &= \frac{2 - n}{4}\,C_{kz}{}^2\,u x^{(1-n)}
\end{aligned}
\right\}
\tag{2}
$$

where the C_k's and n are constants.[1] With these relations $x_k{}^*$ for the stack is then modified to

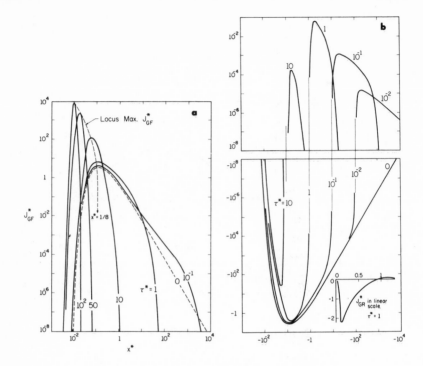

Fig. 5. Ground level flux from a plume of pollutants.
(a) flux of fallout, (b) reflected flux.[8]

$$x_k^* = \frac{C_{kz}^2\, x^{(2-n)}}{4\, h^2} \tag{3}$$

Accordingly, both cases can use the chart of ground-level flux, shown in Fig. 5, which gives

$$
\begin{aligned}
J_{kG}^* &\equiv J_{kGF}^* + \alpha_k\, J_{kGR}^* \\
&= \frac{1}{2}\left(\frac{\tau_k^*\, x_k^* + 1}{x_k^{*2}}\right) \exp\left[-\frac{(\tau_k^*\, x_k^* - 1)^2}{4x_k^*}\right] \\
&\quad + \frac{\alpha_k}{2}\left(\frac{\tau_k^*\, x_k^* - 1}{x_k^{*2}}\right) \exp\left[-\frac{(\tau_k^*\, x_k^* + 1)^2}{4x_k^*}\right]
\end{aligned} \tag{4}
$$

where α_k is the reflection coefficient of the ground; $\alpha_k = 1$ for nonabsorbing ground, $\alpha = 0$ for completely absorbing ground. For constant D_k and actual flux J_{kG} at the ground, J_{kG}^* is defined as

$$J_{kG}^* = \frac{J_{kG}\, h^2}{\pi r_1^2 \rho_{k1} w_1}\left(\frac{4\pi\, uh}{D_{kz}}\right)\left(\frac{D_{ky}}{D_{kz}}\right)^{1/2} \exp\left[\frac{1}{4x_k^*}\frac{y^2}{h^2}\frac{D_{kz}}{D_{ky}}\right] \tag{5}$$

For cases with Sutton diffusivity,

$$J_{kG}* = \frac{J_{kG}}{r_1^2 \rho_{k1} w_1} C_{ky} C_{kz} x^{2-n} \exp\left[\frac{1}{4x_k*} \frac{y^2}{h^2} \frac{C_z^2}{C_y^2}\right] \tag{6}$$

The above correspondence suggests that we should consider using the wealth of stack-plume data[9, 10] and correlations in the estimation of dispersion of pollutants including particulates near an airport. Computer modeling of flight operations and resulting pollutant concentration and flux at an airport and its surrounding area is feasible. This may include all modes of operation of a variety of aircraft over a period of varying meteorological conditions. Such a modeling program would permit optimizing the operation of a given airport for given pollution control regulations, checking the pollution patterns during the design for an expansion or a new airport, and modifiying flight operations according to weather conditions.

It is felt that most estimates concentrate on ground-level flux which in many cases is not as pertinent as ground-level concentration which is given by[8]

$$\rho_k = \rho_{k1} \left(\frac{r_1}{h}\right)^2 \left(\frac{w_1}{u}\right) \sqrt{\frac{D_{ky}}{D_{kz}}} \left(\frac{1}{4x_k*}\right) \exp\left[-\frac{1}{4x_k*} \frac{y^2}{h^2} \frac{D_{kz}}{D_{ky}}\right]$$

$$\cdot \left\{\exp\left[-\frac{1}{4x_k*}\left(\frac{z}{h} + \tau_k* x_k* - 1\right)^2\right]\right.$$

$$\left. + \alpha_k \exp\left[-\frac{1}{4x_k*}\left(\frac{z}{h} + \tau_k* x_k* + 1\right)^2\right]\right\} \tag{7}$$

with $h = x \sin\theta$, $z = z' \cos\theta$ in Fig. 3.

The effect of aircraft on world pollutant flow, high-altitude dispersion of pollutants, condensation trails, carbon dioxide, and carbon particle could be estimated on the same basis.

An example is now given using Fig. 5. For the conditions: $u = 100$ m/sec, $x = 2000$ m, $\theta = 30°$, $D_k = 10^2$ m²/sec, $\tau_k = 10^{-2}$ sec, $g = 9.80$ m/sec², we obtain

$$\tau* = \frac{10^{-2} \cdot 9.8 \cdot 0.866 \cdot 2000 \cdot 0.5}{10^2} = 9.8 \tag{8}$$

$$x* = \frac{10^2}{2000 \cdot 100 \, (0.5)^2}\left[\frac{100t}{2000} - 0.866\right] = 2 \cdot 10^{-3}\,[t - 17.32] \tag{9}$$

Figure 5b shows that the effect of reflection is negligible for this case. Figure 5a shows that fallout begins to be felt at $x* = 10^{-2}$ or at $t = 27.3$ sec after take-off, reaches a maximum at $x* = 10^{-1}$ or $t = 67.3$ sec. and nearly ends at $t = 167$ sec after take-off. For a jet of total exhaust flow of 100 kg/sec with 2% of the contaminant under consideration, $\pi r_1^2 \rho_{1k} w_{1k} = 2$ kg/sec, and the maximum fallout rate at $x* = 0.1$ is $J_{kG}* = 10^2$ or $J_{kG} = 0.016$ kg/m² sec.

However, for $\tau_k = 10^{-3}$ sec (with all other parameters unchanged), fallout may still be felt at $t = 2 \cdot 10^4$ sec, or nearly 6-hr after take-off. It is readily seen that for up to 10 miles from a jetport, even though the number of flights might be reduced to negligible amounts (over say 8 hr of each day), the fallout of species from jet exhaust persists over the entire 24-hr period.

This is true especially for the case of gaseous pollutants. Physical magnitudes are $\tau \simeq 4.5 \cdot 10^{-10}$ sec for SO_2 molecules in air at $300°K$, $\tau \simeq 10^{-9}$ sec for NO, $\tau \simeq 3 \cdot 10^{-6}$ sec for 1 μ water droplets and $3 \cdot 10^{-2}$ for 10 μ fly ash. Hence, gaseous pollutants tend to linger over long periods of time and are more subject to removal by chemical processes and scavenging. In general, $\tau_k \simeq 0$ for all other gases except for heavy molecules or atoms such as uranium.

When a distribution in particle size exists such that the number distribution function at the source is known, an integration procedure can be applied as well as a summation which treats different sizes as different species.[11]

Electric charges on particles and ions contribute additional effect of field forces,[12, 13] but the effect is small at tolerable concentrations of pollutant species.

A. Cruising

The above relations can be extended to treat the effect of high-altitude release of pollutant along a densely traveled air route. Take the simple situation of flight of similar aircraft cruising at an altitude h at a mean spacing L. The accumulation and dispersion of species k in the atmosphere is given by its density[8]

$$\rho_k = \rho_{k1} \frac{w_1 r_1}{4\sqrt{D_{ky}D_{kz}}} \frac{r_1}{L} \exp\left[-\frac{1}{4x_k^*}\frac{y^2}{h^2}\frac{D_{kz}}{D_{ky}}\right]$$

$$\cdot \left\{ \exp\left[-\frac{1}{4x_k^*}\left(\frac{z}{h} + \tau_k^* x_k^* - 1\right)^2\right]\right.$$

$$\left. + \alpha_k \exp\left[-\frac{1}{4x_k^*}\left(\frac{z}{h} + \tau_k^* x_k^* + 1\right)^2\right]\right\} \tag{10}$$

when $\tau_k^* = \tau_k gh/D_{kz}$, $x_k^* = D_{kz}t/h^2$. t is the time, y and z are coordinates parallel and normal to ground, both perpendicular to the direction of flight. For a partially absorbing ground (or ocean), the fallout corresponds to a scavenging mechanism in the atmosphere. The concentration along the flight path will change according to the term within the brackets { } in Eq. (10) for $z/h = 1$:

$$\exp\left[-\frac{\tau_k^2 g^2 t}{4D_{ky}}\right]\left[1 + \alpha_k \exp\left(-\frac{h^2}{D_{kz}t} - \frac{\tau_k gh}{D_{kz}}\right)\right] \tag{10a}$$

Hence, for small $\tau_k{}^*$ $(\tau_k{}^* \to 0)$ and nonabsorbing ground $(\alpha_k \to 1)$, the concentration along the flight path will increase according to

$$1 + \alpha_k \exp\left(- \frac{h^2}{D_{ky}\,t}\right) \tag{10b}$$

For large $\tau_k{}^*$ and completely absorbing ground $(\alpha_k \to 0)$, this concentration will decrease according to $\exp(- \tau_k{}^2 g_k{}^2 t/4D_{ky})$ in Eq. (10a). For nonzero α_k, an increase in ρ_k along the flight path is seen.

An interesting example is that 0.516 lb/min of NO is emitted per plane for a flight lane handling jets of 18,000-lb thrust spaced at 100 km and consuming 260 lb fuel per minute at a specific emission rate of 2 for NO. This flight density will build up to 0.5 ppm NO along the flight path at 10^4 m altitude within 5 years, the NO will reach $0.5/e = 0.18$ ppm within a width of 500 km. Correspondingly, CO_2 will reach 250 ppm under such ideal steady spreading at the flight altitude. These estimates are made with the assumption of no removal from the atmosphere. Removal by physical and chemical processes are considered next.

IV. CHEMICAL AND PHYSICAL CONVERSIONS

Pollutants react chemically with other constituents of the atmosphere, among themselves, and with surfaces of particulates and ground. Pollutants are also affected by the physical processes of absorption, condensation, agglomeration, and scavenging by rain. Extending the concept of multiphase systems,[11] the density or concentration ρ_k of a species k in a principal fluid of density ρ is given by

$$\frac{d\rho_k}{dt} - \rho_k \frac{d\ln\rho}{dt} = -\nabla \cdot \left[\tau_k \rho_k F_k - D_k \Delta \rho_k\right] + \Gamma_k \tag{11}$$

where t is the time, τ_k is the relaxation time for momentum transfer between species k and the principal fluid, F_k is the force per unit mass acting on species k, D_k is its diffusivity, and Γ_k is the rate of generation of k

$$\Gamma_k = m_k \left[- k_1 n_k - \sum k_{2j} n_k \ldots \sum k_s n_k S_s\right] + m_k \sum k_{ij} n_i n_k \ldots \tag{12}$$

where m_k is the mass of a species k molecule, n_k its number density, $\rho_k = n_k m_k$, k_1 is the rate constant for dissociation of k, k_{ij} that for second- or higher-order reactions with j, etc., k_s for adsorption or reaction with surface s of surface area per unit volume S_s, k_{ij} is the rate for yielding species k.

Basic atmospheric processes of SO_2, N_2O, NO, NO_2, with O_2, O_3, and radiation and catalysis are not known in detail.[9] At the natural composition found in the atmosphere the reported half-life of SO_2 is 43 days (corresponds to a reaction rate constant of $2.5 \times 10^{-7} m^3/kg$ mole sec and activation energy of 20 kcal),[9] Note that the reaction rate constants and activation energies of reactions in the atmosphere are not given extensively.[9]

With known rate constants the effect of the chemical process can be computed. Take the example of a second-order reaction in a diffusion process of component k of a pollutant

$$k\rho_k n_j + D_{kz}\frac{\partial^2 \rho_k}{\partial z^2} + D_{ky}\frac{\partial^2 \rho_k}{\partial y^2} = u\frac{\partial \rho_k}{\partial x} \tag{13}$$

for rate constant k and n moles/m³ of jth species. Computation is readily carried out with the pertubation

$$\rho_k = \rho_k{}^{(0)} + (-kn_j)\,\rho_k{}^{(1)} + \ldots \tag{14}$$

$\rho_k{}^{(0)}$ is the density of kth species due to diffusion. It is seen that for small concentrations of ρ_k

$$\rho_k{}^{(1)} = \rho_k{}^{(0)}\,e^{-kn_j\,x/u} \tag{15}$$

and correction in numerical modeling can be carried out. Superposition of reaction rates to physical diffusion and fallout is therefore readily done.

Cases of scavenging and agglomeration can be treated in an analogous manner.[13]

V. PLUME AXIS FROM A STACK

As long as the curvature of the plume path is small, the plume axis (see Fig. 4) as influenced by momentum rise and buoyancy can be computed separately from the plume spread. These calculations are outlined in the following sections.

A. Momentum Rise

The momentun rise can be computed by comparing the correlation of Bosanquet for momentum rise of stack plume[14] to that based on boundary-layer theory, with either Tollmien's diffusivity or Sutton's diffusivity[15] (see Fig. 6). Referring to Fig. 4, the relations of Bosanquet et al., are

$$\frac{\Delta h_{v\;max}}{2r_1} = \frac{4.77(\pi)^{1/2}/2}{(u/w_1)[1 + 0.43(u/w_1)]} \tag{16}$$

while boundary-layer theory gives

$$\frac{\Delta h_{v\;max}}{2r_1} = \frac{3}{32}\frac{r_1 w_1}{D}\left(\frac{w_1}{u}\right)\left\{\frac{w_1}{u}\left(\frac{\pi}{2} - \sin^{-1}\frac{u}{w_1}\right) - \left[1 - \left(\frac{u}{w_1}\right)^2\right]^{1/2}\right\} \tag{17}$$

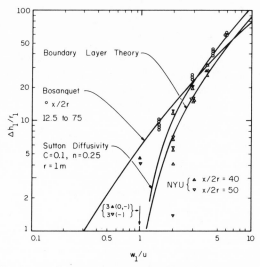

Fig. 6. Maximum momentum rise of stack plume given by various sources. New York University points are from wind tunnel tests and circles are from field measurements.[15]

Bosanquet further gives the path of plume axis as

$$\Delta h_v = \Delta h_{v\,max} \left(1 - 0.8 \frac{\Delta h_{v\,max}}{x}\right) \tag{18}$$

B. Buoyancy Rise

For an ambient temperature T_a, the buoyancy rise Δh_b is given by

$$\frac{\Delta h_b}{2r_1} = 5.00 \, N'_{Gr} \left(\frac{w_1}{u}\right)^3 \left(\frac{T_a}{T_1}\right) Z \tag{19}$$

where

$$N'_{Gr} = \frac{gr_1}{w_1^2} \left(\frac{T_1}{T_a} - 1\right) \tag{20}$$

$$Z = Z(X) \approx 0.29 \, X - (4.8 \cdot 10^{-3}) \, X^2 + 2.71 \cdot 10^{-5} \, X^3 \tag{21}$$

and

$$X = \frac{1}{3.16} \frac{u}{w_1} \left(\frac{T_1}{T_a}\right)^{1/2} \tag{22}$$

C. Buoyancy and Temperature Gradient

The potential temperature gradient $d\Theta/dz$ of ambient atmosphere is given by[14, 16]

$$\frac{d\Theta}{dz} = \frac{dT}{dz} + \Gamma \tag{23}$$

where Γ is the adiabatic temperature gradient (nearly 5.4°F per 1000 ft, or 1°C per 100 m), Θ is then the potential temperature, and $d\Theta/dz > 0$ for stable atmosphere which is the only case which can be correlated.

Bosanquet *et al.*, gave

$$\frac{\Delta h_{gr\ max}}{2r_1} = \frac{6.37\,\pi}{2} \left(\frac{w_1}{u}\right)^3 N'_{Gr}\left(\ln J^2 + \frac{2}{J} - 2\right) \tag{24}$$

where

$$J = \pi^{-1/2} \left(\frac{u}{w_1}\right)^2 \left[0.43\left(\frac{T_1/T_a}{N_{Ri}}\right) - 0.28\,N'_{Gr}{}^{-1}\right] + 1 \tag{25}$$

and

$$N_{Ri} = g\left(\frac{d\Theta}{dz}\right) \frac{T_1}{T_a{}^2} \frac{r_1{}^2}{w_1{}^2} \tag{26}$$

D. Combined Relation

Taking into account the effects of dilution of plume, relative motion, and atmosphere turbulence, Bosanquet gave, for a dilution coefficient C_2 (for which he recommended the value of 0.13)

$$\frac{\Delta h}{2r_1} = \left(\frac{1}{4}\right) C_2{}^{-2} \left(\frac{w_1}{u}\right)^3 N'_{Gr}\left\{f_1(a) + f_{11}(a_0)\right.$$
$$\left. - 0.615 a_0{}^{1/2} \left[\left(\frac{w_1}{u}\right) + 0.57\right]^{-1/2}\right\} \tag{27}$$

where

$$a = \left(\frac{8}{3}\right) C_2{}^2 N'_{Gr}{}^{-2} \left(\frac{u}{w_1}\right)^4 \left[1 + \left(\frac{3}{2}\right)\left(\frac{x}{2r_1}\right)\left(\frac{w_1}{u}\right) N'_{Gr}\right] \tag{28}$$

$$a_0 = \left(\frac{8}{3}\right) C_2{}^2 \left(\frac{u}{w_1}\right)^4 N'_{Gr}{}^{-2} \tag{28}$$

Here N'_{Gr} is based on $(T_1 - T_{a1})$, and T_{a1} is the temperature at which density of stack gas is equal to that of ambient atmosphere. $f_1(a)$ and $f_{11}(a_0)$ are given in two charts. For large a and a_0, Bosanquet gave

$$\left.\begin{array}{l} f_1(a) = \ln a - 0.12 \\ f_{11}(a_0) = 0.311\,a_0{}^{1/2} - \left(\frac{1}{2}\right)\ln a_0 - 1 \end{array}\right\} \tag{29}$$

Provisions are made for

(a) The case of zero buoyancy; and $w_1/u > 0.5$

$$\frac{\Delta h_{\max}}{2r_1} = 6^{-1/2} C_2^{-1} \frac{w_1}{u} \left\{ 1.311 - 0.615 \left[\left(\frac{w_1}{u} \right)^2 + 0.57 \right]^{1/2} \right\} \tag{30}$$

and for small w_1/u,

$$\frac{\Delta h_{\max}}{2r_1} = 6^{-1/2} C_2^{-1} (0.9) \left(\frac{w_1}{u} \right)^{3/2} \tag{31}$$

(b) The existence of the density gradient in a static atmosphere in Eq. (28) is given by

$$a \left(\frac{2r_1}{w_1} \right) \left(\frac{w_1^4}{u^4} \right) \frac{N'_{Gr}}{4C_2} = 1.527 \, (2)^{1/2} \left(\frac{T_1}{T_a} \right) N_{Ri}^{-1/2} \left(\frac{2r_1}{w_1} \right) \tag{32}$$

with the value of 200 sec as the maximum value.

VI. PLUME PATHS FROM AIRCRAFT

To apply the stack plume relation to calculating the plume path of an aircraft, we must recognize the nature of the plume path of the latter. The

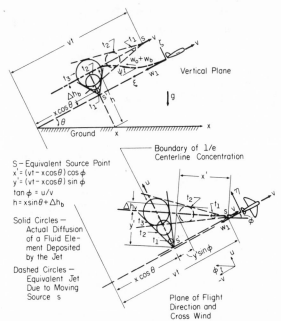

Fig. 7. Plume path and diffusion pattern of an aircraft taking off in a crosswind and the plume is under the influence of buoyancy.

behavior of the plume path and dispersion of pollutant from an aircraft is depicted in Fig. 7.

Figure 7 shows the case of an aircraft at take-off with velocity V, as is shown in Fig. 3, except now we consider the influence of a crosswind at velocity u and buoyancy of the plume represented by the modified Grashof number N'_{Gr} and the Richardson number N_{Ri}.

The influence of buoyancy is obviously in the vertical plane. If there is no crosswind, at a distance from the aircraft, each element of contaminant species will be accelerated to a buoyant velocity w_b until a buoyancy rise Δh_b is reached. At Δh_b the element is left in the atmosphere at zero velocity. This element diffuses in the meantime as shown by the solid circles at time t_1, t_2, t_3 in Fig. 7. These solid circles represent the boundaries of $1/e$ axial concentration of a contaminant. The net effect is identical to the concentration at various locations along the plume axis starting from an equivalent source S, as depicted by circles of dashed lines.

The crosswind constitutes what corresponds to a "momentum rise" Δh_v of the plume in the plane of the flight direction and that of the crosswind (see bottom of Fig. 7). An element of contaminant deposited along the path of an equivalent source will drift windward and diffuse according to the progress represented by solid circles. This is equivalent to a moving source S with plume axis making an angle ϕ with the flight direction.

Such a three-dimensional plume is readily accounted for by extending the case treated for stacks[15] as an illustration. The jet initially at velocity w_1 produces velocity w_0 along the paths according to

$$\frac{w_0}{w_1} = \frac{r_1 w_1}{D} \left[\frac{3}{8} \frac{r_1}{s} + \frac{w_b}{w_1} \right] \tag{33}$$

where

$$\frac{w_b}{w_1} = \frac{N'_{Gr}}{8} \int_{s_0/r_1} \frac{d(s/r_1)}{(s/r_1)(w_0/w_1)} - \frac{N_{Ri}}{8} \int_{s_0/r_1} \frac{\zeta - \zeta_0}{r_1} \frac{d(s/r_1)}{(s/r_1)(w_0/w_1)} \tag{34}$$

and $s_0 = (3/8)(r_1 w_1/D)r_1$ with g in N'_{Gr} and N_{Ri} replaced by $g \cos \theta$, and the coordinates ξ, η, and ζ are given by (see Fig. 7)

$$\left. \begin{aligned} d\eta &= \frac{u}{w_0} ds \\[2mm] d\zeta &= ds \cos \phi \sin \psi = \frac{V}{(u^2 + V^2)^{1/2}} \frac{w_b}{w_0} ds \\[2mm] d\xi^2 &= ds^2 - d\eta^2 - d\zeta^2 = \left[1 - \left(\frac{u}{w_0}\right)^2 - \frac{V}{u^2 + V^2}\left(\frac{w_b}{w_0}\right)^2 \right] ds^2 \end{aligned} \right\} \tag{35}$$

With this information the plume path can be determined numerically. This is how Fig. 6 was obtained.

For determining the fallout flux in Eqs. (4) to (6), the coordinates x, y, and h should be replaced by

$$\left.\begin{array}{l} x' = (Vt - x \cos \theta) \cos \phi \\ y' = (Vt - x \cos \theta) \sin \phi \\ h' = x \sin \theta + \Delta h_b \end{array}\right\} \tag{36}$$

with $\tan \phi = u/V$.

Since these plume paths are characterized by small curvatures, the density distribution of species (even in the part before the jet momentum is dissipated) may be determined by neglecting the curvature,[17] i.e.,

$$\rho_k = \left(\frac{\rho_{k1} w_1 r_1^2}{8 D_k}\right) (1 + 2 N_{Sc}) \left[1 + \frac{3}{64}\left(\frac{r_1 w_1}{D_k}\right)\frac{r^2}{s^2}\right]^{-2 N_{Sc}} \tag{37}$$

where $N_{Sc} = D/D_k$ is the Schmidt number of species k, and D is the turbulent diffusivity in the jet.

VII. WING-TIP VORTICES

Wing-tip vortices from aircraft influence the initial dispersion of pollutants near their source. The wing-tip vortices arise from the circulation γ of a wing, which, for span B of an elliptic wing, is related to the total lift according to the well-known relation:

$$F_L = \left(\frac{\pi}{4}\right) \rho V \gamma B \tag{38}$$

where ρ is the air density and V is the flight speed. For a 747 aircraft, with $F_L \simeq 355$ tons, $B \simeq 210$ ft, $V \simeq 916$ fps, we get $\gamma = 1780$ ft^2/sec. The wing-tip vortex thus gives rise to a tangential velocity of nearly 178 fps at a radius of 10 ft. The above equation also shows that at the lowest flight speed, the vorticity is greatest for a given load. Because of favorable pressure gradients, these vortices would take a mile to decay and the tangential velocity of the solid body core (which is expected to have a radius of less than one foot) which theoretically reach 1780 fps or more[18]. The dispersion of particulates as well as gaseous compounds emitted from the engines is expected to be influenced by this swirling flow. Since the velocity distribution in such a system is known,[19] its contribution to dispersion is readily calculated. The net effect is an increase in apparent diffusivity.

VIII. PLUME VISIBILITY

The visibility of a jet plume or its optical characteristics is of interest because of (1) the desirability of treating smoke plume quantitatively and

(2) the possibility and desirability of monitoring the contribution of an individual aircraft by lidar measurement,[20] or by other optical means. Such knowledge will help in enforcing control guidelines. A large jet plume may look darker even though its pollution per passenger mile may be less than a small plume. Plumes have been studied extensively in relation to stack emission and pollution. A study of stack plumes was presented by Ensor and Pilat[21] extending the results given in the surveys of Robinson[20] and Purdom.[22]

Since the visible plumes from aircraft consist mainly of soot particles, that is, they are black-smoke plumes, Ringelmann numbers provide a good measure of their density. In addition, we shall include "white" plumes in the present discussion to account for the condensation of water vapor, especially when dealing with pollution in the upper atmosphere.

In terms of specific projected particle extinction area S (m^2/g) and particle density ρ_p (g/m^3), the fraction of transmitted light I/I_0 over a path length L (m) is given by

$$\frac{I}{I_0} = \exp\left(-S\rho_p L\right) \tag{39}$$

and S is given by

$$S = \frac{1}{K\bar{\rho}_p} \tag{40}$$

where $\bar{\rho}_p$ is the density of material constituting the particles and K is a light extinction parameter. K is a function of particle size distribution, refraction index, and wavelength of light. Ensor and Pilat gave K in cm^3/m^2 as a function of refractive index of material at a wavelength of 5500 Å (see Fig. 8). The geometric mass mean particle radius $\bar{r}$ in μm and geometric standard deviation σ for the distribution in particle size is given by

$$f(r) = \frac{1}{\sqrt{2\pi}\, r \ln \sigma} \exp\left[-\frac{(\ln r/\bar{r})^2}{2(\ln \sigma)^2}\right] \tag{41}$$

For a jet plume, from Eq. (7), taking the depth at the boundary where the particle density is $1/e$ of that at plume axis, L is roughly $4[D\,x/u]^{1/2}$ for $y = L/2$; x is measured from the equivalent source. It is interesting to note that based on this model and for reference particle density ρ_{p1}, the light transmittance along the axis of the plume is given by

$$\frac{I}{I_0} \simeq \exp\left[-\frac{\sqrt{\pi}}{2}\,\mathrm{erf}\,(1)\,\frac{\rho_{p1}}{\bar{\rho}_p}\left(\frac{r_1{}^4 w_1{}^2}{u\,Dx\,K^2}\right)^{1/2}\right] \tag{42}$$

for either constant diffusivity or Sutton diffusivity.

The magnitude of transmittance is seen in the illustration given by Ensor and Pilat. For a stack radius of 10 m, $\bar{r} = 2\ \mu$m, $\bar{\sigma} = 3$, $\rho_p = 2$ g/cm^3, and index of refraction $m = 1.96 - 0.66i$ for carbon, their chart gave

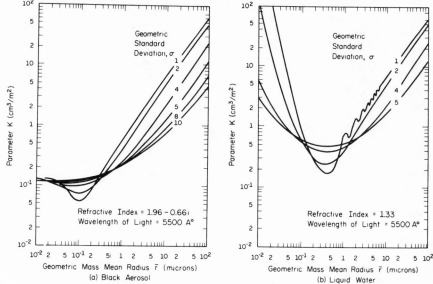

Fig. 8. Extinction parameter K as a function of the log normal size distribution for (a) carbon or black aerosol, and (b) liquid water given by Ensor and Pilat.

$K = 0.6$ cm^3/m^2, leading to $\rho_p = 0.039$ g/m^3 for Ringelmann 1 or 80% transmittance. For a 1-m-diam. plume of soot, from a jet engine with $\bar{r} = 0.05$ μm, $\sigma = 4$, $K = 0.1$, the same transmittance corresponds to $\rho_p = 0.045$ g/cm^3, which is much higher than is usually encountered. Hence, a more rigorous means than the Ringelmann number is desired for evaluation and comparison.

IV. CONCLUDING REMARKS

This survey shows that adequate information exists to develop a computer model for emission and dispersion at an airport and also for distribution of pollutants in the upper atmosphere. Such a model is desirable not only in the planning of a new airport but also to simulate and assist in modifying the flight operations of an existing airport to improve air quality.

NOTATION

a, a_0	plume-rise parameter	dT/dz	temperature gradient
B	wing span	$d\Theta/dz$	potential temperature gradient
C_2	dilution coefficient	D	diffusivity
C_k	constant in Sutton diffusivity	D_k	diffusivity of species k

f	frequency of landing–take-off cycle	r_1	exit orifice radius
F_L	lift force	$r, \bar{r}$	particle radius and mean particle radius
g	gravitational acceleration		
h	height of plume	s	coordinate along plume path
h_s	height of stack	S	specific projected particle extinction area
I	intensity of light		
J_{kG}	ground-level flux	t	time
k_{ij}	reaction rate constant of species i and j	T	temperature
		u	wind speed
L	mean spacing in flight lane	V	flight speed
$\dot{m}_k$	mass rate per unit descent of species k	w_1	velocity at the exit or source
		w_b	buoyancy induced plume velocity
n, C	constant quantities in Sutton diffusivity	W_k	width of the downwind trail
		w_0	velocity along plume axis
N'_{Gr}	modified Grashof number	x, y, z	Cartesian coordinates
N_{Ri}	Richardson number	x', y'	coordinates as defined
N_{Sc}	Schmidt number	Z, X	dimensionless quantities

Greek Letters

α_k	reflection coefficient of ground	ρ	density of atmosphere
β	wing span	ρ_k	concentration of species k of pollutant
γ	circulation		
Γ	adiabatic temperature gradient	ρ_{k1}	density of concentrate of species k at the exit
Δh	plume rise		
θ	angle made by flight direction with ground	σ	geometric standard deviation of particle size
Θ	potential temperature of the atmosphere	τ_k	relaxation time of momentum transfer of species k
ξ, η, ζ	coordinates as defined	ϕ, ψ	angles

Superscript

*	dimensionless quantities as defined

Subscript

1	exist condition	0	axial quantity
a	ambient	p	particles
b	buoyancy	v	velocity effect
k	species k	x, y, z	x, y, z components
max	maximum		

REFERENCES

1. George, R.E., Verssen, J.A., and Chass, R.L., Jet aircraft: A growing pollution source, *J. Air Poll. Assoc.* **19** (1969) 847–855.
2. Bastress, E.K., and Fletcher, R.S., Aircraft engine exhaust emissions, *ASME* Paper No. 69-WA/APC-4, (1969).
3. Heywood, J.B., Fay, J.A., and Linden, L.H., Jet aircraft air pollutant production and dispersion, *AIAA* Paper No. 70-115, AIAA Eight Aerospace Science Meeting, New York, June 1970.

4. Sawyer, R.F., Reducing jet pollution before it becomes serious, *Astronautics and Aeronautics* **8** (4), (April 1970) 62–67.

5. Sawyer, R.F., Applications of aerospace technology to air pollution problems, AIAA Paper No. 70–815, *AIAA Fifth Thermodynamics Conference*, Los Angeles, Cal., June 29–July 1, 1970.

6. Fay, J.A., *Air Pollution from Future Giant Jetports*, Fluid Mechanics Laboratory Publication No. 70–6, May 1970, Massachusetts Institute of Technology, Cambridge, Massachusetts.

7. Nolan, M., *A Survey of Air Pollution in Communities Around the John F. Kennedy International Airport Report*, Robert F. Taft Sanitary Engineering Center, Public Health Service, Cincinnati, Ohio, June 1966.

8. Soo, S.L., Effect of simultaneous diffusion and fallout from stacks and aircraft, *Atmospheric Environ.* **5** (1971) 283–295.

9. Stern, A.C., *Air Pollution* (Second Ed.), Vol. **1**, Academic Press (1968), pp. 126, 149, 179, 254.

10. Thomas, F.W., Carpenter, S.B., and Colbaugh, W.C., IV, Recent results of measurements: Plume rise estimates for electric generating station, *Phil. Trans. Roy. Soc. London*, **A265** (1969) 221–243.

11. Soo. S.L., *Fluid Dynamics of Multiphase Systems*, Blaisdell Publishing Co., Waltham, Massachusetts, (1967), Chapter 6.

12. Lee, Y.N., and Soo, S.L., A study of jets of electrically charged suspensions, *Environ. Sci. Tech. Ind. Eng. Chem.* **4** (1970) 678–685.

13. Soo. S.L., Dynamics of charged suspensions, International Review of Aerosol Physics and Chemistry (G.M. Hidy and J.R. Brock, Eds.), Pergamon Press, 1971, pp. 62–149.

14. Strom, G.H., Atmospheric dispersion and stack effluents, *Air Pollution* (A.C. Stern, Ed.), Second Edition, Vol. 1, Academic Press, New York, 1968, pp. 240–253.

15. Soo, S.L., Modeling and similarity relations of physical and chemical processes, *Pure and Appl. Chem.* **24**, (1970) 721–729.

16. Panofsky, H.A., Air pollution meteorology, *American Scientist* **57** (2) (1969) 269–285.

17. Soo. S. L., Electrohydrodynamic jets, *Developments in Mechanics, Proceedings of the Eleventh Midwestern Mechanics Conference*, Iowa State University, Ames, Iowa, **5**, pp. 215–224, 1969.

18. Soo, S.L., Core velocity distribution in a swirling pipe flow, *J. Appl. Math. Phys.* **ZAMP** (70/1), (1970) 125–129.

19. Rose, W.G., A swirling round turbulent jet, *Trans. ASME, J. Appl. Mech.* (December 1962) 615–625.

20. Robinson, E., Effect on the physical properties of the atmosphere, *Air Pollution*, (A. C. Stern, Ed.), Vol. 1, Part II, Academic Press, New York, (1968), pp. 377.

21. Ensor, D. S., and Pilat, M. J., Calculation of smoke plume opacity from particulate air pollution properties, *J. Air Poll. Control Assoc.* **21** (8), (1971) 496–501.

22. Purdom, P. W., Source monitoring, *Air Pollution* (A. C. Stern, Ed.), Vol. 2, Academic Press, New York, 1968, Chapter 29.

Chapter 8

Instrumentation and Techniques for Measuring Emissions

R. W. Hurn

Bartlesville Energy Research Center
Bureau of Mines, U.S. Department of the Interior
Bartlesville, Oklahoma

I. METHODS FOR MEASURING POLLUTANTS

A. Terms of Expression

In emission measurement, volume concentrations of the several components are characteristically expressed in the following terms (weight measurements are discussed in Section II.B):

Carbon dioxide, carbon monoxide, and oxygen are expressed as percent of the sample volume.

Nitric oxide and nitrogen dioxide as volume parts of NO (or NO_2) per million volume parts of the sample (ppm). The total of NO + NO_2 is designated NO_x.

Hydrocarbon as parts of hydrocarbon per million parts of sample, or, alternatively, as parts of carbon per million parts of sample (ppmC). The latter term is arbitrarily defined as the volume (or mole) concentration of hydrocarbon in the sample multiplied by the average number of carbon atoms per molecule of that hydrocarbon. Thus, 1 ppm propane (C_3H_8) is the equivalent of 3 ppmC hydrocarbon. In the early days of emission measurement, hydrocarbon emissions were measured in terms of the carbon equivalent of hexane, or ppm hexane. Thus, in early usage ppm values were often assumed to be ppm hexane even though not designated as hexane; this usage is ambiguous and should be avoided.

B. Instrumental

1. Flame Ionization for Measurement of Hydrocarbon

a. Principle. An oxygen/hydrogen flame is virtually ion free if the burning mixture contains only pure hydrogen, oxygen, and inert gases. If hydrocarbon is introduced, the ion flux is increased, and only a minute amount of hydrocarbon need be present to produce the effect. Moreover, under appropriate conditions the ion yield is proportional to the amount of hydrocarbon introduced into the flame. This characteristic of the hydrogen flame is now utilized in the flame-ionization measurement of hydrocarbon. Although the phenomenon was first recognized years ago, its practical application was possible only with the advent of high-impedance, fast-response, low-level amplifiers.

The basic elements of a flame-ionization detector (FID) are, as shown in Fig. 1, a burner and ion collector assembly. In practice, sample gas is mixed with hydrogen in the burner assembly and the mixture burned in a diffusion flame. Ions that are produced in the flame move to the negatively polarized collector under the influence of an electrical potential applied between the collector plates. At the negative collector the ions receive, via a current network, electrons that are collected from the flame zone at the positive collector. Thus, a small current (proportional to the amount of hydrocarbon entering the flame) flows between the collector plates. This small current is amplified using a high impedance, direct-current amplifier; the output of which becomes the indication of hydrocarbon present. The instrument is shown schematically in Fig. 2.

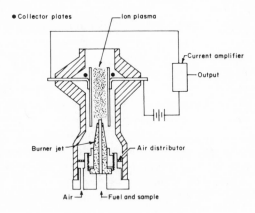

Fig. 1.   FID burner-collector assembly.

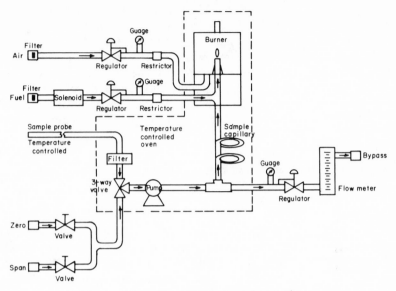

Fig. 2. Schematic of FID hydrocarbon analyzer.

The detector responds to carbon that is linked with hydrogen, as in (1),

$$R-\underset{\displaystyle H}{\overset{\displaystyle H}{\underset{|}{\overset{|}{C}}}}-H \qquad (1)$$

and the response is largely independent of the molecular configuration, i.e., hydrocarbon species. (The species does, however, affect an oxygen synergism discussed later.) Thus, the detector is essentially a carbon atom counter.

The dynamic range of the detector is very broad. A well-designed burner will generate ion current linear with hydrocarbon over a dynamic range of at least $1:10^6$; a comparable dynamic range is achievable in the current amplifier, and FID instruments typically provide linear output over the fuel range covered. A typical instrument may have its most sensitive range set at 0–100 ppmC with a switch selector providing three or more additional ranges, each successively expanded over the next most sensitive by a factor of 5 or more. Thus, such an instrument might have ranges of 0–100, 0–500, 0–2,000, and 0–10,000 ppmC of hydrocarbon.

Characteristics of the FID are improved with most burner designs if instead of using pure hydrogen fuel, the hydrogen is mixed with an inert gas to decrease flame temperature. This mixture of hydrogen and inert gas is referred to as fuel gas or fuel.

b. Operational Characteristics. The FID responds directly to the amount of hydrocarbon entering the flame; therefore, close control of sample flow is required. In general, the sample flow rate is specified at the minimum amount that will give the required sensitivity in any given instrument. Fuel and air flow rates also influence the response characteristics of the detector. Response typically first rises and then falls with increased fuel rate, as shown in Fig. 3. Best operation and minimal error from variation in fuel rate are obtained with fuel rate near the center of the response peak shown in Fig. 3. For maximum sensitivity, the air supplied to the detector must equal at least the stoichiometric requirement of the fuel, but noise and flame-stability problems are associated with too-high air flow.

The required volume rates of instrument gases are quite low. Typical values are: sample, 3–5 ml/min; fuel/gas mixture, 75 ml/min; air, 200 ml/min. Instrument manufacturers will have, to a degree, optimized these operating parameters consistent with instrument range and other instrument characteristics such as the amplification that is provided. However, slight gains in performance, particularly in special applications, can often be had by determining the several response characteristics of a particular instrument and readjusting the operating parameters to optimum for the intended application.

Interferences in hydrocarbon measurement by FID are virtually absent in the usual sense of the term "interference" in that only hydrogen-linked carbon compounds respond directly. However, free oxygen in the sample does affect the hydrocarbon response, and the effect can amount to possibly 20% or greater change in hydrocarbon response, as free oxygen is increased from 0 to 15 or 20%. The actual amount varies widely depending upon burner and collector design (geometry and materials) and upon detector operating parameters. In some cases, the response is augmented by oxygen; in others, response is diminished. Moreover, the effect in any one system varies with hydrocarbon type. With detectors that are most commonly in use, the response is diminished as oxygen increases. The oxygen effect, often called synergism, can be minimized by reducing sample rate to the minimum consistent with the required signal level and by optimizing the design and operating param-

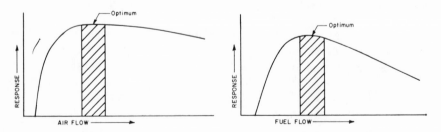

Fig. 3.   Response of flame ionization detector to air rate and to fuel rate.

eters on the basis of experimental trial. For minimal oxygen effect, pure hydrogen should *not* be used as fuel. Either nitrogen or helium may be used as a diluent, but a mixture of helium (60%) with hydrogen (40%) has been found the more advantageous in reducing oxygen synergism. (In addition, the use of the mixture in lieu of pure hydrogen may lessen the degree of nonuniform response to differing types of hydrocarbon that sometimes is encountered when pure hydrogen is used as fuel.) With careful attention to the adjustments for minimal oxygen effect, the hydrocarbon response should change by no more than 2% when oxygen in the sample is increased from 0 to 5%. Commercial instruments will have been designed for minimal oxygen synergism. Nevertheless, each individual instrument should be tested to determine its operating characteristics; corrections for the oxygen that is present can then be neglected or made as appropriate to the accuracy desired.

Hydrocarbon-derived oxygenates, such as aldehydes, alcohols, and similar compounds respond in the FID measurement with a carbon-atom sensitivity somewhat reduced from that of the hydrocarbons. A suitable routine procedure for differentiating hydrocarbon and oxygenates is not available at this time. However, the amount of oxygenated material in combustion products is typically only a small fraction of the hydrocarbon. While this is true of current automobiles, the ratio of oxygenates to hydrocarbon in emissions from future systems cannot be predicted and the oxygenate response in measurement of hydrocarbon by FID should not be neglected.

c. Calibration. The ion production of the hydrogen flame in an FID can be made linear with hydrocarbon concentrations over a very wide dynamic range. Therefore, a well-designed FID will have an output that is linear over the range of concentrations for which it is intended, and one upscale calibration point that is referred to an accurately established zero setting on each range will set the instrument calibration. In practice, it is generally preferable to calibrate each range at two or more upscale concentrations that provide approximately equal increments over the range. Aside from these generalities, calibration requirements will vary, instrument to instrument, depending upon the precision that is built into the instrument's range-changing function and upon other design and operational characteristics. Therefore, the pertinent generalization is that this class of instrument should provide essentially linear output; calibration and readout procedures can take advantage of this characteristic.

d. Application. The FID is insentitive to water vapor and hence sample need not be dried. However, adequate precaution should be taken in introducing sample to ensure that neither liquid nor particulate matter enter the instrument. Either of these may foul the small diameter tubing and orifices that serve critical metering functions. Both fuel and air supplied to the instrument should be filtered, and the gases should be of a purity to give the degree of stability and accuracy that is required. In this regard, it should be noted that hydrocarbon in either the fuel or air supply will produce ion current additional to that produced by hydrocarbon in the sample. The purity that is called for in the fuel and air will depend upon the application, but for precise measurement, these gases must be virtually free of hydrocarbon; such gases are commercially available at modest cost.

Because the detector is sensitive to sample flow, care must be taken to ensure that the design, i.e., specified, sample flow rate is not exceeded. To do so, may move the amplifier/output system into a saturation or nonlinear range.

Typical instruments require operating gases at supply pressures under 50 psi. The gases usually are supplied from high-pressure bottles, although the air supply may be taken from a general air system with appropriate care for drying and purification, as may be necessary. Electrical requirements are nominal.

2. Nondispersive Infrared

a. *Principle.* The nondispersive infrared (NDIR) gas analyzer utilizes as its detection principle the absorption of infrared energy by a quantity of the gas to be analyzed. "Nondispersive" refers to the fact that the energy involved is not resolved into discrete spectral levels but, instead, involves broad but distinctive adsorption regions within the infrared.

Instruments most commonly in use for emission measurement are based upon a principle of differential absorption of energy from two columns of gas — one the gas to be analyzed and the other a gas of invariant composition, free of the component of interest, and relatively nonabsorbing in the IR. These gases are contained in, respectively, a "sample" cell and a "reference" cell. Infrared energy sources are positioned at one end of each cell and energy detectors at the opposite ends. Windows that are transparent to the infrared provide an optical path from the sources, through the cells, and into detectors that are responsive to the energy of wavelengths that are absorbed by the component to be measured. (In the following discussion, the component to be measured is designated "component X." The energy of wavelengths that are absorbed by component X will be referred to as "X-absorbed energy.") In principle, the sources supply a constant infrared energy input to the gas cells, and the detectors respond to the X-absorbed energy that is transmitted through the cells. These amounts of transmitted energy are compared instrumentally, and the difference — attributable to absorption by component X in the sample cell — is translated as concentration of X in the gas sample.

b. *Operational Characteristics.* In the most commonly used NDIR instruments the sample and reference infrared energy detectors are combined in a differential optical detector (Fig. 4). This differential detector has two cavities, essentially identical, that are charged (at subatmospheric pressure) with the gas component to be measured. These cavities are separated by a

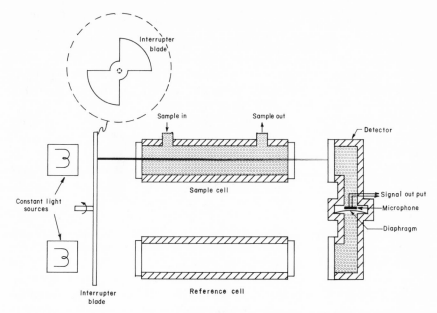

Fig. 4.   Operational Elements of the NDIR analyzer.

thin flexible diaphragm and connected only by a small orifice that permits long-term pressure equalization. By virtue of its charge, the detector absorbs energy in precisely the same spectral bands as are absorbed by the measured component. This characteristic is used in the following way: Gas in the reference half of the detector receives from its source through the reference cell an invariant level of energy of the wavelengths it absorbs; like bands of energy that reach the sample side of the detector are modulated by absorption of energy in the sample cell according to the amount of absorbing gas in the sample. Thus, to the extent that the sample beam is modulated by the absorbing gas, the detector "sees" unequal amounts of energy in its reference and sample sides. To detect and transduce this inequality, the source energy is pulsed by using a rotating beam interrupter. This interruption results in cyclic absorption of energy in the detector cells; which also is to say, cyclic heating of the gases. If there is an energy inequality, the heating effect differs between the sample and reference cavities and the thin diaphragm separating the two is caused to flex as the two gas masses are heated unequally. This flexure is detected and measured by its electrical effects upon a capacitance system, and the resultant output signal is thereby related to the concentration of absorbing gas, component X, in the sample.

The type of detector just described is known as a Luft detector; it has been, and is now, used widely. However, because of its complexity and fragile microphonic character, it may be expected to give way to a more rugged, simple system of detection, for example, one employing solid-state energy detectors.

The amount of X-absorbed energy that is removed from the infrared beam in the sample cell depends upon the concentration of X, its absorbance in the infrared, and upon the depth of the gas column through which the energy beam passes. The depth of gas column obviously is controlled by the sample cell length; this length, therefore, is selected as required to obtain the desired signal output at the concentration to be measured. In instruments commonly used, sample cell lengths vary from about 1/8 to 15 in. Special instruments employ cells of 30 in. or more.

The energy absorption and signal output of NDIR instruments are responsive to the number of molecules of the absorbing gas through which the energy beam passes.* Therefore, sample flow rate can indirectly influence signal output through its effect on pressure (and gas density) in the sample cell. This effect does not compromise instrument accuracy provided it is recognized in the calibration procedure; the primary consideration with respect to sample flow rate, therefore, is to ensure that calibration and sample measurement are made with equal absolute gas pressures in the sample cell. This is done most simply by using identical flow rates for calibration and for sample gases with the exit restriction the same in both cases.

Sample flow rate also affects speed of response of the instruments through an obvious relationship in which the time rate of system volume replacement is dependent upon the gas flow rate.

One of the more serious difficulties, or limitations, in the use of NDIR arises from the fact that most gases absorb in infrared regions in common with other gases. This results in interference from any such gases that are present in the sample; and the fact is that most of the gases to be measured in emissions have absorption in the infrared in common with water and/or one or more of the other gases present. One method for minimizing such interference is to confine the transmitted energy to a wavelength band (or bands) in which absorption is limited to the component of interest. Solid optical filters offer this possibility, and great advances have been made in recent years in producing filters that permit a high degree of selectivity in designing the infrared beam-wave pattern. They are being used to good advantage and with increasing sophistication to provide high specificity in NDIR measurement. Transmission characteristics of the filters obviously will vary with the application, and instrument manufacturers choose filters compatible with the

---

*The partial pressure of the absorbing gas also influences the energy absorption in a phenomenon known as "pressure broadening" — as pressure increases, the absorption bands tend to broaden and definition may tend to deteriorate with parallel loss of interference discrimination. This effect is insignificant with pressure effects attributable to gas flow, but can be quite significant in attempts to increase instrument sensitivity by deliberately pressurizing the sample cell.

energy source and other optical elements with which the filter is mated. Because of the great variability in design and in choice of optical filters, generalities concerning spectral characteristics may be misleading and therefore none are attempted in this review.

In another approach to lessening interference, an "interference filter cell" filled with the interfering gas, or gases, is used in series with the sample cell. This cell, in principle, is a gas optical filter that would remove from the sample beam those wavelengths at which the interfering gases absorb in common with the component of interest. If this removal were complete, appearance of the interfering gases in the sample could then have no effect upon the measurement. Unfortunately, this technique of interference control is only marginally effective in many cases because the interference filter may in fact operate with poor efficiency in removing those wavelengths in which the sample and interfering gases absorb in common. Nevertheless, although interference filter cells may have more or less severe limitations, they sometimes are useful and they are found in practice.

c. Calibration. The fundamental response of the NDIR instrument is related to the concentration of the measured gas in a relationship described by an exponential rather than a linear function. In this relationship the increment of signal per increment of concentration tends to decrease as concentration increases. For very narrow ranges of concentration with only a small fraction of the availble IR energy absorbed in the sample, a linear relationship may be used without serious error. However, in most automotive emission measurements a suitably linear relationship does not hold over a practical choice of ranges.

Two choices are available in interpreting the output. (1) Refer the output to a calibration curve, or alternatively, use an equation that fits the curve and relates concentration to output, or (2) electronically linearize the instrument output so that a simple multiplier relates concentration to output. With the increasing use of automatic data acquisition, and particularly for continuous measurement of emissions, the latter is gaining favor. Expectation is that instruments will become generally available with linearized output.

d. Application. 1. Measurement of Carbon Dioxide. CO_2 absorbs strongly in the infrared, and no other exhaust components interfere at the concentrations that are present. Because of the very strong absorption of CO_2, short sample cells are adequate. In fact cell length must be appropriately restricted to prevent saturation of the signal which results when all, or nearly all, of the available energy is absorbed in the sample cell with only a small amount of CO_2 present. For undiluted exhaust gas a 1/8 in. length of sample cell is typically used; this length may be increased directly as the concentration of CO_2 is decreased in diluted sam ples. An interference filter is not necessary for the CO_2 measurement.

2. Measurement of Carbon Monoxide. Measurement of CO at concentrations of 1.5% or more in exhaust presents no difficulty. Typically, for
this application a 1/4 in. sample cell is used. Water, CO_2, and ethylene
interfere, but the ethylene interference generally is not significant and the water
and CO_2 interferences can be handled satisfactorily using either an optical
filter or an interference filter cell (about 5 in.) filled with CO_2 saturated with
water vapor at about 50°F. Interference problems become seriously troublesome in attempting to measure CO in dilute exhaust at CO concentrations
below about 0.2% — which approaches the concentration level permissible
by U.S. Federal air pollution standards required for 1972 and later. For
measurements below this level, an optical filter is required for satisfactory
discrimination against the interfering gases. In lieu of an optical filter,
interferences can be decreased by continuously flowing through the reference
cell a stream of air taken from the exhaust diluent supply (see Sec. II. A.
2, Constant Volume Sampling). In this arrangement the principal interference,
water vapor from the diluent air, varies equally in the reference and sample
beams, and therefore no differential attributable to background water is
generated. (Recall that the output signal depends upon the differential of
absorption in sample and reference cells.) While this arrangement that com-
pensates background (or diluent) interferences may give useful results, the
interferences from other exhaust components remain serious. These are more
effectively discriminated using an optical filter, and for reliable low-level
measurement such a filter is highly recommended.

3. Measurement of Nitric Oxide. NO absorbs only weakly in the infrared
and, in addition, CO, CO_2, and water vapor interfere seriously. The NO
measurement requires careful attention to these interferences at all levels of
measurement and an optical filter is a virtual necessity for acceptable
discrimination. For measurement at concentrations less than about 300 ppm
NO, sample drying is recommended for satisfactory discrimination of the
strong water vapor interference. This may be done either using an ice bath or
chemical agents. Some of the commonly used driers have been found to
perturb the NO content, and use of such materials should be limited to
nonindicating Drierite or such other agents as the user determines to have
no effect on the NO. (A small amount of indicating Drierite is acceptable as
needed to show progression of the spent agent, but its use should be held to
a minimum.)

Because NO absorbs weakly, the maximum available length of cell is
generally indicated for standard instruments. This requirement for maximum
sample cell length, plus questionable advantage of a gas interference filter,
practically rules out use of an interference filter cell in NO measurement.

Used with care, NDIR is suitable for NO measurement at levels below
50 ppm, but the capability of standard instruments in practical use is reached
at about that level. Further refinements in the adaptation of NDIR to NO

measurement could extend the application, but unless such refinement is brought about, it is expected that the chemiluminescence method will replace NDIR for this measurement.

4. Measurement of Hydrocarbon. Unsaturated carbon–hydrogen linkage does not have an IR absorption that can be used effectively in conjunction with the absorption of saturated hydrocarbon. As a result, the measurement of exhaust hydrocarbon by NDIR has evolved such that it reflects, in general, only the hydrocarbon with saturated carbon–hydrogen linkage. The measurement is, in fact, referred to the IR absorption of hexane, which is the sensitizing gas of the hydrocarbon detector. In the NDIR measurement, olefinic hydrocarbon responds with a sensitivity reduced depending upon the degree of unsaturation in the molecule; aromatic hydrocarbon produces very little response (compared to that of an equal quantity of carbon in a paraffinic hydrocarbon).

As in the case of CO and NO, water interference is serious but manageable at hydrocarbon levels that are typical of those permitted in vehicle exhaust through the early 1970's. However, the NDIR measurement does not offer the requisite sensitivity for measuring the hydrocarbon levels expected in emissions from autos of model years later than the early 1970's. Both because of inadequate sensitivity and nonuniformity of response, the NDIR method will be replaced by FID (discussed in Section I, B, 1). In actuality, by 1971 FID superseded NDIR for virtually all measurements in research and development; by 1973 FID will have largely superseded NDIR in all U.S. emission measurements. Indications from other areas of the world are not so clear, but it is difficult to foresee continued use of NDIR in view of the clear advantages of FID for the hydrocarbon measurement.

5. Chemiluminescence for Measurement of Oxides of Nitrogen. The method of chemiluminescence utilizes the reaction of NO with ozone to produce NO_2 at an excited state. The excited molecule spontaneously relaxes to the unexcited state with the release of a discrete quantity of photo energy. Measurement of this energy provides a measure of the NO_2 and hence the NO involved in the reaction. The method is very sensitive compared with current or anticipated levels of NO in auto emissions at maximum dilutions that are expected in any emission measurement procedure. While the method is specific for NO, NO_2 (and equilibrium N_2O_4) can be converted to NO by a catalytic/thermal process and the total of NO_x can be measured as NO. The development is very recent and details of the instruments that will evolve cannot be reliably predicted. The principle of operation and results with a first-generation instrument have been described.[1,2]

The initial development of the instrument (and instruments currently available commercially) involved reacting the NO with ozone and measuring the luminescence at a pressure of about 5 to 12 torr. A chilled photodetector has been used also to augment sensitivity and stability. Further development is expected to enable the measurement to be made at ambient pressure and at near-ambient temperature. The principal elements of the analyzer probably will be retained in refined instruments, but rapid change from the details of these early instruments is to be expected.

By its nature the method is flow-sensitive — the output signal being directly proportional to the product of sample flow rate and NO concentration. Accurate flow-rate control is therefore essential to accuracy of measurement. The amount of sample that is required is very small, typically less than 60 ml/min at ambient pressure.

The ozone required in the reaction may be generated in an ozononator, furnished as part of the system, using either oxygen or air. Ozone that is excess to the reaction must be destroyed, both to prevent its deleterious effect on downstream components of the system and to avoid its toxicity in the discharge stream. This destruction may be accomplished by passing the ozone over charcoal on which it is irreversibly adsorbed and reconverted to oxygen or combined with other adsorbed organic material. These current practices with respect to supply and disposal of ozone may be expected to change substantially with refinements, but the basic elements of the problem probably will remain inherent to the instrumental principle.

As indicated above, NO_2 in the sample must be reconverted to NO before it will respond in the measurement. This conversion has been approached by heating the sample to force the NO_2:NO equilibrium in the direction of NO; at a temperature of about $600\,^\circ F$ the conversion appears to yield about 90% of the NO_2 as NO. Catalytic action in the process has not been adequately described at this time. One problem in the conversion is possible reduction of the NO by CO that may be present in a sample containing little oxygen. However, this reduction phenomenon has little significance in measuring dilute exhaust containing CO concentrations that are now typical of auto emissions.

If NO and NO_2 are to be determined separately, the NO measurement is made upon sample that bypasses the converter and NO_x is determined upon sample that is passed through the converter prior to measurement. The NO_2 measurement is by difference.

6. Nondispersive Ultraviolet (NDUV) for Measurement of Nitrogen Dioxide. NO_2 absorption in the infrared is inadequate for satisfactory measurement by NDIR. In the ultraviolet, however, NO_2 does absorb sufficiently to enable its measurement by NDUV at concentrations above about 5 ppm. The instrument is similar in principle to the NDIR except that the UV energies are measured using a photocell instead of the Luft detector. Above 200 ppm, accuracy of $\pm 1\%$ of full scale is achievable. Below 200 ppm, accuracy can be within about $\pm 2\%$ of full scale or ± 2 ppm, whichever is the larger uncertainty (but such accuracy requires very careful attention to instrument calibration and alignment).

Inasmuch as the chemiluminescence method for NO_x appears likely to obviate the need for a separate NO_2 measurement, the NDUV method is expected to have little further application in standard emission measurements. Even so, the NDUV instrument should be recognized as useful and reliable for NO_2 measurement at concentrations in exhaust gases as low as 5 ppm. Therefore, in special applications not requiring standard instrumentation, existing NDUV instruments may be used to advantage and should not be assumed to be technically obsolescent.

C. Wet Chemistry for Aldehyde Measurement

Explicit information on the hydrocarbon-derived oxygenated materials (oxygenates) in engine exhaust is inadequate to assess fully the relative importance of the several classes of combustion product in this category. It is known that the oxygenates that typically are present include a broad spectrum of aldehydic and nonaldehydic carbonyls, of which the aldehydes probably are the most important, quantitatively. Analytical measurements have thus far been focused upon the aldehydes and they are the only class of oxygenates for which routine methods are now available. These routine methods, however, are neither rapid nor simple, nor do they provide a measurement of all the aldehydes present in exhaust gas.

A suitable instrumental method is not available, and therefore the aldehydes measurement is thus far confined to wet chemistry; the chromotropic acid procedure is used for formaldehyde and the MBTH procedure for "aldehydes" — a generalization often assumed to mean the total of aldehydes present. The assumption, if made, is erroneous because, as now widely recognized, the method does not include aromatic aldehydes and may measure only partially the C_5 and heavier aldehydes. For a more complete determination of aldehydes, the DNPH method is used. However, the method is tedious and generally unsuitable for other than research use.

1. Chromotropic Acid Method [3,4]

The chromotropic acid method is based on the production of a measurable color resulting from the reaction of formaldehyde with 1,8-dihydroxy-naphthalene-3,6-disulfonic acid (chromotropic acid). In the procedure, a known volume of exhaust is passed through a fritted bubbler containing a 0.1% solution of chromotropic acid in sulfuric acid. A stable color suitable for spectrophotometric measurement at 580 mμ is produced upon heating the solution for 30 min at 100°C. The method is relatively free from interferences. The most serious interference (large amounts of aromatic hydrocarbons in the sample) can be decreased greatly by using 1% sodium bisulfite as the absorbing solution. [3]

2. MBTH Method[5]

The MBTH method depends on the production of a blue dye resulting from a series of reactions involving an aldehyde, 3-methyl-2-benzothiazolone hydrazone (MBTH) and an oxidizing agent. A known volume of exhaust is pulled through an impinger containing 10 ml of a 0.4% aqueous solution of 3-methyl-2-benzothiazolone hydrazone hydrochloride. After a 1-hr waiting period, the solution is rinsed quantitatively into a 100-ml volumetric flask with 25 ml of a 1% solution of ferric chloride in water. After another waiting period of 20 min, the mixture is diluted with acetone to 100 ml. The absorbance at 635 mμ is then determined by comparison with a blank and the aldehyde concentration is read from an appropriate calibration curve.

3. Colorimetric DNPH Method[6]

The DNPH method is suitable for the determination of a broad range of aldehyde and ketone compounds. In this method, a known volume of exhaust is pulled through two scrubbers in series containing the 2,4-dinitrophenyl-hydrazone (DNPH) reagent.

The carbonyls react with the DNPH to form a precipitate which is filtered, washed, dissolved, and combined with a hexane extract of the original filtrate. The combined solution is then diluted to an appropriate concentration. An aliquot is then diluted with pyridine and KOH in water and methanol. The absorbance of the resulting red solution is read at 440 mμ and compared with a suitable calibration curve.

II. AUTOMOTIVE EMISSIONS MEASUREMENT

A. Sampling Methods

1. Total Sample Collection

The most straightforward method of exhaust gas sampling entails capture and collection of the total quantity of exhaust gas that is produced during a period of test. Although the method is simple in principle, it does pose substantial operational problems in containing and handling the large volumes of gas that are involved. Other deficiencies relate to residual contamination, to problems in maintaining the sample integrity during collection, and to the time required in the batch process by which the sample is collected and measurements are made upon it. Such measurement involves the determination both of pollutant concentrations and of gas volumes.

In the procedure, a plastic bag is used to collect the sample which must be cooled in order to maintain an acceptably low temperature for the plastic film. Cooling also serves to remove some of the water vapor. For measurement of pollutant concentrations, the bag is thoroughly agitated and sample is drawn from the total volume of gas that was produced during the test. After concentration measurements are completed, the bag is emptied through a suitable gas-volume measuring system to determine total sample volume. Because of its basic simplicity, this method of bag collection will probably be continued for some applications. However, in standardized testing it is being replaced by the constant-volume sampling system next described.

2. Constant-Volume Sampling (CVS)

This method was termed "variable-dilution sampling" in an early description[7] and was later termed "constant-volume sampling" (CVS). In this method the full stream of exhaust being produced is continuously directed into the intake of an air pump that, operating under the conditions prescribed, has a constant discharge rate. A free supply of diluent gas, typically air, is also continuously available at the pump intake. As required to satisfy pump demand, the pump draws diluent from this supply in an amount such that the total flow of exhaust plus diluent is at all times a constant. The capacity of the air pump must be something greater than the maximum instantaneous exhaust flow but should not greatly exceed the exhaust maximum or the resulting overall dilution may be undesirably high. A simplified system is shown schematically in Fig. 5.

It can be shown that a sample drawn at any constant-volume rate from the exhaust/diluent stream described above contains a true compositional aliquot of the total volume of exhaust that was produced during the sampling interval. This will be true irrespective of variations in the exhaust flow. In brief then, the CVS method requires only (1) that the exhaust be mixed in a

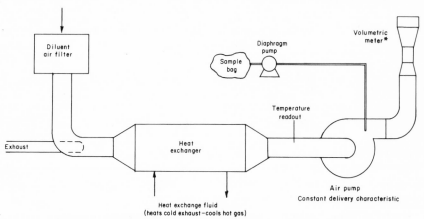

Fig. 5. Simplified schematic of constant-volume sampling system.
*If a positive displacement pump is used, a revolution counter serves to register diluted exhaust volume.

flowing stream with diluent continuously varied in amount as required to maintain a constant flow rate of exhaust plus diluent, and (2) that the dilute stream be sampled at a constant rate for the period of the test. The resulting sample is a diluted aliquot of the total volume of exhaust produced during the interval so sampled.

To simplify explanation, the discussion above referred to a requirement that the flow of exhaust plus diluent be maintained at a constant-volume rate. Actually, to achieve true proportionality in the sampling, the flow of exhaust plus diluent must be maintained at constant *mass* rate. To achieve this in a system that is volume regulated, the gas must be handled at essentially constant temperature and pressure. For this reason, the exhaust gas is cooled prior to introduction to the system, and preferably, both exhaust and diluent should be brought to a common temperature before reaching the exhaust/diluent pump. Alternatively, the exhaust and diluent can be mixed ahead of the pump and the mixture controlled at a level temperature. Variations from this "level" temperature should be held within about $\pm 10\,°F$ if proportioning errors are to remain negligible. Another requirement in operating the system is that the exhaust sample be diluted enough to reduce the moisture content of the mixed sample to a dew point well below the operating temperature of the sampling system. Another obvious problem is diluent air contamination, which is difficult to differentiate from pollutants contributed by the exhaust. In principle, corrections can be made for diluent contamination if the contamination levels are determined by separate measurements upon the diluent; in practice, it is preferable to obtain clean diluent air. Chilling the diluent to about $32\,°F$ or below and passing it over charcoal at an equivalent temperature is an excellent means for reducing both moisture and organic contamination.

The exhaust diluent pump can be any that has a suitable constant-delivery characteristic. Positive displacement pumps of the Roots type are used extensively, but a centrifugal blower can be used satisfactorily in an appropriately designed system. Suitability of a centrifugal blower hinges upon a system design that avoids pressure differentials of magnitude sufficient to cause significant variations in the pump delivery.

The CVS method is inherently well suited to continuous measurement of pollutants and to real-time automatic computation of pollutant weight. In addition, dilution of the exhaust with suitably conditioned diluent air serves to arrest interaction of exhaust products and, as mentioned above, prevents condensation of water vapor from the exhaust sample. Prevention of water condensation is essential if oxygenates are to be measured and is desirable as a general aid in minimizing adsorption–desorption phenomena in the sampling system.

As previously described, the sample upon which emission measurements are made is withdrawn at constant rate from the dilute exhaust stream. Measurements may be made continuously on the sample stream, or, alternatively, sample may be collected over the test period and measurements made upon the composite sample. If measured continuously, the continuous signals are integrated and time-averaged to obtain measurements of average concentrations of the pollutants in the dilute mixture. These concentration data are combined with total gas-volume data to yield measurements of the respective pollutants in whatever terms are desired.

One of the more advantageous features of the CVS system is the ease and reliability with which the required gas-volume measurements are made. If a positive displacement type of pump is used, its delivery can be calibrated against pump revolutions. This revolution count, easily obtained, then provides an accurate measurement of pumped volume.

If a centrifugal pump or blower is used, the discharge rate is easily and accurately monitored by use of a calibrated orifice, air nozzle, or venturi. The venturi is recommended because its characteristic of pressure recovery lessens total system pressure drop. Because the discharge rate remains essentially constant over the sampling period, a continuous reading is not required, but is useful in detecting operating problems that may cause unacceptably large variations in the normally constant delivery.

A CVS system is required in the Federal test procedure for measuring emissions from light-duty vehicles beginning with the 1972 model year. Briefly, that Federal procedure calls for the following equipment items:

(1) A positive displacement unit to pump the dilute exhaust mixture.
(2) Dilution air filters, both charcoal and particulate.
(3) A heat exchanger to maintain the exhaust/diluent mixture within $\pm$ 10°F of a set point (unspecified) during the entire period of a test.
(4) Temperature sensors, pressure gauges, and other flow-regulating and measuring accessories required to measure the amount of gas mixture handled during the test.

The system is described in detail by the U.S. Environmental Protection Agency.[8]

3. Variable-Rate Proportional Sampling

In variable-rate sampling (VRS), sample is withdrawn directly from the exhaust stream at a point well within the exhaust system where back-contamination (or dilution) with air is negligible. At all times, the sample withdrawal rate has a fixed proportionality to the exhaust flow rate; as exhaust flow changes, the sampling rate is automatically varied accordingly. In this manner the total volume of sample collected will be a fixed fraction of the total volume of exhaust produced during the sampling period and will be a true aliquot of that volume. Instrumentally, VRS is accomplished using

(1) An air rate sensor to provide signal proportional to engine air-intake rate.

(2) A sample flow sensor to provide signal proportional to sample withdrawal rate.

(3) A control valve that regulates the rate of sample flow.

(4) An electrical network to compare air intake and sample rates, and to generate an error signal whenever sample and air rates are incorrectly proportioned.

(5) A servo system, operating on the error signal, to correctly position the sample-rate control valve.

(6) A sample collection reservoir, typically a plastic film bag that provides variable capacity at ambient pressure.

In the VRS system that has been used most widely,[9] the intake-air rate is sensed using a laminar airflow element installed as a part of the engine air cleaner; sample flow is sensed using a laminar flow element made of small-diameter tubing. Pressure differentials across the respective flow elements are converted to electrical signals using sensitive capacitance-type pressure transducers, and these electrical signals are continuously and automatically compared to generate an error signal that is the primary input to the sample-rate regulating system. The sample-rate regulating valve typically is a screw-type metering valve controlled by a servo motor that operates to maintain a set balance between the air intake and the sample-rate signals.

Water in the exhaust sample is either removed in a water trap ahead of the proportional sampler, or, alternatively, is maintained as vapor through the sampling system by heating the entire sample-handling system to a temperature above the dew point, e.g., 150°F.

Early experience with the proportional sampler showed that the composition of exhaust gas tended to change during and after its collection. The change was found to result from reaction of the products with residual oxygen and parallel or subsequent reactions involving the oxides of nitrogen. The rate of such reaction is markedly slowed if the exhaust sample is mixed with dry nitrogen as it is collected; in practice, this is done by precharging the sample collection bag with dry nitrogen in an amount equal to four to six times the volume of the exhaust sample that will be collected. Such dilution serves both to inhibit reaction of the products and to lower the vapor pressure of the water in the sample to below the dew point. Typically, the sample is collected at ambient temperature.

Begun in development in about 1960, by 1965 VRS systems had been refined to provide high reliability in accurately proportioning sample acquisition even during sharp engine transients. However, problems in monitoring engine air-intake rate and the overall complexity of the sampling procedure and associated instrumentation served to limit use of the procedure largely to research operations. Although little used today, the VRS remains as a reliable and practical method of sampling exhaust gas in cases in which bag collection may be inappropriate and the CVS system either inappropriate or unavailable. The method is adaptable for use in road testing and has been used quite satisfactorily in that application.

B. Quantifying Emissions

The measurement of a gaseous pollutant in exhaust almost invariably is made in terms of concentration relative to the total of the mixture — generally volume concentration is used. Also, contamination levels of gaseous atmospheric pollutants are logically considered in terms of relative concentrations. Against this background, there evolved the practice in which emission measurements were based on relative rather than absolute concentration or weight values. Thus, government standards — first the State of California and later U.S. Federal standards — expressed permissible emissions in terms of concentration relative to other gases in the exhaust gas; e.g., 275 ppm hydrocarbon and 1.5% CO.

Since air that is excess to the combustion requirement dilutes the combustion products, it became the practice in measurements upon automobiles to refer concentration measurements to the oxygen-free combustion product volume. This is done with relatively little error for dry gasoline exhaust by assuming that the yield of carbon dioxide and carbon monoxide from complete combustion of a stoichiometric mixture will approximate 15 volume % of the total combustion product. Historically, therefore, emission concentrations have been "corrected" to a standard 15% for the total of [CO + CO_2 + unburned carbon atom], each expressed as percentage. If, for example, the measured CO, CO_2, and hydrocarbon are, respectively, 0.5%, 9%, and 300 ppmC, the correction factor would be 15 divided by [0.5 + 9.0 + 0.03], or 1.57. The "corrected" values of the emissions would be, respectively, 0.8% CO, 14.2% CO_2, and 470 ppmC hydrocarbon. Details of this correction are given in Ref. 10. A subsequent revision[11] was intended to make the correction more equitable in its treatment of exhaust from lean-combustion processes. Differences between the two corrections are not large.

In brief, any percentage (or ppm) measurement should be interpreted with consideration for the overall composition of the measured sample. Thus, for example, turbine exhaust may appear to have very low pollutant levels because the combustion product at the point of discharge from the engine is, in effect, diluted with 3 to 10 equivalent volumes of air. When corrected for this dilution, the levels may not appear so drastically low.

Although relative concentration is the historical basis of reference for emission measurements, the significance of relative concentration, per se, in an exhaust stream is lost entirely when the pollutant is transferred to the atmosphere. In the context of consequence to pollution, the cumulative total of pollutants from all sources is the only measurement that is relatable to an atmospheric effect. Here, then, the significant quantity is *amount* because an absolute value is required to answer the question, "How much pollutant is added to the atmospheric burden?" For this reason, the basis for emission measurements more recently has become weight, and relative concentration values are only one factor in the final expression.

Emission weight is readily calculated from concentration and gas-flow data, using either molecular volumes and molecular weight data or, more directly, gas volumes and specific weights. Note that for calculating weights of emission from diluted samples with the total volume of dilute gas being known, as in the CVS procedure, concentration data are applied directly to the total volume of dilute gas being measured. No reference to actual exhaust-gas flow is needed, nor is there need for reference to the concentration of pollutants as they actually appear in the undiluted exhaust. These calculation procedures are straightforward and readily deduced; precise details may also be obtained from applicable regulatory rules if such are involved in the testing (see the following section on "Test Procedures").

A method of emission measurement that is yet little used, but which offers attractive advantages in some applications, involves calculations based on the carbon balance between fuel burned and the exhaust products. No gas volume measurement is needed. Briefly, the method requires (1) a measurement of weight of fuel burned during the test interval; (2) relative concentrations of CO, CO_2, and hydrocarbon in the exhaust; and (3) for refined calculation, an estimate of the carbon/hydrogen ratio in the fuel. Given these data, the weight of any component containing carbon is computed on the basis of its carbon weight as a fraction of the total carbon in the combustion products. The total carbon in the product obviously remains equal to that in fuel burned; the value is determined from fuel weight and from its carbon/hydrogen ratio. The weight of noncarbon emission, e.g., NO or NO_2, is determined on the basis of its volume concentration relative to a prominent carbon compound, with appropriate correction for molecular weight.

The following equations may be used to estimate weights of emissions using the fuel and concentration data as discussed above. More nearly exact values may be obtained from equations presented in Ref. 12, but the approximations generally will give results within $\pm 2\%$ of values obtained by the more rigorous treatment:

$$
\begin{aligned}
W_{CO} &= 2 \times 10^{-4}\,(CO) \div Q \\
W_{HC} &= 1 \times 10^{-4}\,(C) \div Q \\
W_{NO}\ (\text{as } NO_2) &= 3.29 \times 10^{-4}\,(NO) \div Q \\
W_{NO_2} &= 3.29 \times 10^{-4}\,(NO_2) \div Q
\end{aligned}
\tag{2}
$$

where

W_x = emission rate of component X, lb/lb of fuel
Q = 10^{-4} (CO + C) + CO_2
CO = concentration of CO in exhaust, ppm
CO_2 = concentration of CO_2 in exhaust, %
C = concentration of HC in exhaust, ppmC
NO = concentration of NO in exhaust, ppm
NO_2 = concentration of NO_2 in exhaust, ppm

Choice of this method for quantifying emissions will depend upon the ease and accuracy with which fuel consumption can be measured and upon the precision of the emission measurement that provides a carbon balance. Its chief advantage lies in avoiding the problems in gas-flow measurement. When such problems are particularly troublesome, or when the measurement of gas flow is inordinately expensive, the carbon-balance/fuel-weight approach to emission quantification may be highly advantageous. An excellent discussion of this fuel-based mass measurement procedure is given by Stivender.[13]

C. Test Procedures

Procedures to be used in measuring emissions from vehicles have been set forth in governmental regulations in Japan, the United States, and Europe. Such regulations may differentiate classes of vehicles, prescribing for each class a test procedure that is compatible with the vehicular system and that reflects both the pattern of the vehicle's use and the more prominent pollution problems associated with that particular class of vehicle.

Current regulations reflect only a progression toward controls that will encompass ever-broadening aspects of vehicle pollution. In this situation the test procedures undergo rapid evolution in response to advances in technology and to changes in regulatory requirements. Thus, ever-broader categories of vehicles are covered, additional pollutants become subject to control, and the officially prescribed test routines are modified at relatively frequent intervals. Such a condition is expected to prevail through the 1970 decade, during which period the needs and concepts of regulatory practice and the design of power systems may alter markedly. This discussion of test procedures is therefore concerned only with generalities; specific provisions and details applicable at any given point in time should be obtained from the appropriate regulatory agency.

The driving cycle in test procedures for light-duty vehicles is designed to approximate the severity and time sequence of speed and power demands in typical, metropolitan traffic. The first such cycle was the "seven-mode cycle" adopted by the State of California in 1966 and later specified in U.S. Federal test procedures applicable to 1967 model automobiles. This cycle required the automobile to be driven through a progression of idle, acceleration, cruise, and deceleration modes. The driving time of a cycle was 137 sec, and, with a minor variation following the first cycle, the cycle was repeated nine times for a total test time of about 21 min. As first used in emission measurement, portions of the cycle were disregarded in calculating emissions, i.e., either measurements representing those portions were not used, or exhaust from those portions was excluded from the composite sample.

A close approximation to the California seven-mode cycle was used in U.S. procedures until a revised driving cycle was incorporated in procedures to be used with 1972 and later model automobiles.[14] That driving cycle is essentially a reproduction of an actual time–speed history of an automobile driven over a 7.5-mile "typical" metropolitan route. It requires 22 min and 51 sec in the driving.

As a matter of practical necessity, "driving" of a test is done on a chassis dynamometer. Power absorption and inertial effects of the dynamometer are set to approximate the power requirements and inertia of a vehicle driven on the road; these dynamometer characteristics are adjusted to vehicle weight.

Emission measurements either are made continuously during the tests or are made upon bag samples collected during the tests. Although the bag-collection procedure has been the one used most widely to date, continuous measurement offers an advantage in automatic data acquisition and processing. For this reason, continuous real-time measurement should be considered for new or updated equipment.

Vehicle emissions are strongly influenced by engine operating temperature and parallel choke action. Therefore, emissions during and following engine starting may be markedly different from stabilized emissions; in like manner, cold- and hot-start emissions may differ significantly. For this reason, test procedures contain specific provisions for the measurement of "cold start," "hot start," and "stabilized" emissions. Because these provisions are the subject of continuing study and undergo frequent revision, current regulations should be consulted for precise details.

The CVS system of exhaust measurement is specified in the latest U.S. procedure. Also specified are NDIR for CO and CO_2 measurement, FID for hydrocarbon, and chemiluminescence for NO and NO_2. Emissions are expressed in terms of weight (grams) of the pollutant per mile driven. Complete details of the test procedure including calculations are given in the document that sets forth exhaust emission standards for 1973 and later model automobiles.[8]

Heavy-duty vehicles, including trucks, buses, and heavy-utility or special-purpose vehicles, have thus far been largely exempt from regulation. However, some state regulations are in effect and U.S. regulations have been proposed. None as yet include a driving cycle comparable to that for the light-duty vehicles. Instead, the trend is to base emission tests upon a series of measurements during a procedure in which speed and load are changed in step fashion; emissions are measured during the stabilized portion of each step. Thirteen such steps or modes are prescribed in the test procedure for new diesel engines for use in California after 1972. Twenty-one stepped modes, which would apply to both diesel and gasoline-powered vehicles,[15] are described in the U. S. proposal for testing heavy-duty vehicles.

Control of fuel-system evaporative losses is an integral part of current U.S. auto pollution control technology. As such, U.S. procedures for emission measurement include specifications for measuring these losses.[14] Evaporative losses are measured only during a "soak" period, i.e., with the vehicle inoperative, because it may be assumed that the vapor-recovery system, by design, ingests all vapor evolved during the driving cycle and continuously directs it into the engine intake for consumption in the combustion. Two basically different approaches to measurement are used. One, in effect, closes all vapor escape from the fuel system (including the carburetor) except for an opening into a charcoal-packed, vapor-collection canister, During a prescribed test, vapors from the fuel system are thus collected and weighed. In the other approach, the vehicle is in a vapor-tight enclosure during the period of measurement. Purge air circulated across the vehicle entrains the evaporative-loss vapors which subsequently are measured in the discharge stream of the purge air. To establish appropriate vehicle and fuel-system temperatures, and to establish appropriate initial conditions in the vehicle vapor-recovery system, the automobile is operated through a prescribed routine of soak and run periods in advance of the measurement period.

III. MEASUREMENT OF EMISSIONS FROM DIESEL AND TURBINE ENGINES

A. Carbon Monoxide and Carbon Dioxide by NDIR

Both turbine and diesel engines typically operate at idle and low power with air–fuel ratio as lean or leaner than 150:1. The excess air in the exhaust under these conditions serves to dilute the combustion products and poses problems in measuring the pollutants that are at relatively low concentrations. However, the sensitivity problems are similar to those found in comparable measurements on auto exhaust sampled using a CVS system (in which the exhaust may be diluted as much as 10:1). Therefore, the discussion (Section B.2) of auto exhaust measurement by NDIR is largely applicable to diesel and turbine measurement.

CO_2 concentrations in these exhausts typically range between 1.5 and 12%; CO between 200 and 2000 ppm. The CO_2 measurement presents no problem; the CO measurement does require careful attention to interferences. With that attention, and particularly if an optical interference filter is used, the CO measurement can be made with little difficulty. While adequate sensitivity is not a problem, the wide range of concentrations to be covered calls for multirange instruments to provide requisite accuracy and acceptable readability at different concentration levels.

For diesel emission measurement, the following are recommended by the Society of Automotive Engineers (SAE)[16]:

Component	Instrument range
CO_2	0–5% and 0–16%
CO	0–0.1%, 0–0.5%, and 0–2.5%
NO	0–1500 ppm and 0–6000 ppm

For turbine emission measurement, the SAE recommends[12]:

Component	Instrument range
CO_2	0–2%, 0–5%, and 0–15%
CO	0–100 ppm, 0–500 ppm, 0–0.25% and 0–2%*
NO	0–50 ppm, 0–200 ppm, and 0–1000 ppm
NO_2	0–50 ppm, 0–200 ppm, and 0–500 ppm

B. Hydrocarbon by FID

Hydrocarbon is measured using flame ionization in a system that utilizes heated components throughout the sample flow path. This heated system (approximately 375°F) is needed to prevent condensation or excessive adsorption of the hydrocarbon. (In diesel or turbine exhaust, the hydrocarbon is much heavier, i.e., of higher molecular weight, and much less volatile than the hydrocarbon in exhaust produced from gasoline.) One note of caution is appropriate in connection with the use of the heated sampling and analytical system — if the temperature is too high, hydrocarbon is lost in the system by oxidation, polymerization, or other reactions. Experience has shown that to avoid such losses, sample-handling temperature should not exceed about 395°F. This loss phenomenon was experimentally studied and reported by Pearsall.[17] Acceptable practice is explicitly described in Ref. 16 applicable to diesel exhaust and Ref. 12 applicable to turbine exhaust.

C. Oxides of Nitrogen

Concentration levels of NO_x in turbine and diesel exhaust range between about 20 and 1500 ppm. Until very recently, wet-chemical methods primarily were used for the measurements because the lower levels were outside the analytical range of instruments suitable for use in engine testing. Recently,

---

*Afterburning engines only.

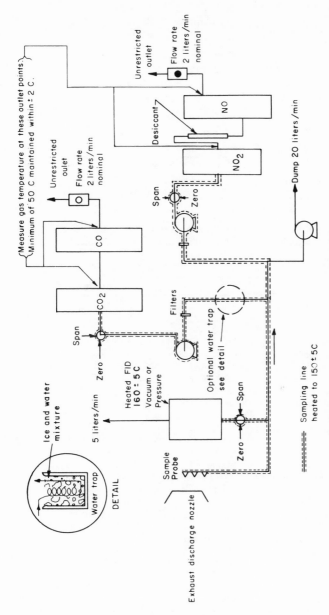

Fig. 6. Schematic of instrumentation for diesel and turbine measurement.

more-sensitive instruments have largely supplanted the wet methods, but wet chemistry is useful in some applications.

The chemical method most prominently used is the phenoldisulfonic acid procedure[18]; a modified Saltzman method[19] has also been used satisfactorily. Both are colorimetric methods requiring nothing more than a standard laboratory colorimeter as instrumental equipment.

As indicated above, instrumental methods are (since about 1970) used more generally and are recommended. NDIR is suitable for NO; NDUV suitable for NO_2. A procedure in which these instruments are used is described in the SAE publications.[12,16] Chemiluminescence promises to be an excellent method for either NO or NO_2 or for the total NO_x, but some further development and refinement of procedures in the use of chemiluminescence are needed before a procedure is described explicitly.

An instrumental package for measurement of each of the pollutants discussed above is shown schematically in Fig. 6. While not portable in the sense of being hand transportable, the instrument package can be mounted complete on a wheeled dolly suitable for use in engine test cells of typical size and configuration.

REFERENCES

1. Stuhl, F., and Niki, H., *An Optical Detection Method for NO in the Range of 10^{-2} to 10^3 ppm by the Chemiluminescent Reaction of NO with O_3*, Ford Motor Company, Scientific Research Staff Report, March 23, 1970, Detroit, Mich.

2. Niki, H., Warnick, A., and Lord, R. R., *An Ozone–NO Chemiluminescence Method for NO Analysis in Piston and Turbine Engines,* SAE Paper 710072, January 1971, Society of Automotive Engineers, New York, N.Y.

3. Altshuller, A.P., Miller, D.L., and Sleva, S.F., Determination of formaldehyde in gas mixtures by the chromotropic acid method, *Anal. Chem.* **33** (April 1961) 621–625.

4. Bricker, C. E., and Johnson, H. R., Spectrophotometric method for determining formaldehyde, *Ind. Eng. Chem., Anal. Ed.* **17** (June 1945) 400–402.

5. Sawicki, E., Hauser, T. R., Stanley, T. W., and Elbert, W., The 3-methyl-2-benzothiazolone hydrazone test, *Anal. Chem.* **33** (January 1961) 93–96.

6. Papa, L.J., Colorimetric determination of carbonyl compounds in automotive exhaust as 2,4-dinitrophenylhydrazones, *Environ. Sci. Tech.* **3** (April 1969) 397–398.

7. Hurn, R.W., Chase, J. O., and Fleming, R. D., Collecting representative exhaust gas samples, *Proceedings of the Tenth National Analysis Instrumentation Symposium, Analysis Instrumentation* — 1964, Instrument Society of America, 1964, pp. 279–285.

8. Environmental Protection Agency, *Exhaust Emission Standards and Test Procedures,* 35 FR 219, Part II, Nov. 10, 1970, as amended by 36 FR 128, Part II, July 2, 1971, pp. 12658–12661.

9. Smith, R., Rose, Jr., A. H., and Kruse, R., An auto-exhaust proportional sampler, *Int. J. Air Water Poll.* **8** (September 1964) 427–440.

10. U.S. Department of Health, Education, and Welfare, *Control of Air Pollution from New Motor Vehicle Engines,* 31 FR 61, Part II, March 30, 1966.

11. U.S. Department of Health, Education, and Welfare,33 FR 108, Part II, June 4, 1968, p. 8314.

12. *Procedure for the Continuous Sampling and Measurement of Gaseous Emissions from Aircraft Turbine Engines,* ARP–1256, October 1, 1971, Society of Automotive Engineers, Inc., New York, N.Y.

13. Stivender, Donald L., Development of a fuel-based mass emission measurement procedure, *SAE Paper 710604,* June 1971, Society of Automotive Engineers, Inc., New York, N.Y.

14. U.S. Department of Health, Education and Welfare, *Control of Air Pollution from New Motor Vehicle Engines,* 35 FR 219, Part II, November 10, 1970, pp. 17288–17313.

15. Environmental Protection Agency, *Exhaust Emission Standards and Test Procedures,* 36 FR 193, Part II, October 5, 1971, p. 19404.

16. Society of Automotive Engineers, *Measurement of Carbon Dioxide, Carbon Monoxide, and Oxides of Nitrogen in Diesel Exhaust,* J177, June 1970, Society of Automotive Engineers, Inc., New York, N.Y.

17. Pearsall, H. W., Measuring the total hydrocarbons in diesel exhaust, *SAE Paper 670089,* January 1967, Society of Automotive Engineers, Inc., New York, N.Y.

18. Beatty, Robert L., Berger, L.B., and Schrenk, H. H., *Determination of the Oxides of Nitrogen by the Phenoldisulfonic Acid Method,* Bureau of Mines Rept. of Investigations 3687, 1943, Pittsburgh, Pa.

19. Saltzman, B. E., *Determination of Nitrogen Dioxide and Nitric Oxide by the Saltzman Method,* Selected Methods for the Measurement of Air Pollutants, Public Health Service Publication No. 999–AP–11, U.S. Dept. of Health, Education, and Welfare, Cincinnati, Ohio, May 1965, pp. C1–C7.

Chapter 9

Direct-Sampling Studies of Combustion Processes

E. L. Knuth

School of Engineering and Applied Science
University of California
Los Angeles, California

I. INTRODUCTION

As has been emphasized in the preceding chapters, long-standing considerations of engine performance have been supplemented now by considerations of engine emissions. These considerations intensify the need for more detailed information on combustion processes. Information of interest includes effects of design modifications, fuel changes, and operating conditions on spatial distributions of chemical species in reciprocating-engine and gas-turbine combustion chambers and on temporal variations in reciprocating-engine combustion chambers.

Specifications for the ideal instrumentation for combustion studies might include the following broad requirements: (a) monitor all chemical species, (b) function in temperature, pressure, and composition ranges typical of combustion processes, (c) resolve spatial and temporal measurements small in comparison with system dimensions and cycle times, and (d) monitor without disturbing the (nonequilibrium) thermodynamic state of the system. Different existing instrumentation systems satisfy to various degrees these several requirements.

Optical spectroscopy has been used to follow the production and/or consumption of individual species in several studies of combustion processes in reciprocating engines. In their pioneering studies, Newhall and Starkman[1] monitored infrared emission from NO in order to determine the NO concentration as a function of time during the expansion stroke. More recently, Lavoie et al.[2] monitored infrared emissions characteristic of the two recombination continua $CO + O$ and $NO + O$, also in order to determine the NO concentration. Smith and Starkman[3] used absorption of external radiation by OH to determine its concentration. Optical spectroscopy has been used also in studies of flames in burners, combustion in pressure vessels, reactions in shock tubes, etc. It satisfies requirement d most closely and requirement a least closely. In a typical application, only one or two species (or reactions) are monitored, sometimes over a limited temperature range.

In an alternative technique, mixture samples are withdrawn from the combustion chamber for analysis under more convenient circumstances, perhaps by a mass spectrometer. An example of a system for sampling from steady-state reactive systems (e.g., from either rocket-motor or gas-turbine combustion chambers) is the system developed by Lewis and Harrison[4] and built around a water-cooled sampling probe. Systems for sampling from reciprocating-engine combustion chambers are exemplified by the system used by Daniel[5] and incorporating a timed poppet-type sampling valve. These systems meet requirement (a) relatively closely and requirement (d) relatively poorly. Composition changes due to additional chemical reactions during transport from the combustion chamber to the analytical instrument are inevitable.

The molecular-beam mass-spectrometer sampling system satisfies, for many combustion studies, the listed requirements with better balance. As shown schematically in Fig. 1, this system consists typically of a source (combustion chamber?), source orifice, source chamber, skimmer, collimation chamber, collimating orifice, detection chamber, and mass-spectrometer detector. The source gas (at pressures up to several atmospheres and temperatures up to several thousand °K) expands via the source orifice into the source chamber (with background pressures from 10^{-3} to 10^{-1} torr). The skimmer transfers the core of the free jet into the collimation chamber (with pressures from 10^{-6} to 10^{-4} torr). The collimating orifice passes those molecules flying near the system centerline into the detection chamber. The mass-spectrometer detector generates a signal proportional (by preference) to the species density in the beam at the detector. Equipment for measuring the

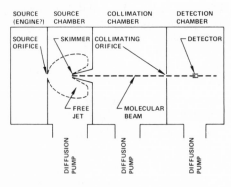

Fig. 1. Schematic diagram of molecular-beam sampling system.

beam speed may be added if one wishes to determine the source temperature from beam measurements.

The applicability of molecular-beam techniques to sampling from high-pressure systems results largely from the rapidity with which the gas temperature decreases during the expansion through the source orifice. This feature is displayed in Fig. 2, where the temperature ratio T/T_0 is plotted as a function of the dimensionless time $a_0 t/d^*$, and in Fig. 3, where $a_0 t/d^*$ is plotted as a function of the dimensionless distance x/d^*. Here T is temperature, a is sound speed, t is flow time (zero at the throat), x is axial distance (zero at the throat), d^* is throat diameter, and subscript o refers to source (stagnation) conditions. For $a_0 = 1000$ m/sec and $d^* = 0.1$ mm, a unit increase in $a_0 t/d^*$ (and a major reduction in T/T_0) is realized in only 10^{-7} sec.

Several investigators have developed molecular-beam mass-spectrometer systems for sampling from high-pressure high-temperature gases. Representative systems include those developed by Foner[6] for studies of free radicals in reactive systems, by Greene and Milne[7] for studies of the thermochemistry

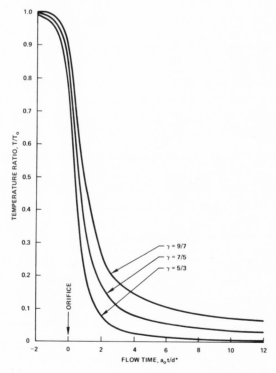

Fig. 2. Temperature ratio as function of dimensionless
flow time for free-jet gas flows from orifice.

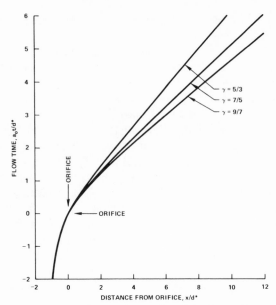

Fig. 3.   Dimensionless time as function of centerline
distance for free-jet gas flows from orifice.

of metal oxides and hydroxides in atmospheric flames, by Sturtevant and
Wang[8] for studies of shock-heated gases, by Dix[9] for studies of arc-heated
gases, by Chang[10] for studies of reactions of alkali metals and air, by
Kahrs[11] for sampling from rocket motors, and by Knuth *et al.*[12] for
sampling from reciprocating-engine combustion chambers.

In this review of molecular-beam sampling of combustion processes,
extensive efforts are made to identify and use convenient dimensionless
parameters. It is hoped that these dimensionless parameters will facilitate
not only applications to new system designs but also correlations of existing
and forthcoming data.

II. A SIMPLIFIED MODEL OF MOLECULAR-BEAM SAMPLING

In order to present the primary features of molecular-beam sampling
without the distractions of secondary (but frequently important) features,
an extremely simplified model is examined here. Procedures for minimizing
and/or handling deviations from this model are considered in subsequent
sections of this chapter.

More specifically, consider a simplified model with the following features:

1. Continuum flow is realized upstream from the skimmer entrance,
    free-molecule flow downstream from this surface.

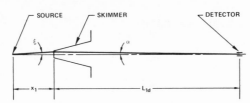

Fig. 4. Geometry and nomenclature used in analysis
of simplified sampling model.

2. The surface of transition from continuum to free-molecule flow is spherical with origin at the source orifice.
3. The distribution of molecular velocities at this transition surface is a Maxwellian distribution superimposed on spherically symmetric radial hydrodynamic velocities.
4. The source gas consists of only one species; its state is steady.
5. Molecules in the three chambers downstream from the source orifice (including molecules which have collided with the skimmer) do not interfere with the free jet and the molecular beam.
6. The detector is located on the molecular-beam centerline.

The corresponding geometry and a convenient nomenclature are indicated in Fig. 4.

The differential flux of molecules per unit solid angle in speed range from u to $u + du$ emanating from a (toroidal) differential area on the transition surface as seen from the detector may be written

$$dJ_d(u, \xi, \alpha) = u\, 2\pi x_1^2 \sin \xi \cos (\xi + \alpha) f_1(u, \xi, \alpha)\, u^2 du\, d\xi \tag{1}$$

where $2\pi x_1^2 \sin \xi \cos (\xi + \alpha)\, d\xi$ is a differential area on the transition surface as seen from the detector, $f_1(u, \xi, \alpha)$ is the number of molecules per unit volume of physical space at the skimmer and per unit volume of velocity space in the vicinity of the angle $\xi + \alpha$ from the direction of the hydrodynamic velocity, and $u^2 du$ is a differential volume in velocity space per unit solid angle. The distribution function may be written

$$f_1(u, \xi, \alpha) = n_1 \left(\frac{m}{2\pi k T_1}\right)^{3/2} \exp \left[-\frac{m(\mathbf{u}_p - \mathbf{U}_1)\cdot(\mathbf{u}_p - \mathbf{U}_1)}{2k T_1}\right] \tag{2}$$

where n_1 is the molecular density at the skimmer, m is the mass per molecule, k is Boltzmann's constant, T_1 is the free-jet static temperature at the skimmer, $\mathbf{U}_1$ is the hydrodynamic velocity at the skimmer, and $\mathbf{u}_p$ is the velocity of a molecule moving toward point P at the detector. The differential molecular density dn_d is related to the differential flux dJ_d by

$$dn_d = \frac{dJ_d}{u L_{1d}^2} \tag{3}$$

Combining Eqs. (1)–(3), one obtains

$$dn_d = \frac{2}{\sqrt{\pi}} \left(\frac{x_1}{L_{1d}}\right)^2 n_1 s^2 \exp\left[-s^2 + 2sS_1 \cos(\xi + \alpha) - S_1{}^2\right]$$
$$\times \cos(\xi + \alpha) \sin\xi\, d\xi\, ds \tag{4}$$

where $s \equiv u/(2kT_1/m)^{1/2}$ and $S_1 \equiv U_1/(2kT_1/m)^{1/2}$ are speed ratios.

Consider now the typical case in which $x_1 \ll L_{1d}$ so that $\cos(\xi + \alpha) = \cos\xi$. Then

$$dn_d = \frac{2}{\sqrt{\pi}} \left(\frac{x_1}{L_{1d}}\right)^2 n_1 s^2 \exp\left[-s^2 + 2sS_1 \cos\xi - S_1{}^2\right]$$
$$\times \cos\xi \sin\xi\, d\xi\, ds \tag{5}$$

Integrating over ξ from 0 to ξ_{max} and over s from 0 to ∞, one obtains for the molecular density n_d at the detector

$$n_d = \frac{2}{\sqrt{\pi}} \left(\frac{x_1}{L_{1d}}\right)^2 \frac{n_1}{4S_1{}^2} \left[(1 - 2S_1{}^2 \cos^2\xi_{max}) \exp(-S_1{}^2 \sin^2\xi_{max}) \right.$$
$$\times \frac{\sqrt{\pi}}{2} (1 + \mathrm{erf}\, S_1 \cos\xi_{max}) + (1 - \cos\xi_{max}) S_1 \exp(-S_1{}^2)$$
$$\left. + (2S_1{}^2 - 1) \frac{\sqrt{\pi}}{2} (1 + \mathrm{erf}\, S_1) \right] \tag{6}$$

where ξ_{max} is the half angle, with vertex at the source orifice, subtended by the skimmer. Equation (6) relates the measured molecular density (n_d) to the gas state at the skimmer entrance (n_1 and S_1) and to the system geometry (x_1/L_{1d} and ξ_{max}). For special cases, it reduces to briefer forms, most of which have appeared in previous literature.

If the speed ratio S_1 is sufficiently large and the subtended half angle ξ_{max} is sufficiently small so that $S_1 \cos\xi_{max} \gg 1$, then Equation (6) reduces to

$$n_d \approx n_1 \left(\frac{x_1}{L_{1d}}\right)^2 \left[1 - \exp(-S_1{}^2 \sin^2\xi_{max}) \cos^2\xi_{max}\right] \tag{7}$$

If, at the other extreme, S_1 vanishes, then

$$n_d = n_1 \frac{A_1}{4\pi L_{1d}{}^2} \tag{8}$$

where A_1 is the cross-sectional area of the skimmer orifice. Equation (8) describes the case of zero hydrodynamic velocity at the skimmer, i.e., an effusive molecular beam with the source located at the skimmer entrance.

Equation (7) has significant implications for selection of the skimmer-orifice diameter. It indicates that n_d increases as the orifice diameter increases (i.e., as ξ_{max} increases), but that already for $S_1 \sin\xi_{max} > 2$,

$$n_d > 0.98 n_1 (x_1/L_{1d})^2 \qquad (9)$$

where $n_1 (x_1/L_{1d})^2$ is the molecular density realized at the detector in spherically symmetric flow with no obstructions. Hence the skimmer orifice need be no larger than

$$\frac{d_1}{x_1} = \frac{4}{S_1} \qquad (10)$$

even if no pumping limitations exist.

If the portion of the free jet intercepted by the skimmer is nearly parallel flow, i.e., if $\xi_{max} \ll \pi/2$, then Eq. (6) reduces to

$$n_d \approx \frac{n_1}{\pi^{3/2}} \frac{A_1}{L_{1d}^2} \left[\sqrt{\pi} \left(S_1^2 + \frac{1}{2} \right) \frac{1 + erf S_1}{2} + \frac{S_1}{2} \exp\left(- S_1^2\right) \right] \qquad (11)$$

If, in addition, $S_1 \gg 1$, then

$$n_d \approx n_1 \frac{A_1}{\pi L_{1d}^2} S_1^2 \qquad (12)$$

whereas, if $S_1 = 0$, then Eq. (8) is recovered again. It is seen that increasing S_1 from 0 to a value much larger than unity increases the molecular density at the detector by the factor $4S_1^2$.

Application of these equations requires relating the gas state (n_1 and S_1) at the skimmer entrance to the gas state in the source. Sherman[13] considered free-jet flows into a vaccum and predicted center-line Mach numbers up to 100 for specific-heat ratios of 5/3, 7/5, and 9/7. For a free jet surrounded by a gas at nonzero pressure, these predictions apply to the region bounded by the orifice and the first shock waves encountered by the jet gas. Knuth[14] examined these predictions and suggested an equation for either interpolating or extrapolating them to other specific-heat ratios. For Mach number M_1 large in comparison with unity,

$$M_1 \approx \left(\frac{2.2}{\sqrt{\gamma (\gamma - 1)}} \right)^{\frac{\gamma-1}{2}} \left(\frac{\gamma + 1}{\gamma - 1} \right)^{\frac{\gamma+1}{4}} \left(\frac{x_1}{d^*} \right)^{\gamma-1} \qquad (13a)$$

$$\approx 3.22 \left(\frac{x_1}{d^*} \right)^{2/3} \qquad \gamma = \frac{5}{3} \qquad (13b)$$

$$\approx 3.64 \left(\frac{x_1}{d^*} \right)^{2/5} \qquad \gamma = \frac{7}{5} \qquad (13c)$$

$$\approx 3.95 \left(\frac{x_1}{d^*} \right)^{2/7} \qquad \gamma = \frac{9}{7} \qquad (13d)$$

where d^* is the source-orifice diameter. Data presented by Ashkenas and Sherman[15] (Fig. 2) indicate that, for values of x_1/d^* of interest here, the

Mach number is independent of the source geomety upstream from the sonic surface. The speed ratio is related to the Mach number by

$$S = \sqrt{\gamma/2}\, M_1 \tag{14}$$

whereas thermodynamic properties are related by isentropic relations, e.g.,

$$\frac{n_0}{n_1} = \left(1 + \frac{\gamma - 1}{2} M_1{}^2\right)^{1/(\gamma - 1)} \tag{15}$$

where subscript o refers to the source conditions. Note that Eqs. (13) and (14) may be combined with Eq. (10) to provide an expression for the maximum useful d_1/d^* as a function of either x_1/d^* or S_1.

As indicated in the preceding section, molecular-beam techniques facilitate also determining the temperature T_0 of the source gas. The most probable speed of the beam molecules may be measured using either time-of-flight techniques or speed filtering. If $x/d^* \gtrsim 4$, then, to good approximation, this most probable speed has reached its maximum value and equals the hydrodynamic speed. Hence, for $x_1/d^* \gtrsim 4$, the temperature T_0 of the source gas may be determined from

$$h_0 \approx \frac{U_1{}^2}{2} \tag{16}$$

where h is specific enthalpy. This method for determining T_0 is particularly useful in cases in which the source-gas temperature is nonuniform — as is the case for gases in reciprocating-engine combustion chambers.

Recall that continuum flow is realized upstream from the skimmer entrance in the model treated here. Under many conditions (including conditions sufficient for negligible skimmer interference at practical values of d_1/d^*), a transition from continuum flow to free-molecule flow occurs upstream from the skimmer. In the region of free-molecule flow, molecular collisions are so infrequent that S remains constant although the density continues to decrease as the inverse square of distance from the source. An analysis of this case, analogous to the analysis given here, indicates that, if $x_1 \ll L_{1d}$, then Eqs. (5)–(16) are still applicable if the subscript 1 in n_1, x_1, S_1, M_1, and U_1 is replaced by the subscript T, referring to the transition surface. Expressions relating x_T, S_T, and $\xi_{\max}$ with source conditions will be given in Eqs. (72), (73), and (77).

III. BOUNDARY-LAYER CONSIDERATIONS

For the simplified model of Section II, it was assumed implicitly that the gas state within the source was uniform except for those state changes associated with acceleration through the orifice. In sampling from combustion chambers, however, nonuniformities (e.g., composition gradients in the vicinity of the chamber wall) exist frequently. Sometimes one wishes to avoid an existing boundary layer (and sample the "bulk" gas); othertimes one wishes to study the boundary layer (i.e., sample the boundary layer). Criteria for realizing, as the case may be, one or the other of these sampling situations would be useful.

Consider the general case in which both bulk gas and boundary-layer gas flow through the source orifice. The sink-type flow upstream from the orifice favors concentration of the bulk gas near the free-jet centerline and the boundary-layer gas near the free-jet periphery. Hence, since the skimmer admits gas from near the free-jet centerline, the sampling process favors sampling of the bulk gas. These considerations motivate

$$d^*/\delta > 1 \tag{17a}$$

as a criterion for sampling bulk gas and

$$d^*/\delta \ll 1 \tag{17b}$$

as a criterion for sampling boundary-layer gas, where δ is the thickness of the boundary layer.

The boundary layer of chief concern here is the composition boundary layer due to quenching of the combustion process by the combustion-chamber wall. Closely related systems have been studied by several investigators. The quenching distance between plane parallel plates was measured for laminar propane flames as a function of pressure, temperature, and air/fuel ratio by Friedman and Johnston.[16] These measurements were extended to higher pressures by Green and Agnew.[17] For a given air/fuel ratio and for the unburned-gas temperature equal to the plate temperature, the data are correlated by an equation of the form

$$\frac{q}{q_r} = \left(\frac{p_r}{p}\right)^a \left(\frac{T_r}{T}\right)^b \tag{18}$$

where q is quench distance, p is pressure, a and b are constants, and subscript r refers to a reference condition. For quenching by a single wall, Daniel[18] suggests dividing the parallel-wall quench distance by 3 whereas Ellenberger and Bowlus[19] suggest dividing by 2.5. Neglecting the difference between the unburned-gas temperature and the wall temperature, Daniel uses $q_r = 0.0031$ in., $p_r = 287$ psia, $T_r = 540°R$, $a = 0.52$, and $b = 0.56$ for a stoichiometric mixture. In the absence of more detailed information, one might use, in Eq. (18), the mean of the unburned-gas temperature and the wall temperature.

After the flame has moved past a given wall surface, the unburned fuel and oxygen begin diffusing from the quench region into the bulk gas. By the beginning of the exhaust stroke, the boundary-layer thickness δ will have grown to approximately

$$\delta \approx 2 \, (D/S)^{1/2} \qquad (19)$$

where D is the mean diffusion coefficient and S is the engine speed in revolutions per unit time. This thickness is frequently an order of magnitude greater than the original quench distance q.

If the fraction of the boundary-layer gases exhausted to the atmosphere is proportional to the boundary-layer thickness, then the above-mentioned dependence of boundary-layer thickness on engine speed would contribute to the observed dependence of unburned-fuel emissions on engine speed. For example, data presented by both Daniel[18] (Fig. 3) and Pinkerton[20] (Figs. 3–4) indicate that the unburned-fuel emission is proportional to the inverse square root of engine speed. It is suggested that Daniel's comment "Increasing the engine speed probably decreases the number of moles of unburned fuel since there is less time for the gases to move into and out of the crevices" be supplemented by a similar statement for diffusion from the quench layer, and that a possible dependence on engine speed of unburned fuel due to quench be examined in future analyses of the type carried out by Daniel.

The sampling technique used by Daniel[5] handles the boundary layer somewhat differently. In that technique, a poppet valve is installed in the combustion-chamber wall. It is timed to open for a small fraction of the cycle and is operated over many consecutive cycles. Concentrations are measured using a gas chromatograph. Boundary-layer gases dominate the sample during the early part of the valve open time — the boundary-layer and bulk-gas proportions in the measured sample are determined by the lift of the valve and by the length of the valve open time. In the sampling technique described in the present paper, boundary-layer gases are concentrated in the free-jet periphery — the boundary-layer and bulk-gas proportions in the measured sample are determined by the ratio of the sampling-orifice diameter and the boundary-layer thickness.

Measurements of Young et al.[21] are available for comparison with the above-mentioned boundary-layer considerations. For a typical set of operating conditions, the calculated quench thickness q was about $0.2 \, d^*$; the boundary-layer thickness δ, estimated using Eq. (19), grew to about $3d^*$ by the beginning of the exhaust stroke. During this time, the measured mole fraction of O_2 increased from about 0.004 to about 0.011. Since the mole fraction of O_2 in the quench layer is of the order of 0.2, these measurements substantiate the criterion for sampling bulk gas, namely Eq. (17a). Note that Eq. (17a) is a less restrictive criterion than the more conservative criterion suggested by Sturtevant.[22]

An orifice diameter smaller than indicated by Eq. (17a) may be used if it is located at the apex of a cone protruding into the combustion chamber. Greene and Milne[7] find that, if the internal half angle of this cone exceeds 45°, then the flow at the free-jet centerline is undisturbed.

IV. CHEMICAL RELAXATIONS

A major consideration in the design of the sampling system is the requirement that the sampling process either (a) disturb the composition of the sample negligibly or (b) change the composition only in a manner which can be predicted. Possible effects of chemical relaxations in the free jet are considered in the present section. Other possible sources of composition disturbances (skimmer interference, mass diffusions, background scattering, etc.) are considered in later sections.

The discussion of chemical relaxations is facilitated by use of the relaxation time — the time required for the deviation from equilibrium to reduce to $1/e$ its initial value if the existing thermodynamic constraints were maintained. In order to identify the appropriate relaxation time and clarify the relation of the relaxation process to kinetic-energy changes, examine the case of a free jet with negligible effects of molecular transports, external fields, and thermal radiation. Then

$$\frac{D}{Dt}\left(h + \frac{U^2}{2}\right) = 0 \tag{20}$$

where h is the enthalpy (including chemical enthalpy) per unit mass and $D(\ )/Dt$ is the hydrodynamic derivative. Divide h into chemical enthalpy h^* and sensible enthalpy h_s

$$h = h^* + h_s \tag{21}$$

Then one may write

$$\frac{D}{Dt}\left(h_s + \frac{U^2}{2}\right) + \frac{Dh^*}{Dt} = 0 \tag{22}$$

At any given point during the expansion, the gas is relaxing toward the equilibrium state associated with the local kinetic energy $U^2/2$, i.e., at constant enthalpy and pressure. Designate this equilibrium state by subscript e. Then

$$h = h_s + h^* = h_{se} + h_e^* \tag{23}$$

If the deviation from equilibrium is not too great, then the rate of change of chemical enthalpy is proportional to its deviation from its equilibrium value and inversely proportional to the relaxation time, i.e.,

$$\frac{Dh^*}{Dt} = -\frac{h^* - h_e^*}{\tau_{h,p}} \tag{24}$$

where $\tau_{h,p}$ is the relaxation time at constant enthalpy and pressure. Combining Eq. (22)–(24), one may write

$$\frac{Dh_s}{Dt} = -\frac{h_s - h_{se}}{\tau_{h,p}} - \frac{D}{Dt}\left(\frac{U^2}{2}\right) \tag{25}$$

Equation (24) states that h^* (and the chemical composition) changes only as a consequence of the relaxation process whereas Eq. (25) states that h_s changes as a consequence of both the relaxation process and the fluid acceleration. Note that the enthalpies h_e^* and h_{se} are not constants but vary according to

$$\frac{D}{Dt}\left(h_e^* + h_{se} + \frac{U^2}{2}\right) = 0 \tag{26}$$

which is a form of the energy-conservation equation.

The importance of identifying and using the appropriate relaxation time does not appear to be generally appreciated. Calculation of a needed relaxation time from a given relaxation time is facilitated by

$$\frac{\tau_{x,y}}{\tau_{z,y}} = \frac{\left(\dfrac{\partial z}{\partial x}\right)_y^0}{\left(\dfrac{\partial z}{\partial x}\right)_y^\infty} \tag{27}$$

where the superscripts 0 and ∞ refer respectively to the equilibrium and frozen processes. See, e.g., Eq. (39) of the excellent discussion of relaxation phenomena by Bauer.[23] Hence the needed relaxation time $\tau_{h,p}$ and the more common relaxation time $\tau_{T,p}$ are related by

$$\frac{\tau_{T,p}}{\tau_{h,p}} = \frac{c_p^0}{c_p^\infty} \tag{28}$$

where c_p is the heat capacity at constant pressure.

As an example, consider the important case of a single effective reaction in a mixture of thermally perfect gases. Then

$$\frac{c_p^0}{c_p^\infty} = 1 + \frac{(\Delta H/RT)^2}{\sum_\alpha \nu_\alpha^2/x_\alpha - (\Delta\nu)^2} \frac{R/W}{\sum_\alpha c_\alpha c_{p\alpha}} \tag{29}$$

where ΔH is the enthalpy increase for the reaction, R is the gas constant, W is the molecular weight of the mixture, ν_α is the stoichiometric coefficient for species α (positive for products, negative for reactants, zero for inert species), x_α is the mole fraction of species α, $\Delta\nu$ is the mole increase for the reaction, and c_α is the mass fraction of species α. Note that, if inert species are present in large mole fractions, then x_α is small for $\nu_\alpha \approx 0$ and $\tau_{T,p} \approx$

$\tau_{h,p}$. On the other hand, as a specific example of a case in which the difference between two relaxation times is significant, note that, for oxygen (no inert species) at 4000 °K and at a pressure such that it is 50% dissociated, $\tau_{T,p} \approx 21\tau_{h,p}$.

If the relaxation time is determined from the reaction-rate coefficients, then care must be exercised here also in selecting the appropriate relationship. As an example, consider again the important case of a single effective reaction in a mixture of thermally perfect gases. Then

$$\frac{1}{\tau_{T,\rho}} = k_f \prod_\alpha \left(\frac{\rho_\alpha}{W_\alpha}\right)^{\nu_\alpha'} \sum_\alpha \frac{\nu_\alpha^2}{x_\alpha} \frac{W}{\rho} \tag{30}$$

$$= k_r \prod_\alpha \left(\frac{\rho_\alpha}{W_\alpha}\right)^{\nu_\alpha''} \sum_\alpha \frac{\nu_\alpha^2}{x_\alpha} \frac{W}{\rho} \tag{31}$$

where k_f and k_r are respectively the forward and backward reaction-rate coefficients, ρ_α is the mass density of species α, W_α is the mass per mole of species α, ν_α' and ν_α'' are respectively the reactant and product coefficients in the reaction equation, and ρ is the mass density of the mixture. See, e.g., Eq. (123) of Bauer.[23] Other relaxation times may be determined from $\tau_{T,\rho}$ by use of Eq. (27).

In many sampling-orifice flows, the chemical reactions are so fast in the region of higher pressures and temperatures ($\tau_{h,p}$ is small in comparison with the time required for the gas to pass through this region) that the mixture composition is essentially in equilibrium in this region, whereas the chemical reactions are so slow in the region of lower pressures and temperatures ($\tau_{h,p}$ is large in comparison with the time required for the gas to pass through this region) that composition changes are negligible in this region. It is convenient to approximate the transition from the equilibrium-composition region to the frozen-composition region by a model containing several "freezing surfaces," one surface for each chemical reaction. (Cf. the discussion of multiple freezing points by McIntyre and Leslie[24] and the review by Bray.[25]) For a given reaction, the terms "partial-equilibrium flow" and "partial-frozen flow" will be used to identify the flows upstream and downstream, respectively, from its freezing surface. In the region of partial-equilibrium flow, another reaction may be frozen whereas in the region of partial-frozen flow, still another reaction may be in equilibrium. A convenient freezing-point criterion is motivated by an examination of $D\tau_{h,p}/Dt$, where $\tau_{h,p}$ is now the relaxation time for the reaction of immediate interest. If this reaction is in equilibrium in some upstream region and is frozen in a downstream region, then, at some surface in the flow, $D\tau_{h,p}/Dt$ must be of order unity. Hence the freezing-point criterion

$$\frac{D\tau_{h,p}}{Dt} = C \tag{32}$$

where C is a constant of the order of unity and $\tau_{h,p}$ is calculated using the partial-equilibrium-flow solution, is motivated. A freezing-point criterion of this form was suggested first by Phinney[26] in his Eq. (29).

This freezing-point criterion may be written alternatively in terms of derivatives of either (a) temperature, (b) extent of reaction, or (c) sometimes mass fraction. Transformation to these other forms are facilitated by writing $\tau_{h,p}$ in the form

$$\frac{1}{\tau_{h,p}} = k_f \prod_\alpha \left(\frac{\rho_\alpha}{W_\alpha}\right)^{\nu_\alpha'} \sum_\alpha \frac{\nu_\alpha^2}{x_\alpha} \frac{W}{\rho} \frac{\tau_{T,p}}{\tau_{h,p}} \tag{33}$$

where, from Eq. (27),

$$\frac{\tau_{T,p}}{\tau_{h,p}} = \frac{\displaystyle\sum_\alpha \frac{\nu_\alpha^2}{x_\alpha} - (\Delta\nu)^2 + \left(\frac{\Delta H}{RT}\right)^2 \frac{R}{W} \Big/ \sum_\alpha c_\alpha c_{p_\alpha}}{\displaystyle\sum_\alpha \frac{\nu_\alpha^2}{x_\alpha}} \tag{34}$$

Consider the case in which the reaction-rate coefficients may be written as Arrhenius equations, i.e.,

$$k_f = A' \exp\left(- E'/RT\right) \tag{35}$$

$$k_r = A'' \exp\left(- E''/RT\right) \tag{36}$$

where the pre-exponential factors A' and A'' and the activation energies E' and E'' are constants. If $\sum_\alpha \nu_\alpha^2/x_\alpha \gg |\Delta U/RT| \gg 1$, where ΔU is the internal-energy increase for the reaction, then the variation of $\tau_{h,p}$ is dominated by the variation of $k_f \prod_\alpha (\rho_\alpha/W_\alpha)^{\nu_\alpha} (\sum_\alpha \nu_\alpha^2/x_\alpha) W$ with temperature. The dominant terms are

$$d \ln \frac{1}{\tau_{h,p}} \approx \left[\frac{E'}{RT} - K' \frac{\Delta U}{RT}\right] d \ln T$$

$$\approx \left[\frac{E''}{RT} - K'' \frac{\Delta U}{RT}\right] d \ln T \tag{37}$$

where

$$K' \equiv -\frac{\displaystyle\sum_\alpha \frac{\nu_\alpha' \nu_\alpha}{x_\alpha}}{\displaystyle\sum_\alpha \frac{\nu_\alpha^2}{x_\alpha}} + \frac{\displaystyle\sum_\alpha \frac{\nu_\alpha^3}{x_\alpha^2}}{\left(\displaystyle\sum_\alpha \frac{\nu_\alpha^2}{x_\alpha}\right)^2} \tag{38}$$

$$K'' \equiv -\frac{\displaystyle\sum_\alpha \frac{\nu_\alpha'' \nu_\alpha}{x_\alpha}}{\displaystyle\sum_\alpha \frac{\nu_\alpha^2}{x_\alpha}} + \frac{\displaystyle\sum_\alpha \frac{\nu_\alpha^3}{x_\alpha^2}}{\left(\displaystyle\sum_\alpha \frac{\nu_\alpha^2}{x_\alpha}\right)^2} \tag{39}$$

(The equality of the two coefficients of $d \ln T$ is consistent with $E' - E'' \approx \Delta U$.) Hence Eq. (32) may be transformed into either

$$\frac{DT}{Dt} \approx -C \frac{RT}{E' + K'(E'' - E')} \frac{T}{\tau_{h,p}} \qquad (40)$$

or

$$\frac{D\xi}{Dt} \approx -C \frac{\sum\limits_{\alpha} W_{\alpha} \nu_{\alpha}'}{W} \frac{1}{\sum\limits_{\alpha} \frac{\nu_{\alpha}^2}{x_{\alpha}}} \frac{E' - E''}{E' + K'(E'' - E')} \frac{1}{\tau_{h,p}} \qquad (41)$$

where ξ is the extent of reaction, defined in terms of the mass-fraction change due only to the reaction of immediate interest, by

$$d\xi \equiv \frac{\sum\limits_{\alpha} W_{\alpha} \nu_{\alpha}'}{W_{\alpha} \nu_{\alpha}} dc_{\alpha} \qquad (42)$$

As was the case for Eq. (32), the left-hand sides of Eqs. (40)–(41) are evaluated for the reaction of immediate interest in equilibrium. The factor

$$\frac{W}{\sum\limits_{\alpha} W_{\alpha} \nu_{\alpha}'} \sum\limits_{\alpha} \frac{\nu_{\alpha}^2}{x_{\alpha}} \frac{RT}{E' - E''} \approx \frac{d \ln T}{d\xi} \qquad (43)$$

is a measure of the coupling between the gas-expansion process and the chemical-relaxation process.

For the special case in which only one chemical reaction occurs in the flow, Eq. (32) may be transformed also into

$$\frac{\rho Dc_{\alpha}}{Dt} \approx -C \frac{\tau_{T,\rho}}{\tau_{h,p}} \frac{E' - E''}{E' + K'(E'' - E')} W_{\alpha} \nu_{\alpha} k_f \underset{\alpha}{\pi} \left(\frac{\rho_{\alpha}}{W_{\alpha}}\right)^{\nu_{\alpha}'} \qquad (44)$$

This form is difficult to extend to the case in which more than one chemical reaction occurs. See, e.g., the results of extensive efforts by Burwell et al.[27] and the review by Bray.[25]

Approximate expressions which are equivalent to

$$\rho \frac{D\xi}{Dt} = -B \left(\sum\limits_{\alpha} W_{\alpha} \nu_{\alpha}'\right) k_f \prod\limits_{\alpha} \left(\frac{\rho_{\alpha}}{W_{\alpha}}\right)^{\nu_{\alpha}'} \qquad (45)$$

where B is of order unity, appear frequently in the literature (e.g., see Ref. 25). This expression is equivalent to Eq. (41) if

$$B \approx \frac{\tau_{T,\rho}}{\tau_{h,p}} \frac{E' - E''}{E' + K'(E'' - E')} C \qquad (46)$$

The factor $\tau_{T,\rho}/\tau_{h,p}$ is approximately unity if the mole fractions of one or more reactive species are small. The other factor (related to the coupling between the chemical relaxation and the gas expansion) is a function of activation energies, stoichiometric coefficients, and composition.

Results of exact calculations indicate the coefficient B in Eq. (45) depends on the nature of the reactive system. For example, McIntyre and Leslie[24] find that, in an expansion of kerosine/air combustion products, ultimate temperatures and compositions are predicted satisfactorily by $B = 2$ for

$$H_2O + N_2 \leftrightarrows H + OH + N_2 \tag{47}$$

and by $B = 0.5$ for

$$H_2 + N_2 \leftrightarrows 2H + N_2 \tag{48}$$

For the given temperatures and compositions, one obtains

Reaction	$\tau_{T,p}/\tau_{h,p}$	$\dfrac{E' - E''}{E' + K'(E'' - E')}$	C
(47)	1.14	2.41	0.7
(48)	1.02	1.05	0.5

where values of C are calculated using Eq. (46). It is seen that C is constant to better approximation than is B.

A more reliable value of C is obtainable perhaps from the exact calculations of Glowacki[28] for the reaction

$$O_2 \leftrightarrows 2O \tag{49}$$

in 30 different nozzle flows of heated air. Values of B which predict satisfactorily the ultimate values of the degree of dissociation (from 0.00462 to 0.827) have been calculated by Phinney.[29] Values of $\tau_{T,p}/\tau_{h,p}$ and of $[E' - E'']/[E' + K'(E'' - E')]$ range respectively from 1.4 to 2.9 and from 1.3 to 2.0. The median value of C for these 30 cases, calculated using Eq. (46), is 0.5, It is suggested that this value be used provisionally in Eqs. (32), (40), (41), and (44).

Dimensionless parameters which facilitate correlating and predicting freezing-point locations would be useful. For identification of these parameters, Eq. (40) is most convenient. In order to emphasize the temperature dependence of $\tau_{h,p}$, note that K' varies only slowly with composition so that it may be approximated by a constant in the integration of Eq. (37). One obtains

$$\frac{1}{\tau_{h,p}} \approx \nu_s \exp\left[-\frac{E' + K'(E'' - E')}{RT}\right] \tag{50}$$

where ν_s is a constant for a given isentrope. Then Eq. (40) may be written

$$\frac{E' + K'(E'' - E')}{RT} \exp\left[\frac{E' + K'(E'' - E')}{RT}\right]\frac{DT}{Dt}$$
$$\approx -C\nu_s T \tag{51}$$

This form of the freezing-point criterion emphasizes that freezing is dominated by the exponential dependence of $\tau_{h,p}$ on temperature and that the dominant dimensionless parameter is the thermodynamic parameter

$$\frac{E' + K'\,(E'' - E')}{RT} \tag{52}$$

This parameter explains the observation by Bray[30] that, for a given gas system, nozzle geometry, and isentrope, the thermodynamic state at the freezing point is essentially independent of the stagnation enthalpy.

The other important dimensionless parameter is the scaling parameter, which is motivated by referring the hydrodynamic velocity to the stagnation-condition sound speed a_0 and the axial distance to the throat diameter d^*. One obtains

$$\frac{E' + K'\,(E'' - E')}{RT}\,\exp\left[\frac{E' + K'\,(E'' - E')}{RT}\right]\frac{u}{a_0}\frac{dT}{d(x/d^*)} \approx -\,CT\,\frac{\nu_s d^*}{a_0} \tag{51a}$$

The dimensionless scaling parameter is seen to be

$$\frac{\nu_s d^*}{a_0} \tag{53}$$

As shown by Ring and Johnson,[31] this parameter has a secondary effect on the thermodynamic state at the freezing point.

As an indicator of early freezing, consider conditions which lead to freezing upstream from the sampling orifice. At the orifice, $T/T_0 = 2/(\gamma + 1)$ and $d \ln T/d\,(x/d^*) \approx -4\,(\gamma - 1)/(\gamma + 1)$. Hence freezing occurs upstream from the orifice if

$$\frac{\nu_s d^*}{a_0} \lesssim 4\,(\gamma - 1)\,\frac{E' + K'\,(E'' - E')}{RT_0}\,\exp\left[\frac{\gamma + 1}{2}\,\frac{E' + K'\,(E'' - E')}{RT_0}\right] \tag{54}$$

This equation provides a quick estimate of the maximum value of the preferred sampling-orifice diameter.

In summary, Eq. (32) is suggested as an improved criterion for freezing of a chemical reaction in a sampling jet. Alternative formulations (involving derivatives of either temperature, or extent of reaction, or mass fraction) are provided by Eqs. (40), (41), and (44). A provisional value of the coefficient C appears to be about 0.5. Indicators of the thermodynamic state at the freezing point are the dominant parameter $[E' + K'\,(E'' - E')]/RT$ and the secondary parameter $\nu_s d^*/a_0$. Equation (54) provides a quick estimate of the maximum value of the preferred sampling-orifice diameter.

V. SPECIES CONDENSATIONS

Species condensations in expanding fluid flows have been observed for many years in supersonic wind tunnels[32] and since 1956[33] in supersonic molecular beams. If condensations were to occur in sampling studies, interpretations of measurements would be complicated. The relevant information which is reviewed briefly here indicates that the probability of significant condensations in molecular-beam sampling from combustion systems is small.

Interesting thermodynamic aspects of condensations in expanding fluid flows are brought out by combining the Clausius–Clapeyron equation for species A

$$d \ln p_{A \text{ sat}} = \frac{\Delta h_A}{p_{A \text{ sat}} \Delta v_A} d \ln T$$

with the isentropic relation for a mixture of thermally perfect gases

$$d \ln p = \frac{\gamma}{\gamma - 1} d \ln T$$

to obtain

$$\frac{d \ln p_{A \text{ sat}}}{d \ln p} = \frac{\gamma - 1}{\gamma} \frac{\Delta h_A}{p_{A \text{ sat}} \Delta v_A} \tag{55}$$

where $P_{A\text{sat}}$ is the saturation vapor pressure, Δh is the enthalpy increase for vaporization, and Δv is the volume increase for vaporization of species A at temperature T. As an example of the relative behavior of $P_{A\text{sat}}$ and p note that, for water vapor in a gas mixture with $\gamma = 7/5$, the right-hand side of Eq. (55) increases from 2.2 at the critical point to 5.7 at the normal freezing point of water. Hence $p_{A\text{sat}}$ always decreases more rapidly than p so that (if the expansion continues indefinitely) the saturation point is reached eventually. However, the relative decrease of $p_{A\text{sat}}$ is smaller at higher temperatures. Furthermore, condensation cannot occur for temperatures above the critical temperature (e.g., $179\,°K$ for NO and $134\,°K$ for CO). Hence, thermodynamic considerations indicate that the probability of condensation occuring is reduced significantly by increasing the source-gas temperature from room temperature to combustion temperatures.

Consider now kinetic aspects of condensation. Unfortunately, although a relatively large number of measurements have been made, the relevant kinetics have been clarified only partially for monatomic gases and very little for polyatomic gases. In a first step toward a framework for correlating some of these data, a dimensionless scaling parameter analogous to Eq. (53) is identified here. Available evidence indicates that condensation is initiated via the formation of dimers in three body collisions, not only for monatomic gases,[34-36] but also for polyatomic gases.[35,37,38] Hence the ratio of the flow time $d*/a_0$ to the mean time τ_0 between consecutive three-body collisions is important to dimer formation. For a single species of hard spheres with diameter σ,

$$\frac{1}{\tau_0} = 3\left(n_0 \frac{4}{3} \pi\sigma^3\right)\left(n_0 \pi\sigma^2 \sqrt{\frac{4}{\pi} \frac{kT_0}{m}}\right)$$

so that the relevant dimensionless scaling parameter may be written

$$\frac{d*}{\tau_0 a_0} = 8 \pi \left(\frac{\pi}{\gamma}\right)^{1/2} n_0^2 d* \sigma^5 \tag{56}$$

Extension to a binary mixture of species A and B yields

$$\frac{d*}{(\tau_A a)_0} = 8\pi \left(\frac{\pi}{\gamma} \frac{m}{m_A}\right)^{1/2} n_{A0} d* \sigma_{AA}^2 \frac{(3n_{A0}\sigma_{AA}^3 + 2n_{B0}\sigma_{AB}^3)}{3} \tag{56a}$$

Note that if either (a) $n_{A0} \ll n_{B0}$ or (b) $3\sigma_{AA}^3 \approx 2\sigma_{AB}^3$, then the right-hand side of Eq. (56a) is proportional to $n_{A0} n_0 d*$; if $n_{A0} \gg n_{B0}$, then Eq. (56) is recovered.

The data most relevant to Eq. (56) is the data for Ar presented by Milne *et al.*[36] in their Figure 2. They find that data for 5 different source pressures are correlated in a plot of excess dimer mole fraction versus $P_0^2 d*$; the excess dimer mole fraction (relative to the equilibrium mole fraction in the source) is approximately a linear function of $P_0^2 d*$. The dimensionless parameter, Eq. (56), is an equally suitable abscissa; then the excess dimer mole fraction is proportional to the ratio of flow time to mean time between consecutive three-body collisions, in agreement with intuitive expectations. Data at additional values of T_0 would be useful in assessing the possible need for an additional temperature-dependent parameter.

The data most relevant to Eq. (56a) is the data for NO–N_2 mixtures presented by Golomb and Good[37] in their Figure 2. They find that data for three different mole fractions are correlated in a plot of NO dimer/monomer ratio versus NO mole fraction; the dimer/monomer ratio is approximately a linear function of NO mole fraction (for fixed p_0 and T_0). Since

$$2 \sigma_{NO\text{-}N_2}^3 \approx 3 \sigma_{NO\text{-}NO}^3$$

the dimensionless parameter, Eq. (56a), is an equally suitable abscissa.

A different scaling parameter, namely $p_0 d*$, has been suggested by Bier and Hagena[39] to correlate condensation effects which occur subsequent to dimer formation. The parameter given in Eq. (56) is not applicable to these subsequent condensation effects.

The smaller the flow time, relative to the time between three-body collisions, the smaller is the dimer production. Hence, according to Eq. (56),

Table I. Measured Dimer-Monomer Ratios

Species	$\dfrac{\text{Dimer}}{\text{Monomer}}$	$\dfrac{d^*}{\tau_0 a_0}$	Ref.
N_2	10^{-5}	5	40
O_2	10^{-5}	5	40
CO_2 in N_2	10^{-2}	30	41
NO in N_2	10^{-2}	50	37

dimer production may be evaluated by making measurements for two values of d^*. In a case in which dimer production is significant and undesirable, one might make measurements for several values of d^* and extrapolate results to $d^* = 0$.

The measured values (all for $T_0 \approx 300\,°K$), shown in Table I, are perhaps most relevant to sampling from combustion chambers. Greene and Milne[42] confirm that this ratio decreases as T_0 increases. Furthermore, Young et al.[21] did not detect significant dimer concentrations in sampling from reciprocating engines. Hence, species condensations are not expected to be significant in molecular-beam sampling studies of combustion processes.

VI. SKIMMER INTERFERENCE

The molecular-beam properties at the mass spectrometer may be affected by skimmer interference with the sampling jet. Typical consequences of this interference are (a) decrease of beam density, (b) increase of velocity-distribution width, (c) decrease of mean velocity, and (d) distortion of beam composition, listed in order of decreasing sensitivity to skimmer interference. Specific examples of the first three consequences are given, e.g., by Bier and Hagena[39,43] and Anderson et al.[34] Distortions of compositions due to skimmer interference have been reported, e.g., by Reis and Fenn,[44] Rothe,[45] and Young et al.[46] Most reported distortions of compositions are due to other causes; they are discussed in Section VII. Sources of skimmer interference, and criteria for minimizing it, are reviewed briefly here. This review emphasizes results obtained since the earlier reviews by Knuth,[47] Anderson et al.,[48,49] and French.[50]

Consider first the effects of the aerodynamic shock structure in the vicinity of the skimmer. A schematic diagram of the most relevant shocks is given in Fig. 5. The indicated barrel shock, stand-off shock, and attached shock are due to free-jet interactions with the background gas, the chamber wall, and the skimmer surface.

The dependence of the shock stand-off distance $(x_w - x_s)/x_w$ on pressure ratio p_0/p_s for several values of nozzle-to-wall distance x_w/d^* is given in Fig. 6,

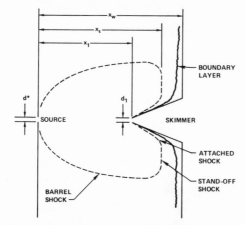

Fig. 5.　Schematic diagram of shock features relevant
to discussion of skimmer interference.

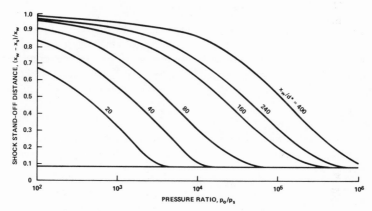

Fig. 6.　Shock stand-off distance as function of pressure ratio for several
values of distance from nozzle to chamber wall.

where p_0 and p_s are the stagnation pressure and the source-chamber background pressure respectively. This figure is a replot of the experimental and analytical results given in Figure 14 of the report by Vick and Andrews.[51] For $x_w/d^* \gtrsim 1.3 \; (p_0/p_s)^{1/2}$, the shock-wave location is independent of the chamber-wall location and is given, on the basis of data by Bier and Schmidt,[52] for all values of γ, by

$$\frac{x_s'}{d^*} = 0.67 \left(\frac{p_0}{p_s}\right)^{1/2} \tag{57}$$

To the approximation that the rate of mass change in the source chamber is small in comparison with mass-flow rates into and out of this chamber, the distance given by Eq. (57) is related to the speed S_s (volume/time) of the source-chamber pumping system by

$$x_s' = 0.75 \left(\frac{1}{\gamma}\frac{kT_0}{m}\right)^{1/4}\left(\frac{\gamma+1}{2}\right)^{\frac{\gamma+1}{4(\gamma-1)}}\left(S_s\frac{m}{kT_s}\right)^{1/2} \tag{58}$$

Note that this distance is independent of source pressure and only a weak function of source temperature. For $x_w/d^* \lesssim 0.3(p_0/p_s)^{1/2}$, the shock-wave location is independent of the pressure ratio and is given, following Roberts,[53] for $\gamma = 7/5$, by

$$x_w - x_s'' = 0.084\, x_w \tag{59}$$

It is recommended that, in order to avoid skimmer interference due to the stand-off shock, the skimmer length satisfy either

$$x_w - x_1 \geqslant x_w - x_s/2 \tag{60a}$$

or

$$x_w - x_1 \geqslant 3\,(x_w - x_s) \tag{60b}$$

whichever is smaller, with x_s taken from Fig. 6. The first criterion is based on the beam-attenuation measurements of Bier and Hagena[43] for $x_w/d^* \gtrsim 1.3$ $(p_0/p_s)^{1/2}$, whereas the second criterion is based on the beam-attenuation measurements of Bossel[54] for $x_w/d^* \lesssim 0.3(p_0/p_s)^{1/2}$. If a high beam density is required, then operation with $x_w/d^* \gtrsim 1.3(p_0/p_s)^{1/2}$ and $x_w - x_1 = x_w - x_s/2$, i.e., with relatively high M_1, is recommended.

Another necessary condition for minimum skimmer interference is attachment of the shock structure to the skimmer lip. The maximum half-apex angle δ_e for attachment of a shock at $M_1 \gg 1$ to a wedge is given by

$$\text{maximum sin } \delta_e = \frac{1}{\gamma} \tag{61}$$

(Sufficiently near the leading edge of a conical skimmer, the flow is wedge-like.) Hence, if the sampled gas consists mostly of diatomic gases, then an external half-apex angle $\delta_e < 45°$ is required. This necessary condition is supported by the studies of Bier and Hagena,[39] which indicate that the beam intensity decreases significantly as δ_e is increased from $43°$ to $52°$.

Additional motivation to reduce δ_e is provided by Oman's prediction[55] that, for attenuations less than 30%, the attenuation due to atoms reflected from the external surface varies exponentially with δ_e. Hence it is recommended that δ_e be made as small as permitted by considerations of the internal half-apex angle δ_i.

Oblique shocks have been observed also inside skimmers by Bier and Hagena[39,43] and by French and McMichael.[56] A sufficient condition for avoiding them is provided perhaps by the rarefaction criterion, involving Knudsen number at the skimmer orifice, given later in this section.

Still another possible source of skimmer interference is collisions of beam molecules with the inner surface of the skimmer. This possible source is avoided if the internal half-apex angle δ_i is sufficiently large. Recall the conflicting requirement of a small value of the external half-apex angle δ_e imposed by considerations of the external flow. Bossel[57] suggests the compromise

$$\delta_i = 3 \sin^{-1} \frac{1}{M_1} \tag{62}$$

This criterion also is supported by the studies of Bier and Hagena,[39] which studies included beam performance measurements for M_1 from 7 to 12 and for $\delta_i = 10°$, $25°$, and $35°$ with δ_e constant. Equation (62) suggests $\delta_i = 25°$ for $M_1 = 7$ and $14°$ for $M_1 = 12$, Bier and Hagena found that the beam performance was about equal for $\delta_i = 25°$ and $35°$ but was dramatically deteriorated for $\delta_i = 10°$. The Molecular-Beam Laboratory, UCLA, uses $\delta_i = 16°$ and $\delta_e = 20°$ in its sampling studies.

Design requirements for the leading edge of the skimmer are not clear. Some investigators expend considerable effort to produce and maintain a sharp leading edge whereas some investigators have found that slightly damaged skimmers perform satisfactorily. A lip radius small in comparison with the skimmer-orifice diameter has been found to perform satisfactorily.

Consider now the orifice size. (Note that none of the possible sources of skimmer interference mentioned so far has placed a limitation on the size of the skimmer orifice.) It is found that if the orifice diameter d_1 is too large, then skimmer interference results. Different criteria for selecting d_1 are used for different design objectives. Different design objectives include (a) maximum beam intensity with negligible skimmer interference and (b) maximum beam intensity without regard for skimmer interference. The first objective is of interest here. Fenn and Deckers[58] found that the fraction of theoretical intensity for relatively large skimmer diameters is correlated by Kn_{01}/M_1 where Kn_{01} is the Knudsen number ($\equiv \lambda_{01}/d_1$) and λ_{01} is the mean-free-path length based on molecular density at the skimmer and on collision cross section at the source temperature. Fisher and Knuth[59] found that, for a conical skimmer, the skimmer-induced decrease in measured Mach number is less than 3% if Kn_{01}/M_1 is greater than about 0.3–0.4.

For cases in which a greater beam density is required and beam density is limited by skimmer interference due to size of the skimmer orifice, one might use a pyramidal skimmer with a rectangular orifice. Since Kn then would be based on the smaller of the two skimmer dimensions, one would realize a greater molecule-flow rate at the skimmer for the same value of Kn. Fisher and Knuth found that, for a pyramidal skimmer, the skimmer-induced decrease in measured Mach number is less than 3% if Kn_{01}/M_1 is greater than about 0.8–1.0.

The use of Kn_{01}/M_1 here and its value at the onset of continuum effects remind one of correlations of rarefied flows around other bodies. Note that

$$\frac{Kn_{01}}{M_1} \equiv \frac{\lambda_{01}}{d_1}\frac{a_1}{U_1} \approx 1.5 \frac{\mu_{01}}{\rho_1 U_1 d_1} \equiv \frac{1.5}{Re_{01}} \tag{63}$$

where μ_{01} is the viscosity coefficient based on the gas temperature at the skimmer and on the collison cross-section at the source temperature and Re is the Reynolds number. (Recall that Re is a measure of the ratio of macroscopic momentum flux to molecular momentum diffusion.) Then note, e.g., that Potter and Miller[60] correlate drag coefficients for a sphere by use of Re and that transition from free-molecule flow to continuum flow begins at $Re \approx 1$.

The important role played by molecules reflected from the skimmer surface has been dramatized by operating skimmers cooled sufficiently to condense the impinging beam molecules. Using CO_2 beams operating initially under skimmer-interaction conditions, both Anderson et al.[34] and Brown and Heald[61] were able to triple the beam intensity by cooling the skimmer to liquid-nitrogen temperatures.

Collisions of beam molecules with other beam molecules have been suggested as forerunners of skimmer interference related to orifice size. Such collisions have been examined by Valleau and Deckers.[62] Their Eq. (4.1) may be put in the form

$$n_d \approx n_1 \frac{A_1}{\pi L_{1d}{}^2} S_1{}^2 \frac{2Kn_1}{2Kn_1 + 1} \tag{12a}$$

It is seen that if $Kn_1/M_1 \gtrsim 1$ and $M_1 \gg 1$, then Eq. (12a) reduces to Eq. (12), in which collisions between beam molecules are neglected.

In summary, skimmer interference may arise from such diverse sources as the Mach disk, shock detachment, the stand-off shock, small internal skimmer angles, a blunt leading adge, and a large skimmer orifice. Fortunately, the several criteria summarized in this section facilitate designs which reduce skimmer interference to acceptable levels.

VII. MASS SEPARATIONS

Mass separations during the transfer of the gas sample from the source to the detector are possible even if mass separations due to skimmer interference are avoided by careful skimmer design. The most important possible separations, namely separations due to pressure diffusions in the free jet and separations due to Mach-number focusing downstream from the skimmer, are discussed in this section. Possible changes in beam composition due to scattering of beam molecules by background molecules are discussed in Section VIII.

Consider first mass separations due to pressure diffusions in the free jet. The pressure gradients relevant to mass separations are those gradients with significant components transverse to streamlines. Such gradients exist (a) within the first three diameters of the free-jet expansion and (b) within the barrel shock which separates the free jet and its surrounding gas. The former gradients lead to relative enrichment of the heavy species at the jet centerline whereas the latter gradients lead to relative enrichment of the light species inside the shock barrel. Criteria for avoiding these separations are of interest in molecular-beam sampling.

Many experimental results which have been reported include effects of both pressure diffusions and Mach-number focusing, and hence do not facilitate identification of the relative contributions to mass separations. Fortunately, however, several studies have been made which obviate Mach-number focusing. These studies include the experimental studies of Becker et al.,[63,64] Waterman et al.,[65,66] Rothe,[45] Abuaf et al.,[67] Anderson,[68] Wang and Bauer,[69] and Sebacher et al.,[70] and the analytical studies of Zigan,[71] Sherman,[72] and Mikami and Takashima.[73] The possibility of skimmer interference was also completely avoided (by absence of a skimmer) except in the studies by Becker et al. and by Waterman et al.

A convenient choice of dimensionless parameters for consideration of pressure diffusion near the source orifice follows from Sherman's analysis. Neglecting second-order terms in $1/Re_0$, one may rearrange Sherman's result to obtain

$$\frac{x_H}{x_L} \frac{x_{L0}}{x_{H0}} = 1 + \frac{C}{Re_0\, Sc_0} \left(\frac{m_H - m_L}{m_0} \frac{\gamma}{\gamma - 1} - \alpha_0 \right) F(\gamma, x/d^*) \qquad (64)$$

where C is the viscosity-temperature constant defined by

$$\frac{\mu}{\mu_0} \equiv C\, \frac{T}{T_0} \qquad (65)$$

Re_0 ($\equiv \rho_0 a_0 d^*/\mu_0$) is Reynolds number based on stagnation conditions, Sc_0 ($= \mu_0/\rho_0 D_0$) is Schmidt number, D is binary diffusion coefficient, α is thermal diffusion factor, $F(\gamma, x/d^*)$ is (for given γ) a monotonically increasing function of x/d^*, and subscripts H and L refer respectively to heavy and light species. The function F has reached its maximum value (for all practical purposes) already at $x/d^* = 3$. Calculations for diatomic gases[72] and calculations and measurements for monatomic gases[45,68] indicate that the maximum value of F is about 10. It is seen that, for the typical case in which

$$\frac{C}{Sc_0} \left(\frac{m_H - m_L}{m} \frac{\gamma}{\gamma - 1} - \alpha_0 \right) = 0\,(1) \qquad (66)$$

mass separation due to pressure diffusion near the source orifice is negligible if $Re_0 \gtrsim 10^3$. Effects of ordinary diffusion (neglected by Sherman) make this criterion conservative for $x/d^* > 3$.

Effects of pressure diffusions in the barrel shock are important in sampling studies if the relative enrichment of light species extends to the jet centerline. This separation has been called "reversed separation" following Waterman et al.[65,66] and "background invasion" by Campargue.[74] These effects have been observed also by Becker et al.[63,75] Reis and Fenn,[44] Rothe,[45] and Sebacher et al.[70,76] Radial composition measurements through the barrel shock by Chow,[77] Bier,[75] Rothe,[45] Sebacher et al.,[70,76] and Schügerl et al.[78,79] indicate clearly the relative enrichment of light species due to pressure diffusion from the relatively high-pressure region outside the barrel shock. The important role of the pressure gradient through the barrel shock suggests that a suitable criterion for no reversed separation effects at the jet centerline might be provided by the criterion for a diffuse shock structure. The shock thickness δ is related to flow parameters by

$$\frac{\delta}{x_s} \sim \frac{1}{Re_0} \left(\frac{P_0}{P_s}\right)^{1/2} \tag{67}$$

A diffuse shock is realized if δ/x_s is sufficiently large. Hence values of the right-hand side of Eq. (67) were calculated for typical experimental conditions.[44,45,65,66,70,74,75,76] Reversed separation does not appear to be realized along the free-jet centerline for

$$\frac{1}{Re_0} \left(\frac{P_0}{P_s}\right)^{1/2} \gtrsim 0.1 \tag{68}$$

This criterion is conservative if the skimmer entrance is located significantly upstream from the Mach disk.

The data on which Eq. (68) is based are measurements of either species densities in the free jet or species fluxes through a sampling probe located in the free jet. The criterion appears to be conservative for measurements by a mass spectrometer located on the beam centerline downstream from the skimmer.

Consider now mass separations due to Mach-number focusing downstream from the skimmer. This phenomenon is clarified by an examination of the case of a binary mixture, nearly parallel flow at the skimmer, and speed ratios at the skimmer large in comparison with unity. Apply Eq. (12) to species A and to species B, take the ratio of the two equations, and rearrange to obtain

$$\left(\frac{x_A}{x_B}\right)_d \left(\frac{x_B}{x_A}\right)_1 \approx \left(\frac{S_A}{S_B}\right)_1^2 \tag{69}$$

This effect of the ratio of speed ratios is called Mach-number focusing. (The term "speed-ratio focusing." would be more precise.) Two limiting cases will be considered.

Consider first the limiting case of continuum flow up to the skimmer. Write Eq. (69) in the form

$$\left(\frac{x_A}{x_B}\right)_d \left(\frac{x_B}{x_A}\right)_1 \approx \left(\frac{U_A}{U_B}\right)_1^2 \left(\frac{T_B}{T_A}\right)_1 \frac{m_A}{m_B} \tag{70}$$

In this case, $T_B = T_A$ and $U_B = U_A$ at the skimmer so that

$$\left(\frac{x_A}{x_B}\right)_d \left(\frac{x_B}{x_A}\right)_1 \approx \frac{m_A}{m_B} \tag{71}$$

Equation (71) is verified by measurements reported by several investigators. It predicts all mass separation effects reported by Greene et al.[80] in their Figure 5–8 except for $p_0 = 5$ torr. (For $p_0 = 5$ torr, continuum flow was not realized at the skimmer. This case will be discussed in a later paragraph.) Equation (71) predicts also the enrichment factor of 20 observed by Klingel-höfer and Lohse[81] for Ar–H_2 mixtures at low values of p_0. (The enrichment factors greater than 20 observed for higher values of p_0 include contributions due to skimmer interference at the first skimmer and Mach-number focusing downstream from the second skimmer. At the first skimmer, $Kn_{01}/M_1 \ll 1$. Measurements made by Y. G. Wang in the UCLA Molecular-Beam Laboratory suggest that Mach-number focusing does not occur at a skimmer under conditions of strong skimmer interference, accompanied perhaps by a continuum expansion within the skimmer.) It predicts also the enrichment factor of 10 for Ar–He mixtures on the beam centerline measured by French and O'Keefe[82] for large values of x/d^*. (For small values of x/d^*, $Kn_{01}/M_1 \ll 1$. The resulting skimmer interference reduces Mach-number focusing, but introduces radial pressure diffusion.) The reader who wishes to examine the above references is alerted to the fact that some, but not all investigators have a ratio of the form $(x_A/x_B)_d/(x_A/x_B)_1$ in mind when they use the term "enrichment factor." The convenience of this form is apparent from Eqs. (64) and (71).

Application of Eqs. (69)–(71) requires a criterion for determining whether the flow up to the skimmer is or is not continuum flow. From measured terminal Mach numbers for monatomic gases and Eq. (13a),

$$\frac{x_T}{d^*} \gtrsim 0.22 \left(\frac{1}{Kn_0}\right)^{0.6} \tag{72}$$

where x_T is the distance beyond which molecular collisions are so infrequent that the speed ratio remains essentially constant. (The molecule density continues to decrease as the inverse square of distance, however.) An analogous criterion for diatomic gases has not been developed yet. Furthermore,

Knuth[47] has shown that the diatomic-gas criterion must include also a function of the collision number for rotational relaxations. Equations (21a), (21b), (23a), and (23b) of Knuth's report[14] suggest that Eq. (72) is a fair approximation also for diatomic gases (such as N_2) with rotational collision numbers from about 5 to 10.

The role of molecular collisions in maintaining $T_L \approx T_H$ and $U_L \approx U_H$ suggests $x_1 \lesssim x_T$ as a criterion for applicability of Eq. (71). Examination of the data by French and O'Keefe,[82] Chang,[10] and Aurich and Schügerl[79] indicates that Eq. (71) is applicable at least up to twice the value of x_T predicted by Eq. (72).

Values of the terminal speed ratio S_T, realized at x_T, are given for monatomic gases by

$$S_T = 1.07 \left(\frac{1}{Kn_0}\right)^{0.4} \tag{73}$$

Equation (73) is useful also for order-of-magnitude estimates when designing for polyatomic gases.

Consider now the limiting case in which transition to free-molecule flow is realized upstream from the skimmer. Figure 5 of Greene *et al.*[80] and Fig. 2 of Aurich and Schügerl[79] indicate that, if x is much larger than x_T, then Eq. (71) is no longer applicable. However, if the flow is nearly parallel upstream from the skimmer and speed ratios at the skimmer are large in comparison with unity, then Eq. (69) is still applicable. An appropriate expression for $(S_A/S_B)_1^2$ is required. In an analysis to be submitted for publication elswhere, the author has examined transition to free-molecule flow upstream from the skimmer for the case of a monatomic trace gas A in a monatomic carrier gas B. It is found that

$$\left(\frac{S_A}{S_B}\right)_T^2 = \left(\frac{m_A}{m_B}\right)^{0.2} \left(\frac{2m_A}{m_A + m_B}\right)^{1.2} \left(\frac{\sigma_{AB}}{\sigma_{BB}}\right)^{1.6} \tag{74}$$

where

$$\sigma_{AB} \approx \frac{\sigma_{AA} + \sigma_{BB}}{2}$$

Substitution into Eq. (69) yields

$$\left(\frac{x_A}{x_B}\right)_d \left(\frac{x_B}{x_A}\right)_1 \approx \left(\frac{m_A}{m_B}\right)^{0.2} \left(\frac{2m_A}{m_A + m_B}\right)^{1.2} \left(\frac{\sigma_{AB}}{\sigma_{BB}}\right)^{1.6} \tag{75}$$

The data of Greene *et al.*,[80] for $p_0 = 5$ torr, are correlated to good approximation by Eq. (75). Note that, for these data, $x_1/d^* = 24$ is much larger than $x_T/d^* \approx 1.3$, where x_T/d^* is calculated using Eq. (72). A quantitative comparison with the data of Aurich and Schügerl[79] is not possible due to the obvious presence of skimmer interference.

Consider finally the case in which collisions involving species A cease to be effective upstream from the skimmer, but B–B collisions are effective up to the skimmer. For nearly parallel flow upstream from the skimmer, speed ratios at the skimmer large in comparison with unity, and a monatomic trace gas A in a monatomic carrier gas B, one obtains

$$\left(\frac{x_A}{x_B}\right)_d \left(\frac{x_B}{x_A}\right)_1 \approx 0.13 \left(\frac{1}{Kn_0}\right)^{0.8} \left(\frac{d^*}{x_1}\right)^{4/3} \left(\frac{m_A}{m_B}\right)^{0.2} \left(\frac{2m_A}{m_A+m_B}\right)^{1.2} \left(\frac{\sigma_{AB}}{\sigma_{BB}}\right)^{1.6}$$

(76)

Equation [76] is applicable in the region

$$1 \leq 0.13 \left(\frac{1}{Kn_0}\right)^{0.8} \left(\frac{d^*}{x_1}\right)^{4/3} \leq \left(\frac{m_A}{m_B}\right)^{0.8} \left(\frac{m_A+m_B}{2m_A}\right)^{1.2} \left(\frac{\sigma_{BB}}{\sigma_{AB}}\right)^{1.6}$$

(Since the thickness of the transition region is nonzero, the factor 0.13 appearing in these two equations is expected to be replaced by a larger factor when more data are available.) Here the enrichment is a function of both source conditions and source–skimmer distance.

If the skimmer is downstream from the transition surface, the source of beam molecules as viewed by the detector (looking through the skimmer orifice) is a portion, with diameter d_1, of the transition surface, located at distance x_T from the source. Then ξ_{max} is given by

$$\sin \xi_{max} \approx \frac{d_1}{2x_T}$$

(77)

with x_T given by Eq. [72]. It is expected that also the factor 0.22 appearing in Eq. [72] will be replaced by a larger factor when more data are available.

In summary, mass separations due to pressure diffusions near the source orifice may be avoided if $Re_0 \gtrsim 10^3$. A criterion for avoiding reversed separation is suggested in Eq. (68). Means for handling the unavoidable Mach-number focusing are provided by Eqs. (69), (71), (75), and (76).

VIII. BACKGROUND SCATTERING

Scattering of beam molecules by background molecules is a problem if it (a) attenuates greatly the beam density and/or (b) alters significantly the relative densities of the several beam species. In principle, this scattering may occur in all vacuum stages of the sampling system. In practice, it is significant frequently in the collimation chamber, sometimes in the source chamber, and seldom in the detection chamber. Hence, expressions useful for either avoiding or handling effects of background scattering are summarized in this section.

The most important factors of background scattering are brought out by an examination of the case of nearly parallel flow at the skimmer and speed ratios at the skimmer large in comparison with unity. Then, for negligible background scattering, the density of species A at the detector is given by Eq. (12), which may be written in the form

$$\frac{n'_{Ad}}{n'_{A1}} \approx \frac{A_1}{\pi L_{1d}{}^2} S_{A1}{}^2 \tag{12}$$

where primes refer to values realized for negligible background scattering. If background scattering in the source chamber and in the collimation chamber are significant, then

$$\frac{n_{Ad}}{n'_{A1}} \approx \frac{A_1}{\pi L_{1d}{}^2} S_{A1}{}^2 \exp\left[-n_s Q_{As} (x_1 - x_m) - n_c Q_{Ac} L_{12}\right] \tag{79}$$

where n_s is the background number density in the source chamber, Q_{As} is the effective cross section for scattering of species A in the source chamber, $x_1 - x_m$ is the path length over which background scattering is important in the source chamber, n_c is the background number density in the collimation chamber, Q_{Ac} is the effective cross-section for scattering of species A in the collimation chamber, and L_{12} is the distance from the skimmer to the collimation orifice. For multicomponent backgrounds, the effective cross sections are defined by

$$n_s Q_{As} \equiv \sum_i n_{is} Q_{Ais} \tag{80}$$

$$n_c Q_{Ac} \equiv \sum_i n_{ic} Q_{Aic} \tag{81}$$

where n_{is} and n_{ic} are the number densities of species i in the source chamber and the collimation chamber respectively. The cross-sections are functions of the molecular species, the system geometry (angular resolution), and relative speed. For source-chamber scattering, measured values include $Q_{Ar-Ar} = 103$ Å^2 for a room-temperature source,[83] $Q_{Ar-Ar} = 11.5$ Å^2 for an arc-heated source,[84] and $Q_{Ar-Ar} = 7.5$ Å^2 and $Q_{Ar-He} = 4.7$ Å^2 for an arc-heated binary-mixture source.[85] For collimating-chamber scattering, typical observed values are $Q_{Ar-Ar} = 90$ Å^2 for an arc-heated beam[84] and $Q_{Ar-Ar} = 110$ Å^2 and $Q_{Ar-He} = 20$ Å^2 for an arc-heated binary-mixture source.[85] LeRoy et al.[86] find that the scattering cross-section varies as the inverse 2/3 power of the relative speed for source temperatures from 290°K to 1160°K.

The required cross sections can be determined for the given system geometry by appropriate differentiations of Eq. (79). Young[85] obtains

$$Q_{Ais} = -kT_s \frac{\partial}{\partial x_1}\left(\frac{\partial}{\partial p_{is}} \ln n_{Ad}\right) \tag{82}$$

$$Q_{Aic} = -\frac{kT_c}{L_{12}} \frac{\partial}{\partial p_{ic}} \ln n_{Ad} \tag{83}$$

Hence the source-chamber scattering cross section is obtained by adding species i to the source chamber for several values of the nozzle-skimmer distance; the collimation-chamber scattering cross section is obtained by adding species i to the collimation chamber.

The required values of x_m also can be determined by differentiation of Eq. (79). One obtains

$$x_1 - x_m = - \frac{kT_s}{Q_{As}} \frac{\partial}{\partial p_s} \ln n_{Ad} \qquad (84)$$

Hence x_m equals the value of x_1 for which $\partial \ln n_{Ad}/\partial p_s$ vanishes. (Note that this definition differs slightly from the definition used by Fenn and Anderson in Ref. 83; they hold p_s constant, vary x_1, and define x_m as the value of x_1 for which n_{Ad} is a maximum.) Values of x_m determined by Fenn and Anderson,[83] Brown and Heald,[61] and Knuth et al.[84] are compared in Figure. 8 of the latter paper. Values determined by Young[85] compare favorably with the values of Ref. 61 and 84. Examination of the data presented in the last three references indicates that, if $x_s \gg x_T$, then $x_m \approx 3x_T$ with x_T given by Eq. (72).

A criterion for negligible altering of the relative densities of the several beam species also follows from Eq. (79). Apply Eq. (79) to species A and B and take the ratio of the two equations to obtain

$$\frac{n_{Ad}/n'_{A1}}{n_{Bd}/n'_{B1}} \approx \left(\frac{S_A}{S_B}\right)^2_1 \exp\ [-n_s(Q_{As} - Q_{Bs})(x_1 - x_m)$$
$$- n_c\,(Q_{Ac} - Q_{Bc})\,L_{12}] \qquad (85)$$

It is seen that the relative densities of A and B are altered negligibly by background scattering if

$$n_s\,(Q_{As} - Q_{Bs})\,(x_1 - x_m) + n_c\,(Q_{Ac} - Q_{Bc})\,L_{12} \ll 1 \qquad (86)$$

If Eq. (86) holds for all species pairs, then Eq. (69) is recovered.

If attenuation of the beam due to background scattering in the source chamber is too great it may be reduced usually by reducing x_1. Attenuation in the collimation chamber is determined (for a given collimation-chamber pumping system) largely by the skimmer-orifice diameter. The considerations of skimmer diameter are simplified (with no loss in generality) by setting $x_1 = x_m$. Then Eq. (79) becomes

$$\frac{n_{Ad}}{n_{A1}} \approx \frac{A_1}{\pi L_{1d}^2}\, S_{A1}^2 \exp\,(-n_c Q_{ac} L_{12}) \qquad (87)$$

The background density n_c is related to the skimmer flow rate and the pumping speed by

$$n_c = \frac{n_1 U_1 A_1}{S_c} \qquad (88)$$

Substituting into Eq. (87) and setting the derivative with respect to A_1 equal to zero, one finds that the number density at the detector n_{Ad} is a maximum if A_1 is selected such that

$$n_c Q_{Ac} L_{12} = 1 \tag{89}$$

in which case

$$\frac{n_{Ad}}{n_{A1}} \approx \frac{A_1}{\pi L_{1d}^2} S_{A1}^2 \frac{1}{e} \tag{90}$$

In a design of a sampling system, one might make L_{1d} as small as is feasible, calculate the maximum allowable value of n_c from Eq. (89), and then the maximum allowable value of A_1 from Eq. (88). Finally, one might check to see if Eq. (86) is satisfied for all species pairs of interest.

In summary, excessive background scattering in the source chamber may be reduced by reducing x_1 toward about 3 x_T, where x_T is given by Eq. (72). The skimmer diameter which maximizes the beam density (taking into account background scattering in the collimation chamber) is specified by Eq. (89). Relative densities of the several beam species are altered negligibly by background scattering if Eq. (86) is satisfied. Which of the latter two equations dominates in the selection of d_1 is determined by the relative values of the several scattering cross sections.

IX. MASS-SPECTRA INTERPRETATIONS

Measuring and interpreting mass spectra are now routine operations in many applications of mass spectrometry. Techniques are described in numerous monographs and texts (e.g., Ref. 87); comprehensive indexes of mass spectra have been compiled (e.g., Ref. 88); and recent advances are reviewed periodically (e.g., Ref. 89). Hence the emphasis here is on those problems of mass spectrometry which are relatively unique to molecular-beam sampling from high-pressure reactive systems. The most important of these problems is perhaps the temperature dependence of the mass spectra.

The fragmentation processes leading to the mass spectra may be modeled by an electron-impact ionization step followed by one or more unimolecular decomposition steps. The several ion-formation probabilities are functions primarily of the vibrational energy of the parent ion immediately after electron impact. This energy is the sum of the initial vibrational energy possessed before electron impact and the excitation vibrational energy received during electron impact. Since the initial energy is a function of the gas temperature, the fragment pattern is temperature-dependent if the initial vibrational energy is a significant fraction of the total vibrational energy. It follows that the temperature dependence of the mass spectrum is weak for simple molecules (N_2, O_2, CO, NO, etc.) near room temperature and strong for complicated molecules (hydrocarbons, etc.) at higher temperatures. Energy considerations dictate that the ion-formation probability decreases with temperature for the parent ion (e.g., $C_3H_8^+$), increases with temperature for a final-product ion (e.g., H^+), and possesses a maximum at some intermediate temperature for all other ions (e.g., $C_2H_5^+$).

For the above model of the fragmentation processes, the total ionization probability (sum of the several ion-production probabilities) is independent of temperature. This feature has been verified experimentally by Cassuto[90] for temperatures up to 300°C and for hydrocarbons up to C_5H_{12}, by Ehrhardt and Osberghaus[91] for temperatures up to 900°C and for hydrocarbons up to C_7H_{16}, and by Komarov and Tikhomirov[92] for temperatures up to 900°C and for hydrocarbons up to C_3H_8.

Measured values of ion-production probabilities are given for C_2H_6, C_3H_8, and C_5H_{12} in Ref. 90, for CH_4 through C_7H_{16} in Ref. 91, for C_2H_4 and C_3H_8 in Ref. 92, and for C_4H_{10} (to 500°C) in Ref. 93. Incidental to their studies of hydrocarbons, Komarov and Tikhomirov[92] find that the ratio of N_2^+ to N^+ decreases only 10% as the gas temperature increases from 150°C to 900°C, confirming the expected weak temperature dependence of the N_2 fragmentation pattern at these temperatures.

Temperature-dependent ion-production probabilities have been predicted by Ehrhardt and Osberghaus[94] and by Komarov and Tikhomirov.[95] Further improvements in these predictions require more complete information concerning the excitation of the vibrational degrees of freedom of the molecule during electron impact.

The mass spectra of condensation products formed from polyatomic gases during free-jet expansions appear to be particularly sensitive to variations in experimental conditions. For example, Milne and Greene[40] find that the dimer mole fraction for NO is doubled if the ionizing-electron energy is decreased from 50 eV to 20 eV, and suggest that perhaps the parent-ion formation probability is relatively low in this case. A low parent-ion formation probability for condensation products might be helpful when sampling from combustion chambers.

It is seen that meaningful interpretations of mass spectra require that the mean vibrational energy of the molecules prior to electron impact be known. Furthermore, in many nozzle flows, the vibrational energy of the gas decreases, particularly during the early portion of the expansion. (Cf. Refs. 25, 93, and 96.) Hence a criterion for predicting the extent of vibrational relaxation in the free jet is required.

Vibrational relaxations are in many respects analogous to chemical relaxations. Hence many of the concepts used in Section IV are applicable here. More specifically, recall the freezing-point criterion

$$\frac{D\tau_{h,p}}{Dt} = C \tag{32}$$

and the ratio of relaxation times

$$\frac{\tau_{T,p}}{\tau_{h,p}} = \frac{c_p{}^0}{c_p{}^\infty} \tag{28}$$

Here $\tau_{h,p}$ and $\tau_{T,p}$ are vibrational-relaxation times, and $c_p{}^0$ and $c_p{}^\infty$ are constant-pressure heat capacities with vibrational degrees of freedom active and inactive respectively. The analog of Eqs. (30) and (31) is

$$\tau_{T,p} = \frac{N_v}{\sqrt{2}\,n\,\pi\sigma^2\,\Omega_\mu}\left(\frac{\pi}{8}\,\frac{m}{kT}\right)^{1/2} \tag{91}$$

where N_v is the collision number for vibrational relaxation and Ω_μ is a factor which handles deviations of viscosity cross sections from hard-sphere cross sections. Neglecting the variation of $c_p{}^0/c_p{}^\infty$ during the expansion, one may write

$$d\ln\frac{1}{\tau_{h,p}} = \left[-\frac{d\ln N_v}{d\ln T} + \frac{d\ln n}{d\ln T} + \frac{d\ln \Omega_\mu}{d\ln T} + \frac{1}{2}\right]d\ln T \tag{92}$$

The temperature dependence of measured values of N_v is in good agreement with the predictions of Landau and Teller (see e.g., Ref. 97), for which

$$\frac{d\ln N_v}{d\ln T} = \frac{1}{6} - \frac{1}{3}\ln N_v \tag{93}$$

For an isentropic expansion

$$\frac{d\ln n}{d\ln T} = \frac{1}{\gamma - 1} \tag{94}$$

The temperature dependence of Ω_μ is given approximately by

$$\frac{d\ln \Omega_\mu}{d\ln T} \approx -\frac{1}{3} \tag{95}$$

Hence the analog of Eq. (37) is

$$d\ln\frac{1}{\tau_{h,p}} \approx \left[\frac{1}{3}\ln N_v + \frac{1}{\gamma - 1}\right]d\ln T \tag{96}$$

Combining Eqs. (32) and (96), one obtains the analog of Eq. (51a), namely

$$\left[\frac{1}{3}\ln N_v + \frac{1}{\gamma - 1}\right]\frac{\tau_{h,p}}{\tau_{h,p0}}\frac{U}{a_0}\frac{dT}{d(x/d^*)} \approx -CT\frac{d^*}{(\tau_{h,p}\,a)_0} \tag{97}$$

The temperature at which Eq. (97) is satisfied fixes the mean vibrational energy of the molecules prior to electron impact and hence (for a given ionizer design) the mass spectra.

   The dimensionless scaling parameter for freezing of vibrational degrees of freedom is seen to be

$$\frac{d*}{(\tau_{h,p}a)_0} \tag{98}$$

Freezing occurs upstream from the source orifice, and the vibrational temperature differs less than about 20% from the source-gas temperature, if

$$\frac{d*}{(\tau_{h,p}a)_0} \lesssim 8\frac{\gamma-1}{\gamma+1}\left[\frac{1}{3}\ln N_v* + \frac{1}{\gamma-1}\right]\frac{\tau_{h,p}*}{\tau_{h,p0}} \tag{99}$$

where the asterisk refers to conditions at the throat of the orifice. Note that the right-hand side is independent of the pressure level and the source-orifice diameter and is only a weak function of the source-gas temperature.

As an example of the application of Eq. (99), consider N_2 at a stagnation temperature of $2600°K$. (Although the vibrational degrees of freedom of N_2 are not excited easily, it is the dominant species in many combustion chambers.) Using the data from Table A.1 of Stevens,[98] one finds that the right-hand side of Eq. (99) is about 50. For the special case in which $p_0 = 25$ atm and $d* = 1$ mm, the left-hand side is about 0.2. Hence, for this case, the vibrational degrees of freedom of N_2 molecules at the mass spectrometer have a mean energy corresponding closely to the source temperature, i.e., closely to $2600°K$.

Another problem of importance in interpretations of mass spectra is that molecules present either as background molecules in the ionizer chamber or as adsorbed molecules on the ionizer surfaces may provide undesired contributions to the measured spectra. For example, Tikhomirov and Komarov[99] find that electrons impacting on adsorbed gases on the walls of the ionizer chamber can produce major contributions to the mass/charge = 19 (fluorine) signal. Greene and Milne[7] suggest that beam modulation be used to discriminate against such effects.

Background and adsorbed gases must be considered also in quantitative calibrations of the mass spectrometer. Young et al.[21] use an effusive source. For in situ calibration, the effusive source is designed with two orifices such that it does not obstruct the supersonic molecular beam. The pressure in the effusive source is determined by metering the mass-flow rate into the source.

In this section, those aspects of mass-spectra interpretations which are particularly important to molecular-beam sampling from continuum sources are considered. Equation (97) facilitates predicting the mean vibrational energy of the molecules entering the mass spectrometer, important to the fragmentation pattern. Attention is drawn also to the need to avoid mass-spectra contributions from background and adsorbed molecules.

X. PRELIMINARY APPLICATIONS

Several different instrumentation systems for studies of combustion processes in internal-combustion engines were compared briefly in Section I. In Section II, an extremely simplified model of the molecular-beam mass-spectrometer sampling system was presented. Procedures for minimizing and/ or handling deviations from this model were considered in Sections III–VIII. Aspects of mass-spectra interpretations which are relatively unique to molec-ular-beam sampling from high-pressure reactive systems were examined in the preceding section. The present section completes the review of this sampl-ing technique by summarizing the highlights of a feasibility study of direct sampling from the combustion chamber of a reciprocating engine.

The relatively complicated problem of direct sampling from reciprocating engines was approached via a series of increasingly complex studies. In the first phase of these studies, the characteristics of a single-species inert supersonic molecular beam with a cycling-pressure source were studied numerically and experimentally.[100] Then several binary inert mixtures of known composition were sampled using the same motored engine used in the first phase.[46] In subsequent tests, qualitative beam densities of N_2, O_2, C_3H_8, H_2O, and CO_2 were measured in sampling from a 2–hp, 4–stroke, single-cylinder engine.[101] As a result of refinements in measuring techniques, the list of detected species was extended then to include NO and CO.[12] In the final phase of the feasibility study, effects of spark timing on the gas com-positions in the engine cylinder were investigated.[21]

Downstream from the sampling orifice, the molecular-beam system used in these studies was a typical supersonic molecular-beam system. The pres-sures in the source, collimation, and detection chambers were from 1×10^{-3} to 7×10^{-3} torr, about 5×10^{-6} torr, and about 5×10^{-7} torr, respectively.

The engine used in Refs. 12, 21, and 101 was a modified Briggs and Stratton Model 60152 engine. Engine speed was controlled by feeding the output from a tachometer back to a servo-load system. The triggering signal for the various electronic devices was provided by a photocell system con-nected to the tachometer shaft. The fuel was propane.

The instrumentation is indicated in Fig. 7. The beam density was meas-ured (one species at a time) by a commerical EAI QUAD 250 mass spectrom-eter. The detector output was amplified by an electron multiplier and a laboratory-built DC amplifier, and then averaged and stored by a Fabri-Tek Model 1062 multichannel signal averager. The chamber pressure was measured by a fast-response Dynascience Model 755–2560 pressure trans-ducer. Its output also was recorded by the signal averager. The information stored in the signal averager was transferred either directly to punched cards

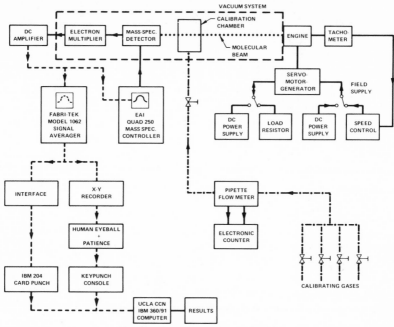

Fig. 7. Instrumentation system.[21]

or temporarily to graph paper. Procedures used in processing the data are described in Ref. 12.

Typical results obtained at an air/fuel ratio of 14.25, an engine speed of 1760 rpm, and a light engine load are presented in Figs. 8–10. The ordinates are proportional to the species mole fraction divided by the nitrogen mole fraction at the detector. Although the peak values are not always shown,

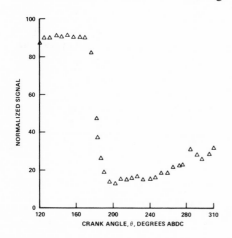

Fig. 8. Relative concentration of C_3H_8 as function of crank angle.

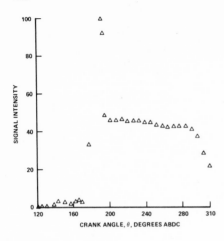

Fig. 9.   Relative concentration of CO as function of crank angle.

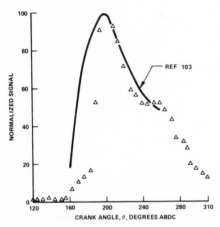

Fig. 10.   Relative concentration of NO as function of crank angle.

the ordinate values have been normalized by dividing them by the peak values and multiplying by 100. The measured peak pressure (323 psia) and calculated peak temperature (2410°R) occurred respectively at 197° and at 201° ABDC.

The several criteria given in earlier sections of this chapter indicate that no significant composition distortions were introduced by the sampling system. More specifically, composition distortions due to chemical relaxations, species condensations, skimmer interference, pressure diffusions, and background scattering are believed to be negligible in Figs. 8–10. The unavoidable species separations due to Mach-number focusing are independent of crank angle and can be handled quantitatively.

The electron-impact fragmentation pattern varies more with temperature variations for C_3H_8 than for CO and NO. For identification of C_3H_8, the

$m/e = 29$ signal was used. Data given in Refs. 91 and 92 indicate that the $C_2H_5^+$ production probability peaks at about 2000°R and is a weak function of temperature in the temperature range of interest here.

The methods of Section III indicate that the sampling is from the bulk gas early in the power stroke ($q \approx 0.2d^*$) and from the outer regions of the boundary layer late in the power stroke ($\delta \approx 3d^*$). The measured increase in C_3H_8 and decreases in CO and NO after about 260° ABDC are consistent with sampling from the boundary layer.

The rapid drop in the relative C_3H_8 concentration between 170° and 195° ABDC indicates that the combustion occurred during this part of the cycle. The propane remaining at 200° ABDC is excess fuel corresponding to the realized equivalence ratio of 1.1.

The decay of the relative CO concentration from its peak value appears to consist of two distinct stages. Both stages may be discussed in terms of the dominant CO-decay reaction, namely,

$$CO + OH \leftrightarrows CO_2 + H \tag{100}$$

Since the relaxation time for this reaction is typically small in comparison with the power-stroke period, the first stage may be a rapid relaxation of the CO concentration into quasi-equilibrium with the existing OH and H concentrations. Since, as shown by Newhall,[102] the ratio of the OH and H concentrations during the power stroke is determined largely by other (relatively slow) reactions, the second stage may be a relatively slow quasi-equilibrium process with the rate limited by the rates of the other OH and H reactions.

The relative NO concentration decays to about 50% of its peak value and then freezes at about 240° ABDC. This decay is greater than measured by Lavoie et al.[2] and by Caretto et al.,[103] but is of the order of magnitude predicted in Fig. 3 of Ref. 103 for the gas which is burned first. Note that the predicted curve included in Fig. 10 also is normalized relative to its peak value.

These preliminary applications confirm the feasibility of using molecular-beam mass-spectrometer sampling systems in quantitative studies of combustion processes. Of greater interest here is that they demonstrate the feasibility of using such systems in quantitative studies of the relatively complicated combustion processes in reciprocating engines. It is suggested that future studies include more precise determinations of sampling-system operating limits within which data may be interpreted most easily. Additional rate data (e.g., relaxation times) and species properties (e.g., fragmentation patterns) are needed. Preparation of this review will be justified if it provides any help in either (a) designing sampling systems, or (b) planning studies involving such systems, or (c) organizing results obtained in such studies.

ACKNOWLEDGMENTS

The studies which led to the present review were made possible by the contributions of all those graduate students who pursued theses in the Molecular-Beam Laboratory, with the capable assistance of Dr. W. S. Young (Postdoctoral Scholar) and Mr. Wayne E. Rodgers (Principal Electronics Technician), and by grants from the National Science Foundation and the Environmental Protection Agency. Dr. W. S. Young and graduate students S. M. Liu, P. K. Sharma, and T. G. Wang read preliminary versions of the manuscript and made valuable suggestions.

REFERENCES

1. Newhall, H.K. and Starkman, E.S., *Direct Spectroscopic Determination of Nitric Oxide in Reciprocating Engine Cylinders,* Paper No. 670122, Society of Automotive Engineers, New York, 1967.
2. Lavoie, G.A., Heywood, J.B., and Keck, J.C., Experimental and theoretical study of nitric oxide formation in internal combustion engines, *Combustion Science and Technology,* Vol. 1, 1970, pp. 313–326.
3. Smith, D.S. and Starkman, E.S., A spectroscopic study of the hydroxyl radical in an internal combustion engine, *Thirteenth Symposium (International) on Combustion,* The Combustion Institute, Pittsburgh, 1971.
4. Lewis, J.D., and Harrison, D., A study of combustion and recombination reactions during the nozzle expansion process of a liquid propellant rocket engine, *Eighth Symposium (International) on Combustion,* The Williams and Wilkins Co., Baltimore, 1962, pp. 366–374.
5. Daniel, W.A., Engine variable effects on exhaust hydrocarbon composition (a single-cylinder engine study with propane as the fuel), *SAE Trans.* **76** (1968) 774–795.
6. Foner, S.N., Mass spectrometry of free radicals, in *Advances in Atomic and Molecular Physics Vol. 2,* (D.R. Bates and I. Estermann, eds.), Academic Press, New York, 1966, pp. 385–461.
7. Greene, F.T., and Milne, T.A., Molecular beam sampling of high temperature systems, AIAA Paper 67–37, *presented at AIAA 5th Aerospace Sciences Meeting,* New York, January 23–26, 1967.
8. Sturtevant, B., and Wang, C.P., Mass spectrometric studies of impurity ionization in shock-heated argon, *Recent Advances in Aerothermochemistry Vol. 2* (I. Glassman, ed.), NATO Advisory Group for Aerospace Research and Development, Paris, 1967, pp. 595–606.
9. Dix, R.E., *Sampling Probe for Instantaneous Mass Spectrometric Analysis of Rarefied High Enthalpy Flow,* AEDC-TR–69–37, ARO Inc., Arnold Air Force Station, Tennessee, April 1969.
10. Chang, J.H., Supersonic molecular beam sampling system for mass spectrometric studies of high-pressure flow systems, AIAA Preprint 69–64, *Presented at the AIAA 7th Aerospace Sciences Meeting,* New York, January 20–22, 1969.
11. Kahrs, J., *Combustion-Gas Sampling System,* AFRPL-TR–70–28, Air Force Rocket Propulsion Laboratory, Edwards, California, March 1970.

12. Young, W.S., Wang, Y.G., Rodgers, W.E., and Knuth, E.L., Molecular-beam sampling of gases in engine cylinders, *Technology Utilization Ideas for the 70's and Beyond,* American Astronautical Society, Tarzana, 1971, pp. 281–289.

13. Sherman, F.S., *Self-Similar Development of Inviscid Hypersonic Free-Jet Flows,* Technical Report: Fluid Mechanics 6–90–63–61, Lockheed Missiles and Space Company, Sunnyvale, California, May 23, 1963.

14. Knuth, E.L., *Rotational and Translational Relaxation Effects in Low-Density Hypersonic Free Jets,* Department of Engineering Report No. 64–53, University of California Los Angeles, Los Angeles, November 1964.

15. Ashkenas, H., and Sherman, F.S., The structure and utilization of supersonic free jets in low density wind tunnels, *Rarefied Gas Dynamics Vol. 2,* (J.H. de Leeuw, ed.), Academic Press, New York, 1966, pp. 84–105.

16. Friedman, R., and Johnston, W.C., The wall-quenching of laminar propane flames as a function of pressure, temperature, and air-fuel ratio, *J. Appl. Phys.* 21 (August 1950) 791–795.

17. Green, K.A., and Agnew, J.T., Quenching distances of propane-air flames in a constant-volume bomb, *Combustion and Flame,* 15 (October 1970) 189–191.

18. Daniel, W.A., Why engine variables affect exhaust hydrocarbon emission, SAE Paper No. 700108, Presented at the *Automotive Engineering Congress,* Detroit, Michigan, January 12–16, 1970.

19. Ellenberger, J.M., and Bowlus, D.A., *Single Wall Quench Distance Measurements,* Presented at the 1971 Technical Session, Central States Section, The Combustion Institute, The University of Michigan, Ann Arbor, March 23–24, 1971.

20. Pinkerton, J.D., *Some Factors Affecting Emissions from Spark Ignition Engines,* Ph.D. Thesis, UCLA School of Engineering and Applied Science, Los Angeles, 1971.

21. Young, W.S., Wang, Y.G., Rodgers, W.E., and Knuth, E.L., *Timing Effects on Gas Compositions in an Engine Cylinder,* Presented at the 1971 Technical Session, Central States Section, The Combustion Institute, The University of Michigan, Ann Arbor, March 23–24, 1971.

22. Sturtevant, B., Application of a magnetic mass spectrometer to ionization studies in impure shock-heated argon, *J. Fluid Mech.* 15 (1966) 641–656.

23. Bauer, H.–J., Phenomenological theory of the relaxation phenomena in gases, in *Physical Accoustics Principles and Methods, Vol. II, Part A, Properties of Gases, Liquids and Solutions* (W.P. Mason, ed.), Academic Press, New York, 1965, pp. 47–131.

24. McIntyre, R.W., and Leslie, R.S.E., Comparative evaluation of several approximate methods of analysis of non-equilibrium flows, in *Recent Advances in Aerothermochemisty Vol. 2,* (I. Glassman, ed.), NATO Advisory Group for Aerospace Research and Development, Paris, 1967, pp. 685–699.

25. Bray, K.N.C., Chemical and vibrational nonequilibrium in nozzle flows, *Nonequilibrium Flows, Part II* (P.P. Wegener, ed.), Marcel Dekker, Inc., New York, 1970, pp. 59–157.

26. Phinney, R., *Mathematical Nature of the Freezing Point in an Expanding Flow,* RM–172, Martin Marietta Co., Baltimore, April 1964.

27. Burwell, W.G., Sarli, V.J., and Zupnik, T.F., Applicability of sudden-freezing criteria in analysis of chemically complex rocket nozzle expansions, in *Recent Advances in Aerothermochemisty Vol. 2,* (I. Glassman, ed.), NATO Advisory Group for Aerospace Research and Development, Paris, 1967, pp. 701–759.

28. Glowacki, W.J., *Effect of Finite Oxygen Recombination Rate on the Flow Conditions in Hypersonic Nozzles,* NOLTR 61–23, September 1961.

29. Phinney, R., *Review of Freezing Point Techniques for Computing Relaxation Flows,* RM–143, Martin Marietta, Baltimore, March 1963.

30. Bray, K.N.C., Simplified sudden-freezing analysis for nonequilibrium nozzle flows, *ARS J.* 31 (June 1961) 831–834.

31. Ring, L.E., and Johnson, P.W., *Correlation and Prediction of Air Nonequilibrium in Nozzles,* AIAA Paper No. 68–378, Presented at the AIAA 3rd Aerodynamic Testing Conference, San Francisco, April 8–10, 1968.

32. Wegener, P.P., Gasdynamics of expansion flows with condensation, and homogeneous nucleation of water vapor, in *Nonequilibrium Flows, Part I* (P.P. Wegener, ed.), Marcel Dekker, Inc., New York, 1970, pp. 163–243.

33. Becker, E.W., Bier, K., and Henkes, W., Strahlen aus Kondensierten Atomen und Molekeln in Hochvakuum, *Z. Physik,* **146** (September 21, 1956) 333–338.

34. Anderson, J.B., Anderes, R.P., Fenn, J.B., and Maise, G., Studies of low density supersonic jets, *Rarefied Gas Dynamics Vol. II,* (J.H. de Leeuw, ed.), Academic Press, New York, 1966, pp. 106–127.

35. Golomb, D., Good, R.E., and Brown, R.F., Dimers and clusters in free jets of argon and nitric oxide, *J. Chem. Phys.* **52** (February 1, 1970) 1545–1551.

36. Milne, T.A., Vandergrift, A.E., and Greene, F.T., Mass-spectometric observations of argon clusters in nozzle beams. II. The kinetics of dimer growth, *J. Chem. Phys.* **52** (February 1, 1970) 1552–1560.

37. Golomb, D., and Good, R.E., Clusters in isentropically expanding nitric oxide and their effect on the chemiluminous NO–O reaction, *J. Chem. Phys.* **49** (November 1, 1968) pp. 4176–4180.

38. Good, R.E., Golomb, D., DelGreco, F.P., Hill, D.W., and Whitfield, D.L., Clusters in nitric oxide jet expansion, *Rarefied Gas Dynamics Vol. 2* (L. Trilling and H.Y. Wachman, eds.), Academic Press, New York, 1969, pp. 1449–1453.

39. Bier, K., and Hagena, O., Optimum conditions for generating supersonic molecular beams, in *Rarefied Gas Dynamics Vol. II,* (J. H. de Leeuw, ed), Academic Press, New York, 1966, 260–278.

40. Milne, T.A., and Greene, F.T., Mass-spectrometric detection of dimers of nitric oxide and other polyatomic molecules, *J. Chem. Phys.* **47** (November 1, 1967) 3668–3669.

41. Milne, T.A., and Greene, F.T., Mass spectrometric observations of argon clusters in nozzle beams. I. General behavior and equilibrium dimer concentrations, *J. Chem. Phys.* **47** (November 15, 1967) 4095–4101.

42. Greene, F.T., and Milne, T.A., Mass spectrometric detection of polymers in supersonic molecular beams, *J. Chem. Phys.* **39** (December 1, 1963) 3150–3151.

43. Bier, K. and Hagena, O., Influence of shock waves on the generation of high-intensity molecular beams by nozzles, *Rarefied Gas Dynamics* (J.A. Laurmann, ed.), Vol. I, Academic Press, New York, 1963, pp. 478–496.

44. Reis, V.H. and Fenn, J.B., Separation of gas mixtures in supersonic jets, *J. Chem. Phys.* **39** (December 15, 1963) 3240–3250.

45. Rothe, D.E., Electron beam studies of the diffusive separation of helium–argon mixtures, *Phys. Fluids* **9** (September 1966) 1643–1658.

46. Young, W.S., Rodgers, W.E., Cullian, C.A., and Knuth, E.L., Molecular–beam sampling of gas mixtures in cycling–pressure sources, *Proceedings of the Seventh International Symposium on Rarefied Gas Dynamics held at Pisa, Italy,* June 29–July 3, 1970.

47. Knuth, E.L., Supersonic Molecular Beams, *App. Mech. Rev.* **17** (October 1964) 751–762.

48. Anderson, J.B., Andres, R.P., and Fenn, J.B., High intensity and high energy molecular beams, in *Advances in Atomic and Molecular Physics* (D.R. Bates and I. Estermann, eds.), Academic Press, New York, 1965, pp. 345–389.

49. Anderson, J.B., Andres, R.P., and Fenn, J.B., Supersonic nozzle beams, *Molecular beams* (J. Ross, ed.), Wiley, New York, 1966, pp. 275–317.

50. French, J.B., Continuum-source molecular beams, *AIAA J.* **3** (June 1965) pp. 993–1000.

51. Vick, A.R., and Andrews, E.H. Jr., *An Investigation of Highly Underexpanded Exhaust Plumes Impinging upon a Perpendicular Flat Surface,* NASA TN D–3269, February 1966, 55 pp.

52. Bier, K. and Schmidt, B., Zur Form der Verdichtungsstösse in frei expandierenden Gasstrahlen, *Z. für angewandte Physik* **13** (November 1961) 493–500.

53. Roberts, L., The action of a hypersonic jet on a dust layer, IAS Paper No. 63–50, *Presented at the IAS 31st Annual Meeting,* New York, January 21–23, 1963.

54. Bossel, U., Skimmer interaction: transition from a 'shock beam' to a supersonic nozzle beam, *Entropie* No. 30 (November-December 1969) 11–15.

55. Oman, R.A., Analysis of a skimmer for a high-intensity molecular beam using a three-fluid model, *The Physics of Fluids,* **6** (July 1963) 1030–1031.

56. French, J.B., and McMichael, G.E., Progress in developing high energy nozzle beams, *Rarefied Gas Dynamics, Vol. II* (C.L. Brundin, ed.), Academic Press, New York, 1967, pp. 1385–1392.

57. Bossel, U., *On the Optimization of Skimmer Geometries,* submitted for publication.

58. Fenn, J.B., and Deckers, J., Molecular beams from nozzle sources, *Rarefied Gas Dynamics, Vol. I* (J.A. Laurmann, ed.), Academic Press, New York, 1963, pp. 497–515.

59. Fisher, S.S., and Knuth, E.L., Properties of low-density freejets measured using molecular-beam techniques, *AIAA J.* **7** (June 1969) 1174–1177.

60. Potter, J.L., and Miller, J.T., Sphere drag and dynamic simulation in near-free-molecular flow, *Rarefied Gas Dynamics Vol. I* (L. Trilling and H.Y. Wachman, eds.), Academic Press, New York, 1969, pp. 723–734.

61. Brown, R.F., and Heald, J.H. Jr., Background gas scattering and skimmer intraction studies using a cryogenically pumped molecular beam generator, *Rarefied Gas Dynamics Vol. II* (C.L. Brundin, ed.), Academic Press, New York, 1967, pp. 1407–1424.

62. Valleau, J.P., and Deckers, J.M., Supersonic molecular beams. II. Theory of the formation of supersonic molecular beams, *Can. J. Chem.* **43** (January 1965) 6–17.

63. Becker, E.W., Bier, K., and Burghoff, H., Die Trenndüse, *Z. Naturforschung* **10a** (July 1955) 565–572.

64. Becker, E.W., Beyrich, W., Bier, K., Burghoff, H., and Zigan, F., Das Trenndüsenverfahren, *Z. Naturforschung* **12a** (August 1957) 607–621.

65. Waterman, P.C., and Stern, S.A., Separation of gas mixtures in a supersonic jet, *J. Chem Phys.* **31** (August 1959) 405–419.

66. Stern, S.A., Waterman, P.C., and Sinclair, T.F., Separation of gas mixtures in a supersonic jet. II. Behavior of helium-argon mixtures and evidence of shock separation, *J. Chem. Phys.* **33** (September 1960) 805–813.

67. Abuaf, N., Anderson, J.B., Andres, R.P., Fenn, J.B., Miller, D.R., Studies of low density supersonic jets, in *Rarefied Gas Dynamics Vol. 2,* (C.L. Brundin, ed.), Academic Press, New York, 1967, pp. 1317–1136.

68. Anderson, J.B., Separation of gas mixtures in free jets, *AIChE.* **13** (November 1967) 1188–1192.

69. Wang, J.C.F., and Bauer, P.H., Measurements of spatial distribution of species in helium argon gas mixtures expanding in supersonic jets, in *Rarefied Gas Dynamics* (L. Trilling and H.Y. Wachman, eds.), Vol. II, Academic Press, New York, 1969, pp. 1009–1013.

70. Sebacher, D.I., Guy, R.W., and Lee, L.P., Diffusive separation in free jets of nitrogen and helium mixtures, *Rarefied Gas Dynamics* (L. Trilling and H.Y. Wachman, eds.), Vol. II, Academic Press, New York, 1969, pp. 931–938.

71. Zigan, F., Gasdynamische Berechnung der Trenndüsenentmischung, *Z. Naturforschung* **17a** (1962) 772–778.

72. Sherman, F.S., Hydrodynamical theory of diffusive separation of mixtures in a free jet, *Phys. Fluids* **8** (May 1965) 773–779.

73. Mikami, H., and Takashima, Y., Separation of gas mixture in an axisymmetric supersonic jet, *Int. J. Heat Mass Transfer* **11** (November 1968) 1597–1610.

74. Campargue, R., Aerodynamic separation effect on gas and isotope mixtures induced by invasion of the free jet shock wave structure, *J. Chem. Phys.* **52** (February 15, 1970) 1795–1802.

75. Bier, K., Umkehrung der Trenndüsen-Entmischung in Überexpandierten Gas-

strahlen, *Z. Naturforschung* **15a** (August 1960) 714–723.

76. Sebacher, D.I., Diffusive separation in shock waves and free jets of nitrogen–helium mixtures, *AIAA.* **6** (January 1968) 51–58.

77. Chow, R.R., *On the Separation Phenomenon of Binary Gas Mixture in an Axisymmetric Jet,* Technical Report HE–150–175, University of California Institute of Engineering Research, Berkeley, November 4, 1959.

78. Schügerl, K., Investigations and applications of supersonic molecular beams, *Rarefied Gas Dynamics* (L. Trilling and H.Y. Wachman, eds.), Vol. 2, Academic Press, New York, 1969, pp. 909–930.

79. Aurich, V., and Schügerl, K., Determination of the radial distributions of the number densitites of the components in supersonic free jets of binary gas mixtures by molecular beam sampling, *Entropie* No. 30 (November–December 1969) 21–24.

80. Greene, F.T., Brewer, J., and Milne, T.A., Mass spectrometric studies of reactions in flames. I. Beam formation and mass dependence in sampling 1-atm gases, *J. Chem. Phys.* **40** (March 15, 1964) 1488–1495.

81. Klingelhöfer, R., and Lohse, P., Production of fast molecular beams using gaseous mixtures, *Phys. Fluids* **7** (March 1964) 379–381.

82. French, J.B., and O'Keefe, D.R., Omegatron studies of a skimmed beam system, *Rarefied Gas Dynamics* (J.H. de Leeuw, ed.), Vol. II, Academic Press, New York, 1966, pp. 299–310.

83. Fenn, J.B., and Anderson, J.B., Background and sampling effects in free jet studies by molecular beam measurements, *Rarefied Gas Dynamics* (J.H. de Leeuw, ed.), Vol. II, Academic Press, New York, 1966, pp. 311–330.

84. Knuth, E.L., Kuluva, N.M., and Callinan, J.P., Densities and speeds in an arc-heated supersonic argon beam, *Entropie* No. 18 (November-December 1967) 38–46.

85. Young, W.S., and Knuth, E.L., A binary-mixture arc-heated supersonic molecular beam, *Entropie* No. 30 (November–December 1969) 25–29.

86. LeRoy R.L., Govers, T.R., and Deckers, J.M., Background scattering of a supersonic free jet: Source temperature dependence, *Can. J. Chem.* **47** (1969) 2305–2306.

87. Beynon, J., *Mass Spectrometry and its Applications to Organic Chemistry,* Elsevier Publishing Co., New York, 1960.

88. Kuentzel, L.E., *Index of Mass Spectral Data,* American Society for Testing and Materials, Philadelphia, 1963.

89. Rinehart, K.L., Jr., and Kinstle, T.H., Mass spectrometry, *Annual Review of Physical Chemistry* (H. Eyring, C.J. Christensen, and H.S. Johnston, eds.), Vol. 19, Annual Reviews, Inc., Palo Alto, 1968, pp. 301–342.

90. Cassuto, A., Variations in mass spectra with the temperature of the ionization chamber between $-150°C$ and $+200°C$, *Advances in Mass Spectrometry, Vol. 2,* Pergamon Press, Oxford, 1963, pp. 296–312.

91. Ehrhardt, H. and Osberghaus, O., Massenspektrometrische Untersuchungen von Kohlenwasserstoffen bei hohen Temperaturen, *Z. Naturforschung* **13a** (1958) 16–21.

92. Komarov, V.N., and Tikhomirov, M.V., Temperature dependence of mass spectra. I. Mass spectra of ethylene and propane, *Russ. J. Phys. Chem.* **40** (December 1966) 1594–1597.

93. Milne, T.A., Beachey, J.E., and Greene, F.T., Study of relaxation in free jets using temperature dependence of n-butane mass spectra, *J. Chem. Phys.* **56** (March 15, 1972) 3007–3013.

94. Ehrhardt, H., and Osberghaus, O., Temperaturabhängigkeit der Massenspektren von Kohlenwasserstoffmolekülen und ihre Bedeutung im Rahmen der statistischen Theorie, *Z. Naturforschung* **15a** (1960) 575–584.

95. Komarov, V.N., and Tikhomirov, M.V., The effect of temperature on the mass spectra of propane and butane, *Russ. J. Phys. Chem.* **40** (August 1966) 1047–1048.

96. Rich, J.W., and Treanor, C.E., Vibrational relaxation in gas -dynamic flows, *Annual Review of Fluid Mechanics, Vol. 2*, Annual Reviews, Inc., Palo Alto, 1970, pp. 355–396.

97. Herzfeld, K.F., and Litovitz, T.A., *Absorption and Dispersion of Ultrasonic Waves,* Academic Press, New York, 1959.

98. Stevens, B., *Collisional Activation in Gases,* Pergamon Press, New York, 1967.

99. Tikhomirov, M.V. and Komarov, V.N., Effect of the surface on the mass spectrum of tetrafluoroethylene and the appearance potential of F^+, *Russ. J. Phys. Chem.* **40** (June 1966) 751–753.

100. Young, W.S., Rodgers, W.E., Cullian, C.A., and Knuth, E.L., Supersonic molecular beams with cycling-pressure sources, *AIAA J.* **9** (Feb. 1971) 323–325.

101. Young, W.S., Rodgers, W.E., Cullian, C.A., Wang, Y.G., and Knuth, E.L., A method for sampling the instantaneous chemical compositions in an internal combustion engine, *Proceedings of the Second International Clean Air Conference* (H.M. Englund and W.T. Beery, eds.), Academic Press, New York, 1971, pp. 418–424.

102. Newhall, H.K., Kinetics of engine-generated nitrogen oxides and carbon monoxide, *Twelfth Symposium (International) on Combustion,* The Combustion Institute, Pittsburgh, 1969, pp. 603–613.

103. Caretto, L.S., Muzio, L.J., Sawyer, R.F., and Starkman, E.S., The role of kinetics in engine emission of nitric oxide, *Presented at the Third Joint Meeting, The American Institute of Chemical Engineers and Instituto Mexicano de Ingienieros Quimicos,* Denver, Colorado, August 30–September 2, 1970.

Index

Acetylene
 combustion of 59, 91–94
 formation of, 18
Activation energy, 52, 55, 82–93, 97, 99,
  100, 154, 216, 217
Additives, effect on emissions, *see* Fuel,
  Tetraethyl lead
Adiabatic temperature gradient, 282
Aerosols, *see* Particulates
After-injection, 224
Agglomeration
 in aircraft emissions fallout, 280
 in particulate formation, 18
Air injection, 70, 71
Aircraft emissions, 78, 122, 267, 268, 270–
  272
 comparison to motor vehicles, 269–271
 components, 268
 diffusion and fallout of, 273–279
 effects of, 267
 emissions index, definition, 270
 emissions of various types of aircraft,
  270–272
 particulates, effects, 268
 reduction techniques, 271
 removal of, 279, 280
 wing-tip vortices, effect of, 285
Aircraft plumes, *see* Plumes
Aldehydes, 67
 effect of load in diesel, 229
 formation in diesel, 234
 measurement, 303, 304
 reactivity, 54
Alkanes, combustion of, 91
Ammonia, production of, 68, 69
Aromatics, 4
 additives, as, 57
Atmosphere, *see also* Fallout, Photochemical
  smog, Visibility
 airborne lead, 6, 186, 202, 206
 aircraft emissions, 278, 279

Atmosphere *(cont'd)*
 pollutants, 37
 removal of pollutants, 38, 279, 280
Automobile Manufacturers Association, 27
Automotive emissions, 2–4, 65, *see also*
  Spark-ignition engines, and specific
  emission of interest
 amounts, 77, 78, 205, 269
 evaporative, 3
 exhaust system, 63–69
 measurement, 28, 29, 185–193, 291,
  304–313
 particulates, 37, 204–206
 polynuclear aromatics, 141–145
 sampling, 304–309
 test procedures 27–29, 113, 114, 311, 312
 warm-up, effect of, 23–26

Batch reactors, 33
Benz(a)anthracene, 141, 142
Benz(a)pyrene, 140–142, 144
Boundary layer effects
 in molecular-beam sampler, 326–329
 in SI engine, 327, 328
Buses, *see also* Diesel
 emissions standards, 312, 313

California Diesel Emissions Standard, 118
California, State Department of Public Health,
  27
California, State Motor Vehicle Pollution
  Control Board, 27
Carbon,
 emissions measurement, 310, 311
 formation in engines, 17–20, 238
 particle size, 18

365